Microbes, Man and Animals

The Natural History of
Microbial Interactions

Microbes, Man and Animals

The Natural History of Microbial Interactions

Alan H. Linton

Department of Microbiology
University of Bristol

with contributions by
Mary P. English
L. W. Greenham
A. E. Jephcott
K. B. Linton
Rosemary Simpson

A Wiley–Interscience Publication

1807 1982

JOHN WILEY & SONS

Chichester · New York · Brisbane · Toronto · Singapore

Copyright © 1982 by John Wiley & Sons Ltd.

Library of Congress Cataloging in Publication Data:
Main entry under title:

Microbes, man, and animals.
  'A Wiley–Interscience publication.'
  Includes index.
  1. Medical microbiology. 2. Veterinary
microbiology. 3. Host–parasite relationships.
I. Linton, Alan Henry. II. English, Mary P.
QR46.M537   616'.01   81-14719

ISBN 0 471 10083 8   AACR2

British Library Cataloguing in Publication Data:
Linton, Alan H.
  Microbes, man and animals.
  1. Medical microbiology
  I. Title
  616'.01   QR46

ISBN 0 471 10083 8

Typeset by Pintail Studios Ltd, Ringwood, Hampshire
Printed in Great Britain by Page Bros. (Norwich) Ltd.

Dedicated
to the memory of the late
Anna Mayr-Harting
teacher, colleague and friend
who inspired this book

Contents

Contributors

MARY P. ENGLISH, M.Sc., D.Sc. *Formerly Consultant Mycologist, Bristol Royal Infirmary and Research Fellow in Mycology, University of Bristol.*

L. W. GREENHAM, B.Sc., B.V.Sc., Ph.D., M.R.C.V.S. *Senior Lecturer in Veterinary Bacteriology and Virology, University of Bristol.*

A. E. JEPHCOTT, M.A., M.B., M.D., Dip.Bact., F.R.C.Path. *Director Public Health Laboratory, Bristol and Clinical Lecturer in Bacteriology (Public Health), University of Bristol.*

A. H. LINTON, Ph.D., D.Sc., F.R.C.Path. *Reader in Veterinary Bacteriology, University of Bristol.*

K. B. LINTON, B.Sc., Ph.D., M.R.C.Path. *Senior Lecturer in Bacteriology, University of Bristol.*

ROSEMARY A. SIMPSON, B.Sc., Ph.D., M.R.C.Path. *Recognized Teacher in Microbiology, University of Bristol.*

Foreword

This book grew out of a course on Pathogenicity and Epidemiology taught to Honours Microbiology students at Bristol by the late Dr Anna Mayr-Harting and myself. Until recently very few texts brought these subjects together in a form readily available to students, and the need prompted us to consider publishing a joint work. Sadly, Dr Harting did not survive long enough to contribute more than a few notes and, in consequence, it has taken much longer to complete than was originally anticipated. The book is now presented in the hope that it will fill a gap in the student literature.

In order to widen the coverage of the subject a number of colleagues have kindly contributed specialist articles. In spite of this the text does not claim to be comprehensive but rather a collection of subjects selected to illustrate principles and the author's own interests. 'I have gathered a bouquet of flowers from other men's gardens: naught but the string is my own' (Montaigne).

The book is aimed at science microbiology undergraduates but since examples are taken from the medical and veterinary fields, both undergraduate and post-graduate students of these professions should also find it useful.

ALAN H. LINTON
1982

Introduction

Infection must be considered as a struggle between two organisms . . .
the parasite and its host. This brings about adaptations on both sides.

METCHNIKOFF, 1891

Micro-organisms are ubiquitous but one of their major habitats is in association
with other living organisms. Animals raised conventionally are exposed to micro-
organisms from the time they leave the sterility of the uterus or egg. Most of these
contaminants are rapidly eliminated from the animal's body but some are able to
colonize the young animal or chick and persist for part or the whole of its life.
These associations are not solely a matter of contamination, i.e. one of mutual
indifference, but involve specific interactions creating the so-called host–parasite
relationship.

Both animals, and the micro-organisms that colonize them, are dynamic
systems capable of independent existence. Gnotobiotic animals (Chapter 2), for
instance, removed aseptically from the uterus of the dam in the late stages of pregnancy, can be raised independent of micro-organisms. Similarly gnotoxenic chicks
can be raised after hatching under sterile conditions. These 'sterile' animals can
survive for long periods of time and grow; a few species have even been
successfully bred under aseptic conditions. They exhibit only minor differences in
physiological function from those raised conventionally.

Micro-organisms, too, can follow an independent existence. The majority of
micro-organisms able to colonize the animal body can be propagated in pure
culture, i.e. axenic culture, on inanimate media, and this can continue indefinitely
under artificial conditions.

It is therefore not unexpected that when the two independent biological systems
live in association with each other they will interact to a greater or lesser degree.
The degree of interaction will vary according to the nature and biological status of
each. At one end of the spectrum are those associations in which only minor
adaptations are experienced by each partner and this results in an ideal state of
parasitism; at the other extreme are associations which damage or even kill one or
other of the partners.

Micro-organisms which damage the host in the course of their association are
recognized as pathogens. Microbial disease is the consequence of damage to the
host resulting from the interaction and a wide range of clinical response is
experienced. The level of the interaction and the extent of the resultant damage

depends on both the virulence of the pathogen and the host's defences. From the standpoint of the pathogen, its ability to colonize and damage the host depends, at least in part, on how well it can adapt its physiology to the host's environment. Adaptations are necessary to overcome, or to come to terms with, the anatomical structures and physiology of the defence mechanisms of the host. If the pathogen can overcome the host defences, the host will be damaged and may not survive; if, on the other hand, the defences overcome the pathogen, the micro-organism may fail to establish itself in the host and die. The sequence of events occurring throughout the interaction is the subject of pathogenesis; the ensuing clinical manifestations constitute the disease syndrome.

A study of pathogenesis therefore must take into account the intrinsic biological properties of both host and parasite, and their responses to each other during interaction. Ideally, since each adapts to the sequence of modifications consequent upon their interaction, a study of the two in association must be attempted (Chapter 3). This approach is more difficult than studying the host and parasite independently but the *in vivo* approach, developed over the last 25 years, has been richly rewarding. It has been possible to identify virulence factors which arise only *in vivo*, to unravel the biochemical basis of certain specific mechanisms and other aspects of the host–parasite interaction.

Most texts are written either from the medical or veterinary standpoints. This book is an attempt to present an integrated account of our knowledge at the present time bringing together information and examples from the medical and veterinary fields. The text is divided into three parts.

Part I is concerned with interactions between the individual host and the parasite. The emphasis throughout the book is on the parasite but it is not possible to treat this in isolation and frequent reference is made to the host's contribution to the interaction. The first chapter considers some of the factors essentially dictated by the host. It includes a brief description of the various host defences and responses to infection, together with a brief consideration of differences in susceptibility of various animal hosts (host-specificity), differences in susceptibility between individuals of the same animal species, and reasons why certain host tissues become preferentially colonized by pathogenic organisms during the course of an infection (tissue-specificity).

It is against this background that the role of the micro-organism is considered *in extenso*. Most micro-organisms are harmless to animals and man: indeed many are highly beneficial, if not essential, to the host or in the natural cycles of nature. The autotrophic micro-organisms which utilize inorganic chemicals as sources of energy and nutrients are completely independent of organic molecules and, therefore, of living hosts for survival, and are never pathogenic. The heterotrophic micro-organisms, in contrast, require organic nutrients of differing degrees of complexity and, among these organisms, all levels of host–parasite interaction are found ranging from complete independence to total dependence upon a living host. This large group of organisms is responsible for degrading dead organic materials in the various cycles of nature. Parasitic micro-organisms, in contrast, have become adapted to grow in association with the living animal or plant and depend to a greater or lesser degree on the living host for essential nutrients. They include non-pathogens or commensals and organisms regularly associated with disease.

Under normal circumstances the healthy animal is in delicate balance with the micro-organisms of its environment, particularly with those which make immediate contact with the exposed surfaces of the body. Usually the two co-exist in equilibrium without causing apparent adverse effects to each other, and possibly benefit one another. Micro-organisms regularly found at the same body site are termed the 'normal' or 'indigenous flora' (Chapter 2). In the healthy individual, by virtue of its defence mechanisms, commensal organisms of the indigenous flora are confined to the external surfaces of the body.

Comparatively few of the large number of parasitic heterotrophs are regularly pathogenic. Parasitism *per se* is not therefore an indication of pathogenicity but the converse is usually true; that is, pathogenic micro-organisms are usually parasitic. Microbial pathogens have unique properties or virulence factors by which they overcome the body defences and actively damage host tissues. The means by which the virulence factors of microbial pathogens can be investigated and quantified are considered in Chapter 3.

Bacterial virulence factors, determined by both *in vitro* and *in vivo* studies, are the subject of Chapter 4. These are considered in the sequence characteristic of an infective process. Those factors which help the organism to establish a primary lodgment in the host are treated first. This is followed by factors which affect the multiplication of the pathogen and its invasion of the tissues leading to the development of overt disease.

In many bacterial diseases the greatest damage to the host is due to the activity of bacterial toxins. The protein toxins (exotoxins) are considered in Chapter 5 and examples of diseases in which exotoxins play an important role are included; bacterial lipopolysaccharide toxins (endotoxins) are the subject of Chapter 6. Part I is concluded by a chapter on viral pathogenicity in which differences between bacterial and viral pathogenicity are emphasized (Chapter 7).

Individuals usually live in groups, herds, or communities of smaller or greater size. This leads to a further level of interaction between infected and susceptible individuals in contact with each other. These community aspects are considered in Part II.

The perpetuation of a microbial species depends upon its ability to move from one individual to another without dying out in the process. Microbes which kill their host run the greater risk unless they can escape from the dead or dying host, survive outside the host, and later gain a foothold in another susceptible one. The manner in which infection is perpetuated in a community and the build-up of reservoirs of infection are considered in Chapter 8. Already differences in susceptibility between individuals have been considered (Chapter 1) but this assumes even greater importance within the community. Apart from differences in immune status, individuals will vary in the degree of exposure to a community infection (the contact probability), in the size of the dose of infection to which they are exposed, in the genetic constitution both of themselves and of the community and various environmental factors which also play an important role.

Outbreaks of infectious disease within a community lead to illness with its accompanying suffering or handicaps and considerable economic losses. Their control is therefore of first importance. Three lines of approach are generally followed: attempts are made to control or eradicate the sources of infection and to limit the spread of infection within the community (Chapter 9), to protect susceptible individuals against infection by artificial immunization (Chapter 10), and prophylactic treatment (Chapter 11). The special case of controlling cross-infections in hospitals is considered in Chapter 12.

Part III is concerned with the subject of 'epidemiology'. From a micro-biologist's point of view epidemiology is much wider than the statistical evaluation of morbidity and mortality rates. It embraces the ecology of the microbial pathogen, its natural history, and the effects produced by the outbreak at local, national, or international levels. In order to bring these various aspects together examples have been selected on the basis of their current interest, their value in illustrating important principles of epidemiological investigation or control, and the way they represent different pathways of infection. A general treatment of human diseases of animal origin—the zoonoses—is first considered (Chapter 13). This is followed by chapters on the epidemiology of foodborne, waterborne, air-borne, contact, vectorborne, and wound infections respectively (Chapters 14, 16, 17, 18, 19, and 20). Other subjects of current interest include brucellosis (Chapter 15), and the epidemiology of mycological diseases (Chapter 21).

PART I

INTERACTIONS BETWEEN THE HOST AND THE PARASITE

Chapter 1

Host defences and responses to micro-organisms

Man and animals are in continuous contact with the micro-organisms of their environment but only certain species are able to colonize the various body sites and, of these, only comparatively few cause overt disease. There are many reasons for these various degrees of interaction. Some are essentially linked with the host; these play a significant role in deciding the nature and extent of the interaction. A summary of the host's contribution to the host–parasite interaction is the subject of this chapter. Others are inherent in the micro-organisms themselves; the contribution made by micro-organisms to the host–parasite interaction is the major theme of the rest of the book.

Different animal species show marked differences in susceptibility to a particular microbial pathogen. Some species succumb to attack while others are partially or totally resistant to the same pathogen. Even within the same animal species individuals may exhibit differences in the degree of response to a pathogen. There are many reasons for these differences, some of which will be considered first.

The healthy animal actively defends itself against microbial attack at different stages in the process of infection. The *non-specific defences* are common to all healthy animals and provide general protection against contamination or assault by micro-organisms. The *specific defences* or *immune responses* are stimulated by the particular pathogen causing the infection; this is a highly efficient host defence. Both types of defence are briefly described.

A. DIFFERENCES IN SUSCEPTIBILITY OF ANIMAL HOSTS TO MICROBIAL PATHOGENS

1. Host-specificity to microbial pathogens

Apart from differences in host response to micro-organisms (see below) the various animal species display either a *natural susceptibility* or an *innate resistance* to the same pathogen; this may be absolute or relative. Certain pathogens exclusively affect man; others affect only animals. Lower animals do not naturally contract syphilis, gonorrhoea, diphtheria, leprosy, typhoid fever, poliomyelitis or measles—infections common to man. Man, on the other hand, is not naturally susceptible to Johne's disease, vibrionic abortion or distemper—diseases found only in various species of animals. These are examples of absolute resistance. In contrast, many pathogens show only relative or very low host-specificity, being able to infect a wide range of animals species; these include the microbial pathogens causing brucellosis, leptospirosis, tuberculosis, anthrax, salmonellosis, tetanus, listeriosis, and psittacosis. Because they are often trans-mitted from animals to man, these are referred to as zoonoses (Chapter 13).

At present, only limited knowledge is available to explain differences in host-specificity. It may be directly related to the ability of the organism to grow at the body temperature of the host. The avian variety of *Mycobacterium tuberculosis*, unlike the mammalian varieties, is able to grow at the higher body temperature of birds. On the other hand, the anthrax bacillus, not being able to grow at the higher body temperature of birds, does not normally cause infection unless the birds are artificially cooled, e.g. by standing in cold water.

Other host-specificities may be linked with exacting nutritional requirements of certain pathogens. This would appear to be the explanation of purine-dependent strains of *Salmonella typhi*, which can only grow in hosts supplying purines. In mice and guinea pigs, which lack this growth factor, purine-requiring strains of *Salm. typhi* are avirulent for these rodents. By injecting purines into these animals in quantities sufficient to satisfy the nutritional requirements of the strains, the organisms prove to be virulent. Naturally occurring purine-dependent mutants of *Salm. typhi*, and of other bacterial species, have been reported (Chapter 4); other examples of host-specificity due to nutrients are considered later.

Another basis of host-susceptibility lies in the ability of the microbe or its products to damage the host. For an animal to be susceptible to a pathogen it must possess a target site for the microbe to attack. For example, injection of the exotoxin of the diphtheria bacillus into rats fails to kill the animal. The unchanged toxin is excreted in the rat's urine, and if a sample of the urine is injected into a guinea pig, the animal dies and lesions typical of those caused by diphtheria toxin (Chapter 5) are produced. This indicates that a sensitive target site is absent from the rat but present in the guinea pig, and also that the toxin excreted by the rat is chemically unmodified.

2. Differences in susceptibility of individuals within the same animal species

Healthy individuals are more likely to resist infection than those suffering with disease, injury or undergoing certain forms of treatment. An attempt is made in Table 1.1 to evaluate the order of host defence efficiency based on the health status of the individual. This is a reasonable ranking order but should not be interpreted too literally.

Apart from health status, individuals within a community vary in their response

Table 1.1   The efficiency of host defences in relation to health

Host state	Host defence efficiency
Healthy adult	High
Adults with local defence impairment, e.g. wounds, especially with tissue damage, skin diseases, foreign body (e.g. obstructed bladder)	↑
Uncontrolled diabetes	
Newborn infant	
The elderly	
General impairment, e.g. malignant disease of reticuloendothelial cells, leukaemia, etc., and by immunosuppressive drugs and corticosteroids	↓ Low

to the same infectious agent. In those diseases in which a well-balanced host–parasite relationship has been established the majority of individuals will be infected subclinically and exhibit no symptoms of illness. Of those which do fall ill most will exhibit the normal range of clinical illness characteristic of the infecting agent. Mild clinical symptoms may indicate greater resistance to infection and, in most instances, this will be due to immunity, or differences in the virulence of the organism (Chapter 3). By prior exposure to the same infective agent, or as a consequence of artificial immunization, active immunity will have been stimulated, thereby providing subsequent protection (Chapter 10).

There are many reasons why individuals of the same animals species may exhibit greater susceptibility than the majority to the same infective agent. In the normal, healthy animal this may be linked with age, sex, or both. Where the individual's health is impaired this is often accompanied by reduced resistance to infection, frequently due to deficiency in the host defences. Many factors can bring this about such as injury (accidental, surgical, or due to insidious damage as may be caused by smoking which leads to chronic bronchitis), malnutrition, environmental stress, organic disease, and various drugs used in therapy. Their effects on host defences are summarized in Table 1.2. Host resistance may be reduced, not only to regular pathogens, but also to organisms not normally pathogenic, and infections by these 'opportunist' organisms are being encountered with increasing frequency (Chapter 2). Selected factors which influence host resistance will be considered.

(a) Age as a factor in disease

Infectious diseases are often more severe at the extremes of life, in the very young, and in the elderly (Table 1.3). The newborn animal is particularly prone to infection, especially if it has not received protective passive immunity from its mother. Many neonatal diseases are alimentary in nature. In the early days of life the gut,

Table 1.2   Normal host defences and factors predisposing to opportunist infections (modified from Klainer and Beisel, 1969, and reproduced by permission)

Altered host defence	Predisposing factors	Possible mechanism
Protection afforded by normal flora	1. Burns, trauma, other infection	May alter normal skin flora by changing skin ecology and physiochemical properties
	2. Surgery	Preoperative or prophylactic antibiotics may alter normal flora
	3. Hospitalization	May colonize host with new or resistant organisms
	4. Antiobiotics	See Table 1.4
Anatomic barriers and secretions	1. Burns, trauma, bites, surgery, other infection, inflammatory diseases	See above; also direct penetration or other disruption of integrity provide new portals of entry for micro-organisms
	2. Extremes of age	May alter physiological defences, e.g. loss of gastric acidity as a protective mechanism against ingested organisms or toxins; depressed cough reflex, deficient ciliary action in respiratory tract, defective clearing mechanism in lung predispose to pulmonary infection
	3. Foreign body, prostheses	May act as nidus for infection, provide new entry portals, or cause obstruction with stasis and infection
	4. Diagnostic procedures	May provide new portals of entry for micro-organisms
	5. Urinary tract and intravenous catheters	May provide new portals of entry; may result in obstruction, stasis; or may act as nidus for infection
	6. Antimetabolites, irradiation	See Table 1.4
	7. Local ischaemia	May alter permeability of skin or mucous membranes to produce new portals of entry
Inflammatory response	1. Diabetes mellitus, renal failure	Accompanying acidosis results in sluggish polymorphonuclear response of reduced intensity, defective leucocyte function, ineffective phagocytosis and lack of fibro-blastic proliferation
	2. Diseases of hematopoietic system	Infiltration of bone marrow may result in deficient and defective granulocytic pool
	3. Antimetabolites, irradiation, corti-costeroids, other drugs	See Table 1.4

Table 1.2. (*continued*)

Altered host defence	Predisposing factors	Possible mechanism
Reticuloendothelial (RE) system	1. Diseases involving the lymphoid or RE systems, e.g. lymphoma, reticulum cell sarcoma	Depress RE system function
	2. Antimetabolites, X-irradiation, corticosteroids	See Table 1.4
Immune response	1. Disease of the lymphoid tissues and RES, e.g. multiple myeloma chronic lymphatic leukaemia	Decrease in normal immunoglobulins, delayed and defective antibody response to antigenic stimuli, and/or production of varying amounts of abnormal immunoglobulins
	2. Hodgkin's disease	Depression of delayed hypersensitivity early, lymphopenia later
	3. Dysproteinaemias	Nature of defect reflects factor(s) which are lacking
	4. Extremes of age	During first 3–6 months dependent on maternal antibody; possible decrease in immune response in elderly
	5. Debilitating diseases, e.g. liver disease, renal failure	May affect immunity via defects in protein synthesis and cell division
	6. Antimetabolites, irradiation, corticosteroids	See Table 1.4

which is free of bacteria at the time of birth, becomes progressively colonized with micro-organisms from the birth canal and, later, from food and the immediate environment. During this period pathogens have a greater opportunity to become established and produce disease. Infections with enteropathogenic *Escherichia coli* or *Salmonella* (e.g. *Salm. dublin*), frequently occur in calves in the first weeks of life. Neonatal diseases of babies may result from infections with human enteropathogenic serotypes of *E. coli* or *Pseudomonas aeruginosa*. Similarly, gut infections of lambs with *Clostridium perfringens* type B, produce lamb dysentery (Chapter 5).

The transfer of passive immunity from mother to offspring is of great importance in protecting the individual in the early months of life until the young animal or baby becomes immunologically competent. In some species transfer of immunoglobulins (see later) occurs wholly or largely *in utero* (man, rats, guinea

Table 1.3.   Age as a factor in infection

Disease	Agent	Host	Age
White scours, neonatal diarrhoea	*Escherichia coli* (specific serotypes)	Cattle, humans	Very young calves, babies
Lamb dysentery	*Clostridium perfringens* type B	Sheep	Lambs
Meningitis	*Neisseria meningitidis, Haemophilus influenzae, Streptococcus pyogenes, Streptococcus pneumoniae*	} Humans	Infants
Whooping cough	*Bordetella pertussis*		
Scarlet fever	*Streptococcus pyogenes*	Humans	Infants
Diphtheria	*Corynebacterium diphtheriae*	Humans	
Measles	Measles virus	Humans	
Mumps	Mumps virus	Humans	Mainly infants and children
Poliomyelitis	Poliovirus	Humans*	
Chickenpox	Varicella	Humans	
Johne's disease	*Mycobacterium johnei*	Cattle, sheep	Adults (infection in early life)
Brucellosis	*Brucella abortus*	Cattle	Adults
Tuberculosis	*Mycobacterium tuberculosis*	Man, cattle, pigs, etc.	All ages but particularly adults
Shingles	*Herpes zoster*	Man	Adults

*Changing age patterns have been experienced over a number of years (p. 142).

pigs, mice, rabbits). About 5–10 per cent of total IgG (immunoglobulin G) crosses the placenta in dogs and cats. In other species (horses, cattle, sheep, goats, pigs, dogs, cats) the immunoglobulins are largely present in the colostrum—the first milk—and ingested during the early hours of life. Protection is limited to infections by strains against which the mother was immune. The importance of colostrum in protection against disease is well illustrated by the problem of scouring in calves. Animals which have received their mother's colostrum may become infected with pathogenic strains of *E. coli* which usually produce scouring only; calves similarly infected, but which have not received colostrum, frequently develop a colisepticaemia, often proving fatal.

Diseases introduced by the respiratory route may occur in children who lack immunity. These include diphtheria, scarlet fever, measles, mumps, and varicella (chickenpox). Some diseases, more prevalent in childhood, may be linked with anatomical ease of infection, as in meningitis which may be caused by a variety of infectious agents including bacteria and viruses (e.g. pneumococci, *Haemophilus influenzae, Neisseria meningitidis*, Coxsackie viruses). Initially an infection sets up a rhinopharyngitis. The nasopharynx is closely connected with the subarachnoid space by means of the 15th and 16th branches of the olefactory nerve which pierces the cribiform plate of the ethmoid. Infection may therefore pass from the nasopharynx *via* the nerve sheath space to the subarachnoid space. Another route of infection, also common in children, is from a middle ear infection by direct extension.

It is rare for an individual infected early in life to contract the same disease a second time; this is due to the immunity previously established. In the elderly, however, certain diseases may develop due to a breakdown of the immune status, particularly cell-mediated immunity, as occurs in, e.g. shingles, or due to anatomical changes. For instance, enlargement of the prostate gland in men leads to urinary stasis which predisposes to infection. Chronic infections, such as tuberculosis, are more frequently seen in the elderly.

(b) *Sex as a factor in disease*

Some diseases of adults are essentially linked with the development of sex organs. In these the dual factors of sex and age render an animal susceptible to specific infections. Hence mastitis and abortion are essentially associated with the adult female and orchitis with the adult male. Calves are relatively insusceptible to infections with *Brucella abortus*, but they can excrete and spread infection. The infection is, however, self-limiting in the calf.

The same organism is responsible in the adult for infections of the placenta leading to contagious abortion in the pregnant cow and infections of the testicles, seminal vesicles, and epididymis in the bull. Apart from tissue susceptibilities of the sex organs (see later), the possible effects of sex hormones on infection must be considered. A seasonal recurrence of pyometra with *Escherichia coli* in maiden bitches has been associated with the oestrous cycle. It is probable that *Corynebacterium pyogenes* endometritis in the cow may have a similar aetiology.

(c) *'Stress' as a factor in disease*

'Stress' is a complex of different factors. Although not well understood it has a very real influence on health. In domestic animals exposure of an animal to undue exertion, shock, transportation, change of environment, climatic change, and other less well-defined environmental factors, increase the incidence of infectious disease and may cause latent infections to become active.

Clinical experience demonstrates the value of rest in the treatment of disease. Fatigue stress, whether nervous in origin or due to muscular activity, increases susceptibility to infection. Rats subjected to fatigue have shown a higher mortality when infected orally with *Salmonella enteritidis* compared with controls. Similarly rats carrying latent salmonella infection may develop full clinical disease when subjected to fatigue. Calves transported from one centre to another often show clinical salmonellosis on arrival due to the flare-up of a carrier state of infection due to the stress of travel. It is well known that the rate of passage of micro-organisms through the gut is greatly increased under stress, and alimentary infections—contracted during transport or time of holding in the abattoir lairage—are excreted much more rapidly than normal. In time of stress the output of cortisone from the adrenal cortex is increased; this suppresses the inflammatory processes of the body and the effect may be deleterious to the host. Frequently animals in transit develop a bronchopneumonia, often due to a primary infection with a parainfluenzavirus followed by secondary invaders, such as *Pasteurella multocida*. This disease often occurs as the consequence of transport stress and is frequently called 'transit fever', 'shipping fever', or 'shipping pneumonia'.

Sudden and prolonged change in temperature, humidity, and radiation also play their part. For instance, excessive exposure to the ultraviolet rays in sunlight has been shown to activate latent tuberculosis. It has been observed that substances present in serum able to destroy bacteria and viruses (e.g. properidin and complement, see later) disappear from the serum following γ-radiation. This is followed by invasion of the blood stream and organs by commensal bacteria from the gut and may be one factor in producing symptoms typical of radiation sickness.

Sudden changes of climate predispose to infectious diseases. The seasonal incidence of respiratory diseases in man is well known. In sheep there is a seasonal incidence of braxy—a clostridial invasion of the abomasal wall leading to high mortality—this coincides with cold and frosty weather and may be linked with the eating of frosted herbage. Pasteur and his colleagues found that natural resistance to anthrax could be overcome by immersing fowls up to their thighs in cold water. Low temperature lowers the phagocytic response (see later) and the anthrax bacilli proliferate.

(d) *The effect of diet, malnutrition, and husbandry on host resistance*

It is extremely difficult to assess the importance of these factors, particularly in human populations where experimentation is forbidden on ethical grounds; there is also the difficulty of considering diet in isolation from other debilitating factors in the environment. However, the frequent association of famine with 'pestilence' has been recognized throughout the history of the human race. Today in underdeveloped countries, where animal protein is in very short supply, ill health is prevalent, and this is accompanied by high levels of infectious disease, lower manpower efficiency, and high infant mortality rates. Diseases, such as tuberculosis, are often prevalent.

A diet of one kind of food over a long period of time, such as one consisting solely of fruit and vegetables, often predisposes to a number of infections. This is particularly evident in certain African tribes. Infectious disease may be linked with vitamin, as well as protein, deficiencies. In contrast, sudden changes of diet or excessive eating may play a role in stimulating latent infections to become active. A number of clostridial diseases of sheep are associated with changes of pasture feeding and it is among the 'greedy eaters' that infection is most prevalent. This may be due to packing of food in the various stomachs of the sheep resulting in anaerobiosis which encourages the growth of clostridia. It is well recognized that overfed babies are less healthy than those underfed. Overfeeding results in obesity and deposition of fats in tissues, each of which may predispose to vascular disease in later life and reduction in the expectation of life.

(e) *Intercurrent disease as a factor in host susceptibility*

The normal defences of the host body are frequently impaired by organic disease (Table 1.2). Infectious diseases are common in patients suffering from debilitating diseases, such as leukaemia, multiple myeloma and Hodgkin's disease. Conditions such as diabetic acidosis, renal and heart failure are often complicated by infections with opportunist organisms (Chapter 2). Frequently the inflammatory response is delayed or suppressed, or the patient lacks a normal response to the antigenic stimulus of the infection either by failing to respond at the normal rate or in an abnormal production of immunoglobulins.

(f) *Therapy as a factor in host susceptibility*

Certain modern therapeutic procedures often used in the management of the diseases mentioned above can render an individual more susceptible to infection. Under these conditions not only pathogens but organisms of the normal flora (Chapter 2) and 'non-pathogens' in the host's environment may be able to initiate infection. These man-imposed effects are termed iatrogenic diseases, and are summarized in Table 1.4. Corticosteroids, for instance, suppress all phases of the inflammatory response and thereby reduce the efficacy of this important non-specific defence. Adrenocorticol hormones are known to suppress the formation and activity of interferon. The various immunosuppressive drugs used in surgery to prevent transplant rejection, and the cytotoxic drugs and γ-rays used to destroy selectively cancerous cells, all depress bone marrow function (lymphopoiesis) and that of the reticuloendothelial system. Cytotoxic drugs also injure rapidly growing tissue cells, such as those of the intestinal mucosa, thereby creating new portals of entry for opportunist infections.

Despite scrupulous aseptic techniques, opportunist organisms frequently localize and thrive on diseased endocardium, endothelium and on intravascular foreign implantations, such as cardiac valve replacement and vascular prothesis. Another problem of increasing importance is bacteraemia, leading to life-threatening bloodborne diseases in patients with defective host defences. Infections may be introduced by intravenous catheters, the administration of

Table 1.4. Therapy which may reduce the efficiency of host defences (after Klainer and Beisel, 1969 and reproduced by permission)

Agent	Mechanism
Corticosteroids	1. Suppress the inflammatory response 2. Depress antibody formation 3. Suppress formation and activity of interferon 4. Others not yet characterized which allow 'non-pathogens' to multiply *in vivo*
Irradiation	1. Depresses bone marrow to reduce granulocyte and macrophage production 2. Suppresses the reticuloendothelial system 3. Depresses antibody formation in part by inhibiting the induction phase of the immune response and by reducing the mass of antibody-forming tissue 4. Injures other rapidly growing cells, e.g. mucosa of gastrointestinal tract, to produce ulceration, bleeding, and subsequent portals of entry 5. May produce chromosomal changes in leucocytes 6. May modify preformed protein so as to alter its specificity and role in immunity
Cytotoxic drugs (antimetabolites)	1–4. As listed under 'Irradiation'
Antibiotics	1. Alter the normal microbial flora of the various body sites (Chapter 2) 2. May result in selection of resistant species or strains 3. Encourage growth of some fungi 4. May result in quantitative changes in serum immunoglobulins

intravenous solutions or the self-administration of drugs, particularly by addicts using non-sterile syringes.

Despite the life-saving and economic value of antibiotics these too have their own serious disadvantages. They suppress the normal bacterial flora (Chapter 2) which reduces competition between the normal flora and the infecting pathogens thereby facilitating infections, and also reduce vitamin synthesis. Tetracycline, in particular, is reputed to stimulate the growth and virulence of *Candida albicans*. Other disadvantages are considered elsewhere (Chapter 2).

3. Range of disease caused by the same microbial pathogen

In one animal species the same range of disease is invariably produced by a particular pathogenic microbe. With some pathogens this is relatively narrow; with others a wide range is experienced. In man, for instance, *Staphylococcus aureus (pyogenes)* produces well-defined abscesses, either at the site of primary lodgement of infection (Chapter 4), or in various organs of the body (kidney, brain, etc.) to which the organism may have been carried in the blood stream from the initial focus of infection. *Streptococcus pyogenes*, too, may remain localized at the site of infection, as in tonsillitis and puerperal fever, or spread through organized tissues, as in scarlet fever, erysipelas, and cellulitis.

Some microbial pathogens produce disease differing both in form and severity in different animal species. Anthrax, for instance, is chiefly a disease of herbivores (cattle, sheep, horses, etc.) but can also affect man. Anthrax infections usually take the form of a peracute or acute, fatal septicaemia in cattle; a less acute but sometimes fatal invasion of the pharynx in pigs and dogs; and a localized lesion (malignant pustule), formed at the site of entry, in man. Birds are generally insusceptible, probably as a result of their higher body temperature.

In certain host–parasitic relationships a pathogen may show a wide range of disease within the same animal species. These differences are most probably due to variation in virulence of the organism (Chapter 3) but may also be due to differences in immunity of the host (Chapter 8). *Erysipelothrix rhusiopathiae* is an important pathogen in porcine pathology. In the pig this organism may cause a peracute or acute septicaemia, or a mild urticaria characterized by diamond-shaped red patches of haemorrhagic necrosis in the skin (diamond skin disease). Chronic forms of the disease, also experienced, are characterized either by arthritic joint infections or progressive endocarditis in which large vegetations develop on the heart valves and seriously impair their function.

From these examples two observations may be made: firstly many pathogens are relatively host–specific; and secondly, within an animal host, many pathogens attack particular tissues or organs and are therefore tissue-specific. These points will now be considered.

4. Tissue-specificity of microbial pathogens

In many microbial diseases the pathogen exhibits maximum growth in particular organs or tissues with corresponding degrees of damage. The primary site of infection may show only minimal pathological changes but spread of the pathogen by lymph or blood is followed by characteristic lesions developing in tissues remote from the portal of entry. Examples of these tissue-specificities are given in Table 1.5.

A number of factors may contribute to tissue-specificity; they include the route by which the pathogen enters the host body and the differences between tissues in inhibiting or supporting multiplication of the pathogen.

(a) *The route by which the pathogen enters the body*

Sources of infection may be divided broadly into two; namely, exogenous and endogenous. In the former the pathogen enters from outside the body either directly by cross-infection from a diseased animal, e.g. by inhalation, or indirectly,

Table 1.5. Examples of tissue specificities by different microbial pathogens in man and animals

Pathogen	Host	Tissue	Disease
BACTERIA			
Corynebacterium diphtheria	Man	Throat	Diphtheria
Proteus spp.	Man	Kidney medulla	
Neisseria gonorrhoeae	Man	Epithelium of urino-genital tract	Gonorrhoea
Shigella spp.	Man	Mucosa of large intestine	Dysentery
Mycobacterium johnei	Cattle, sheep	Intestinal mucosa	Johne's disease
Corynebacterium renale	Cow	Kidney medulla	
Brucella abortus	Cow	Placenta	Abortion
	Bull	Testis	Orchitis
Fusiformis nodosa	Sheep	Hoof tissue	Foot rot
Leptospira ictero-haemorrhagiae	Man, rat, etc.	Kidney	Leptospirosis
VIRUSES			
Poxviruses			
variola virus	Man	Skin	Smallpox
vaccinia virus	Cattle	Skin	Cowpox
varicella virus	Man	Skin	Chickenpox
pig pox	Pig	Skin	Pigpox
sheep pox	Sheep	Skin	Sheep-pox
Foot and mouth virus	Cattle	Mouth and feet	Foot and mouth disease
Encephalitis	Man, horses, etc.	CNS	
Rabies virus	Dog, man, vampire bat	CNS	Rabies
Louping ill virus	Sheep		Louping ill
Poliovirus	Man	Gut epithelium and CNS	Poliomyelitis

CNS = Central nervous system.

e.g. by ingestion of contaminated food or water, or by contact with a contaminated object. Endogenous infections arise by pathogens already present in the individual, such as *Staphylococcus aureus* carried in the nose or on the skin, or pneumococci, resident in the oral cavity. A predisposing cause often initiates infection by an endogenous microbial pathogen; this may be àn abrasion or burn on the skin or a malfunctioning sweat gland (see later).

Usually pathogens are unable to establish themselves in the host unless they are introduced via a specific route. The portal of entry is therefore of considerable importance in the initial stages of infection (Table 8.4). Spores of *Clostridium tetani* may be carried in the animal gut without tetanus developing: if, however, they are introduced into the tissues during surgery of the gut, or in a wound caused by an instrument contaminated with faecal excreta containing the spores, they may germinate under anaerobic conditions and produce the neurotoxin which causes symptoms of tetanus to develop. In contrast, natural infections of man by enteric pathogens, such as *Salmonella typhi*, *Vibrio cholerae*, and *Shigella dysenteriae*, are initiated only after ingestion.

The major portal of entry for the same pathogen may sometimes vary between animal hosts. In cattle, anthrax bacilli usually infect via the gut when ingested in compounded feed incorporating contaminated bone meal; in man, the spores are usually introduced through an abrasion on the face, neck or arms of porters, butchers or dockers carrying imported, contaminated animal hides over their shoulders. Before the treatment of wools with disinfectants prior to being sorted fatal cases of human anthrax (woolsorter's disease) resulted from the inhalation of spores.

The site of primary lodgement (Chapter 4) is largely determined therefore by the portal of entry. By virtue of their biological characters some diseases remain localized near to the point of entry into the host. Thus staphylococci, leprosy bacilli, and dermatophyte fungi localize in the skin; respiratory pathogens in the lungs and upper respiratory tract; and enteric pathogens in the gut. Pathogens which lead a parasitic existence in the nasopharynx enter the tissues by the respiratory route, and consequently show a similar distribution of lesions. *Corynebacterium diphtheria* and *Streptococcus pyogenes* (Lancefield type A) lodge and multiply in the human tonsil, although streptococci tend to spread more extensively. Other pathogens (pneumococci, *Haemophilus influenzae*, *Staphylococcus aureus*, etc.) spread to the lungs and cause pneumonia. Some pathogens (pneumococci, *Neisseria meningitidis*, *H. influenzae*, etc.) spread from the nasopharynx to the meninges of the brain, giving rise to meningitis.

Dependent both on the nature of the pathogen and the local anatomy, particularly the direction of lymph flow draining the area, many pathogens are carried from the primary site to remote parts of the body where they localize and multiply. The usual site of the primary lesion (chancre) of syphilis is determined by the venereal mode of infection but the sites of secondary and tertiary stages, so highly characteristic of the disease, are found in the mouth, throat and skin, in the genital and anal regions, and in the formation of gumma in arteries, brain, liver, and bones. In syphilis, therefore, the ultimate location of diseased tissues bears no relation to the site of primary infection. Many other examples could be quoted to illustrate this point. It must be concluded, therefore, that apart from infections which characteristically remain localized, the route of entry does not play a major role in tissue-specificity.

(b) *Differences between tissues*

Certain tissues are known to be inhibitory to micro-organisms. Microbial pathogens reaching these will not be able to multiply and consequently the tissues will not become colonized. Other tissues, not exhibiting inhibitory properties, may become colonized and support the growth of the microbial pathogen. These tissue differences may play a part in the phenomenon of tissue-specificity.

The other possible basis for specificity is the presence of growth metabolites in certain tissues which support multiplication of the pathogen. Two examples have been widely investigated. The first is concerned with infections of the kidney medulla in man by *Proteus* spp. and in cattle by *Corynebacterium renale*; the second is of infection by *Brucella abortus* involving the foetal tissues of susceptible hosts. These two examples form models for future work and will be considered in some detail.

(i) *Infections which localize in kidney tissue.* Many species of bacteria are known to infect kidney tissue of both man and animals. These are chiefly Gram-negative rods including *Proteus* spp., *Pseudomonas* spp., and *Escherichia coli* but also a few Gram-positive bacteria—namely enterococci and *C. renale*. Within this range of organisms some produce infections which are of relatively short duration, self-limiting, and rapidly heal. Others, caused by *C. renale* and *Proteus* spp. (particularly *P. mirabilis*) produce more severe pyelonephritis resulting in irreversible damage to the kidney medulla in cows and man respectively. Investigations have shown that the greater severity of these infections is due to the ability of both organisms to produce copious quantities of the enzyme urease.

C. renale is a specific cause of pyelonephritis in adult, female cattle. Infected animals pass bloodstained, alkaline urine in which many pus cells and the causal organism may be demonstrated. Both kidneys may be enlarged, the ureters dilated, the walls thickened, pus and debris are present in the urine due to necrosis of the kidney medulla, where the primary focus of infection occurs. A cystitis, thought to be due to alkaline irritation, frequently accompanies this condition.

C. renale was found to infect the kidneys of 70 per cent of mice inoculated by the intravenous route and the organism was regularly found to multiply in the kidney tissue but not in the lungs and spleen; other corynebacteria caused kidney infection in only 14 per cent of challenged mice. Biochemical studies showed that urea, present in the kidney tissue and in urine, was the substrate utilized by the organism. *C. renale* produces copious quantities of urease which splits urea to release large amounts of ammonia, which renders the urine strongly alkaline and irritant. It would appear that the ability to produce urease is probably the most important single factor in the persistence of this infection in cattle. However, this does not entirely explain the host-specificity of the strain, the fact that female and not male cattle are affected, or why other corynebacteria (e.g. *C. ovis*), which also produce urease, do not infect kidney tissue.

The ability to produce urease is also considered to be the principal reason why *P. mirabilis*, in contrast to the non-urease producing *E. coli*, is able to grow better in kidney cell tissue cultures, and produce more severe lesions both in experimental rats and in the natural disease occurring in man (pyelonephritis). In experimental animals, using fluroescent microscopy, it has been shown that *P. mirabilis* grows intracellularly and invades the kidney tubules. Tissue-specificity and toxicity is thus attributed to the ability of these organisms to utilize urea as a specific metabolic substrate in the kidney.

(ii) *Infections which localize in the placenta/foetal tissues.* Brucellosis is a disease presenting a range of features of great interest in the study of pathogenicity. It is primarily an acute disease of ungulates (including the cow, sheep, goat, and pig). Within these domestic animals, both male and female, there is a distinct tissue-specificity. The pregnant female is the most vulnerable; infection results in an acute placentitis with subsequent abortion. In males, infection is localized in the genitalia, particularly in the seminal vesicles. Infected males may transmit the organism venereally to susceptible females at coitus. In non-ungulate animals (including man) the infection takes a chronic course being chiefly confined to the lymphatic system and causing periodic fever (hence the name undulant fever in man) and Type 4 hypersensitivity (Table 1.8). The latter may be stimulated by the intracellular nature of brucellae infections. Apart from these various forms of host- and tissue-specificity there is a close relationship between animal host and microbial species. Thus *Brucella abortus* is chiefly found in cattle, *Br. melitensis* in goats and *Br. suis* in pigs. Each of these species may infect other hosts, e.g. man, but the principal reservoirs of infection are in the primary hosts.

The biochemical basis for the host- and tissue-specificity has now been established (for a useful summary see Keppie, 1964). Based on the observation that the acute disease occurred in the female, pregnant cow, and that the main pathological changes were located in the lumen of the uterus, the distributions of the pathogen in the various tissues and organs of the cow were investigated. Pregnant cows were inoculated intravenously with a virulent strain of *Br. abortus* and 3–4 weeks later, when the animals were about to abort, were sacrificed. Quantitative estimates of the numbers of organisms in each tissue were determined. The total yield of organisms in each pregnant cow ranged from 0.3 to 1.5×10^{14} but 90 per cent of the organisms were found on the foetal side of the placental barrier (see Table 1.6). Most of the infection was in the foetal tissues and this correlated with the pathology of the disease which indicated a placentitis within the uterus.

The tissue-specificity within the foetal tissues was thought to be due to the presence of a metabolite for brucellae in these tissues. This was supported by the finding that sterile saline extracts of foetal tissues increased the yield of organisms when added to 'standard' media in the laboratory. Parallel studies also showed that a substance fractionated from the allantoic and amniotic fluids of a cow about to abort also stimulated the intracellular growth of brucellae in bovine leucocytes. The fraction was dialysable, therefore of small molecular weight, and subsequently shown to be the carbohydrate erythritol. The optical isomer, D-threitol, was inactivated at ten times the concentration at which erythritol was present in foetal tissues.

The tissues which were shown to have brucella growth-stimulating properties

Table 1.6. Distribution of *Brucella abortus* in the tissue of experimentally infected pregnant cows when abortion was imminent (after Keppie, 1964, and reproduced by permission of Cambridge University Press)

Tissue	Approx. total no. ($\times 10^{12}$) of *Br. abortus* per tissue of cows					Approximate percentage
	I	II	III	IV	V	
Foetal tissue						
Cotyledon (placenta)	63	90	140	23	27	64
Allantoic and amniotic fluid	0.4	0.9	3.7	7.9	1.2	27
Chorion	5.6	2.1	9.9	0.6	3.3	9
Spleen, kidney, lung, brain	In no cow did the number exceed 0.1×10^{12}					
Adult tissue						
Caruncle (placenta)*	4.4	11	13	2.4	5.1	
Uterine mucosa	—	—	3.0	2.4	0.4	
Spleen, lung, iliac gland, kidney, udder, thoracic gland, liver, muscle	In no cow did the number exceed 0.1×10^{12}					

*It was impossible to separate completely all the foetal tissues from the maternal caruncle; this is the reason for the apparent but low infection of the maternal attachments.

were also those which had significant levels of erythritol and these were limited to the foetal tissues which were most prone to heavy infection. Further support for the growth-stimulating properties of erythritol was demonstrated by injecting infected newborn calves for 1 week with erythritol; this enhanced spleen counts by up to 30,000-fold. Erythritol was also demonstrated in the bovine seminal vesicles and to a lesser extent in the testes—sites where infection becomes localized in the bull; it was absent in the same tissues from insusceptible animal hosts.

Parallel observations were made with *Br. melitensis* and *Br. suis* and the conclusion was reached that erythritol is without doubt the cause of host- and tissue-specificity of the three important species of brucellae in their respective hosts and explains the localized nature of the characteristic placentitis or orchitis in the respective sexes. Further, erythritol was shown to be metabolized by fully virulent, and not by attenuated, strains, and therefore is directly related to pathogenicity in susceptible hosts.

B. GENERAL DEFENCES OF THE HOST

From considering the innate susceptibility or resistance of an animal to infection we turn to the means by which the body defends itself against microbial attack. Two broad types of host defences are recognized—the general or non-specific defences and the immune or specific defences.

Non-specific, or *general defences*, are of two kinds: the preventive defences, which rely on the structural integrity and proper physiological functioning of the body tissues; and the inflammatory response of the tissues to the stimulus resulting from injury caused by the infecting micro-organism.

1. Anatomical defences

The structural integrity of the body surfaces forms an initial barrier to colonization by micro-organisms and their subsequent penetration into the subdermal tissues. Although some organisms are introduced into the deeper tissues by the bite of a vector or through a wound, the majority of infectious agents impinge on the skin or the mucous membranes of the oral cavity, the upper respiratory, alimentary, and urogenital tracts. Where these structures are intact and function adequately they usually form an effective barrier to initial lodgement of micro-organisms.

(a) *Skin*

The skin is not only a physical barrier but a living tissue actively secreting bactericidal substances, such as lactic and fatty acids, from sweat and sebaceous glands and other sites. The surface cells are being constantly replaced. A break in the integrity of the skin surface, as caused by minor abrasions, may allow access of highly invasive organisms (e.g. leptospirae, pathogenic staphylococci and anthrax bacilli) with which the host may be in contact. A non-functional sweat gland may become colonized by *Staphylcoccus aureus* leading to the formation of a local abscess. Debilitation of the biological activity of the skin, e.g. by burns or inflammation (Table 1.2), may lead to infection by opportunist organisms (Chapter 2), such as yeasts and *Pseudomonas aeruginosa*.

(b) *Nose, nasopharynx, and respiratory tract*

Micro-organisms are inhaled through the nose in droplets of moisture or on dried dust particles of organic origin. Most of these are filtered out by inhalation through the tortuous passages ('turbinate baffles') of the anterior nares and fine hairs with which they are lined. Those which pass this barrier may stick to the mucous surfaces of the nasopharynx or trachea and be actively removed to the mouth by the ciliated epithelium lining this region. Organisms thus trapped are swallowed and destroyed in the stomach. Mucous secretions contain substances which can compete with cell surface receptors for viral neuraminidase and thereby inhibit penetration of certain viruses into cells.

(c) *Mouth, stomach, and intestinal tract*

Micro-organisms entering by the oral route, possibly more than any other, have to compete with a large and well-adapated normal flora (Chapter 2) of the mouth and of the intestinal tract. Those swallowed in food or drink upon reaching the stomach are immediately confronted with acid secretions and the various gastric juices. Many are destroyed at this stage. Those which survive and pass out of the stomach must compete with the ever-increasing mixed aerobic and anaerobic flora of the small and large intestine. Prior to the establishment of the normal alimentary flora (Chapter 2), the neonate is more susceptible to alimentary infections than adults, particularly with specific serotypes of *Escherichia coli*, *Salmonella* spp. and, in lambs, *Clostridium perfringens* type B.

(d) *Urogenital system*

The flushing mechanisms of sterile urine from the normal kidney usually maintain the bladder free of micro-organisms. This natural defence may be bypassed where organisms from the external surface are introduced mechanically into the bladder by the insertion of a catheter. They may enter also between an indwelling catheter and the wall of the urethra. If they become established and multiply, they may cause cystitis. Infections of the genitalia, such as gonorrhoea and syphilis in man, and vibriosis and brucellosis in cattle, are often introduced venereally.

(e) *Conjunctiva*

Usually the surface of the eye is free of micro-organisms by virtue of the continuous flushing of the surfaces with tears from the lachrymal glands. The enzyme lysozyme, present in these secretions, is able to destroy Gram-positive bacteria. Lysozyme is a low molecular weight, basic protein which, as a mucolytic enzyme, attacks bacterial cell wall mucopeptide structures. Despite this protective mechanism infection can be initiated by accidental splashing of pathogens into the eye, either in the laboratory or as a natural infection—as sometimes occurs in stockmen handling live animals infected with *Brucella abortus* (Chapter 15). Indeed the term 'inocula' derives from this route of infection (in-ocula).

2. Tissue bactericides

One of the features of certain virulent organisms is their ability not only to colonize one or other of the body surfaces, but also to breach the preventive defences and invade the tissues. They then encounter a further line of non-specific defences in the tissue bactericides present in body fluids and organized tissues.

(a) *Bactericides in body fluids*

Extracellular tissues fluids, especially blood plasma, contain bactericides which play a distinct role in tissue defence. These bactericides are unrelated to each other either chemically or biologically. The presence of lysozyme in lachrymal secretions has already been mentioned but this enzyme is also present in most body secretions (except cerebrospinal fluid, sweat and urine) and enhances the activity of complement (see later).

Bactericides in blood, plasma, or serum can be readily demonstrated by their inhibitory effects on the growth of micro-organisms. In diagnostic bacteriology this factor must be taken into account when attempting to isolate pathogens from the blood of a patient. The blood must be diluted with culture broth, beyond the inhibitory level (usually ten-fold) to allow growth to occur.

At least two groups of bactericidal substances in blood have been characterized. Complement, a non-specific component of normal serum, combines with micro-organisms coated with specific antibody (see below) and, in the case of Gram-negative bacteria, brings about their death or lysis. Its most important role is therefore in the immune animal. A second group of serum components, found in some animal hosts, are the β-lysins. They are relatively thermostable (i.e. they are not inactivated by temperatures of 56–60°C after 30–40 minutes). In the serum of rats β-lysins have been shown to be anthracidal, i.e. bactericidal to anthrax bacilli.

(b) Bactericides in organized tissues

The organized tissues of the body are frequently found to be inhibitory to microbial multiplication. Extracts of spleen, brain, kidney, thymus, and pancreas are actively bactericidal against a wide range of organisms including *Staphylococcus aureus*, *Bacillus anthracis*. *Escherichia coli*, and *Streptococcus pyogenes*. The bactericidal agents have been shown to be basic proteins or polypeptides. Their presence in tissues may account, at least in part, for the resistance of particular organs to colonization by infecting organisms. On the other hand their absence in other organs may account for the ability of some pathogens to colonize them. These properties account, at least in part, for the tissue-specificity of some microbial pathogens.

3. The inflammatory response

The entry of an injurious agent, either infectious or non-infectious, into the tissues of a living host produces an inflammatory response (Figure 1.1). Inflammation is a dynamic process. It may range from an acute exudative inflammation, where exudate and cells collect in injured tissues to defend against further damage, to chronic productive inflammation, where repair and inflammation occur simultaneously. In acute inflammation the local area becomes red, swollen, and painful. This is due to vascular changes and leakages of fluid and cells from the blood vessels. The principal vascular changes are increased blood flow and permeability of the capillary walls allowing leucocytes and, where much damage has occurred, erythrocytes, to pass from the vessels into the tissue spaces. These

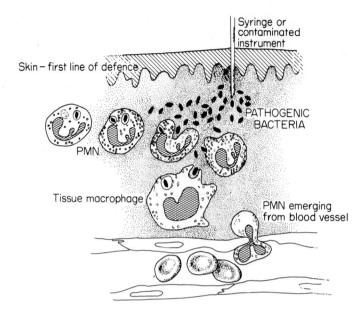

Figure 1.1. Schematic representation of phagocytosis by polymorphonuclear leucocytes (PMN) and tissue macrophages following penetration of the skin and introduction of pathogenic
bacteria into deeper tissues. The PMNs are more efficient in
phagocytosis than the macrophages. Note the PMNs are mobilized
into tissues from blood vessels during the inflammatory response
(after Bellanti, 1971 and reproduced with permission)

changes are brought about by a variety of substances including proteases (e.g.
plasmin, an enzyme-lysing fibrin), polypeptides (e.g. leukotaxine and bradykinin,
causing increased permeability) and pharmacologically active amines (e.g.
histamine, often associated with hypersensitive and allergic inflammatory reactions). The exudate which leaves the blood vessels as a consequence of increased
permeability has a high protein content resembling plasma, contains fibrinogen
which forms fibrin, and many cells. The first to appear, and the most dominant,
are neutrophils—later macrophages become increasingly involved.

Neutrophils or polymorphonuclear leucocytes circulate in the blood stream of
normal animals and are found only in small numbers in connective tissue spaces.
In an acute inflammatory response they actively migrate through the endothelial
cell junctions in large numbers as part of the inflammatory exudate. As long as
infection persists they continue to migrate to the focus of infection and ingest or
phagocytose the foreign agents. Neutrophils, engorged with bacteria, usually die
and form pus.

Macrophages are cells of the reticuloendothelial system, and readily
phagocytose foreign particles, such as bacteria. In connective tissue they are

called histiocytes (or resting wandering cells) and in the blood stream, monocytes. In addition to phagocytosing infective agents they also process the antigenic components of the infective agents and convey them to lymphoid tissue where they initiate an immune response in the immunocytes (see later).

Both exudate and its cellular components are important in overcoming an infection.

In addition to bactericidal properties, the exudate dilutes toxins produced by the invading bacteria, carries to the site of infection antibodies which neutralize the toxins or combine with bacteria rendering them more susceptible for phagocytosis, and supplies plasma factors (e.g. complement) which take part in bactericidal processes and promote the flow of blood cells to the injured site. The main function of the leucocytes in the inflammatory process is to phagocytose the foreign bodies (e.g. pathogenic bacteria) and to clear the wound of dead tissue debris, both being prerequisites of the healing process. The cellular responses involved in inflammation come within the discipline of cell pathology and one of the excellent texts listed at the end of the chapter should be consulted for a fuller account. Of all the defence mechanisms of the animal host the inflammatory response is, without doubt, the most important mechanism for dealing with a microbial infection. However, as indicated below, the animal may fail to survive unless the specific mechanisms come to its aid.

C. SPECIFIC HOST DEFENCE MECHANISMS (ACQUIRED IMMUNITY)

The non-specific defence mechanisms, so far considered, play an important role in protecting the host against an infectious disease. Often, however, they are insufficient to protect the host against overt pathogens. Other defences are essential to enable the host to overcome the infecting parasite. These are the immune responses of the host to the infecting agent. Children, for instance, suffering with a deficiency of the immunological responses (e.g. agammaglobulinaemia, i.e. an inability to produce antibody) die prematurely from an infection and rarely reach adult life. It may be concluded, therefore, that whilst pathogenic microbes may overcome the non-specific defences, they are usually susceptible to the specific immune responses once these have developed.

The first requirement of the immunological system is to be able to recognize substances (antigens) which are foreign to the host. Heterologous antigens (i.e. not-self), in contrast to antigenic components present in the hosts tissues (self), provoke a response by the immune system. In addition to being foreign, antigens are substances of variable molecular weight, and consist of protein, polysaccharide, or lipid, or combinations of these. Some compounds of these chemical classes are good antigens, notably those of high molecular weight; others are poor. Many substances, like blood serum, tissue transplants and micro-organisms, possess antigenic determinants. Microbial cells, and many of their products, are usually strongly antigenic. By virtue of its complex structure a single micro-organism consists of multiple antigens. The host will respond to each antigen able to come into intimate contact with its immune system. For this to occur the antigens must be exposed at the microbial surface (e.g. capsular, flagellar and somatic antigens) or be able to diffuse away from it (e.g. exotoxins).

Antigens stimulate the immune system by producing humoral antibody, cell-mediated immunity or both. The responses are associated with two subpopulations of lymphocytes derived from primitive bone marrow cells (Figure 1.2). Lymphocytes responsible for humoral or antibody-mediated immunity are processed by lymphoid tissue. In chicks this tissue is present in the bursa of Fabricius near the anus, while in mammals it is thought that lymphoid tissue also associated with the gut, e.g. Peyer's patches, is responsible. Lymphocytes processed by these tissues are termed B lymphocytes, since the early work on chicks demonstrated the involvement of the bursa. Under antigenic stimulus B lymphocytes transform, first into large blast cells, then into smaller ones, and finally into antibody-secreting plasmacytes. Plasmacytes synthesize large amounts of specific immunoglobulin which corresponds stereochemically to the stimulating antigen; the antibody is usually released into the body fluids. Some B lymphocytes retain an immunological memory of the stimulating antigen by which the same antigen is recognized when subsequently introduced into the host, and a rapid synthesis of the specific antibody ensues—the anamnestic response.

Lymphocytes responsible for a cell-mediated response are processed by the thymus gland. Thymus-derived or T lymphocytes also cooperate with B lymphocytes in the production of antibody. Under antigenic stimulus T lymphocytes transform into immune intact cells; these cells are committed to recognize the specific antigen. They have a protective role in recognizing and eliminating somatic cells whose surface antigens are foreign to the host. They include tissue transplants, which may be rejected by 'killer' lymphocytes; tumour cells, which arise by cellular mutation and are foreign to the host in that they differ from the normal cells; and macrophages which are parasitized by micro-organisms, as often occurs in chronic infections. Since sensitized whole cells are involved in this type of reaction, the protection given to the host is termed cell-mediated immunity. It provides the host with a form of immunological surveillance.

Which of the two types of immune response occurs to a microbial infection is decided, in part, by the nature of the stimulating antigens. Microbial antigens which stimulate B lymphocytes include bacterial structures (such as pneumoccocal polysaccharides) and bacterial products (such as exotoxins). Those which stimulate T lymphocytes are often microbial pathogens causing chronic infections, particularly those involving intracellular parasitism, such as viruses, salmonellae, brucellae and mycobacteria. Many microbes stimulate both antibody and, to a lesser degree, cell-mediated immunity. The presence of adjuvants (Chapter 10) enhances both types of response.

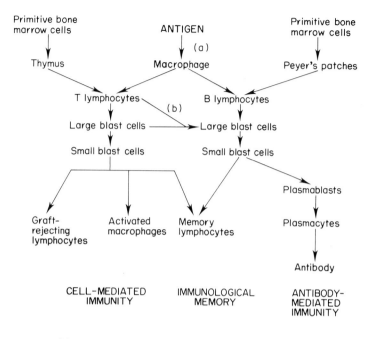

(a) Preparation of large antigen molecule by macrophage RNA.

(b) Cooperative activity.

Figure 1.2. Cellular involvement in the immune responses (modified from Allan, 1980)

1. Humoral antibody

Immunoglobulins (γ-globulins) consist of a group of serum proteins which can be separated from other serum proteins by their distinctive ability to migrate in an electric field. A number of γ-globulins are recognized in man—at least five are identified, namely IgM, IgG, IgA, IgD, and IgE. Apart from physical and chemical differences they exhibit unique biological activities (Table 1.7) and their synthesis proceeds at different stages and at different rates throughout the course of an infection.

Immunoglobulin M (IgM) is the first immunoglobulin to appear in the blood stream during the course of an infection. It is a much larger molecule than IgG and, as a consequence, is confined largely to the blood stream and cannot traverse the placenta barrier to the foetus *in utero*. It is important in giving the host protection against bloodborne infections—IgM deficiency is often associated with septicaemia. Of all the immunoglobulins it is the most efficient in agglutinating particulate antigens. This renders antigens more susceptible to phagocytosis by

Table 1.7. Characteristics of various immunoglobulins

Symbol	IgM	IgG	IgA	IgE
Molecular weight	900,000	160,000	160,000 +	Reagins
Time of production in primary infection	Early	Late	?	?
Major component after secondary stimulation	–	+	–	–
Transport across placenta	–	+	–	–
Function	Agglutinins, opsonins, lysins, complement-fixing (+++)	Precipitins, antitoxins, complement-fixing (+)	Secretory at mucous membranes	Reagins
Fix to skin or other cells	+	–	–	+++

+++ = all IgE is cell-fixed.

macrophages (see below). In the presence of complement agglutinated, whole cells are damaged, killed or lysed (cytolysis or, in the special case of red blood cells, haemolysis).

Immunoglobulin G (IgG) has a relatively small molecular size and can cross the placental barrier from mother to foetus in humans and to a lesser degree in dogs and cats, thereby giving protection to the newborn against infections to which the mother was immune. The half-life of IgG in the newborn is of the order of 21–23 days and the titre of maternal IgG falls off over the first 3 months of life, but by the end of this period the infant is synthesizing its own IgG which compensates for loss of maternal antibody. For this reason it is unwise to vaccinate before 12 weeks of age, e.g. puppies with distemper vaccine. IgG can also diffuse out from the blood stream into extravascular spaces and is therefore the immunoglobulin most commonly present in these sites. Large amounts of IgG are produced rapidly following a second or subsequent stimulation by the same antigen.

Immunoglobulin A (IgA), unlike other immunoglobulins, is synthesized locally, being secreted in the seromucous secretions of the body including saliva, tears, nasal fluids, colostrum, and secretions of the lung and gastrointestinal tract. IgA immunoglobulins which bathe the gastrointestinal surfaces are frequently termed 'copro-antibodies'. They act as a protective coating by protecting the mucous surfaces against microbial adhesion.

Certain plasma cells synthesize immunoglobulin E (IgE). The majority of IgE is bound to tissue cells, particularly mast cells, and very little is detectable in serum. Upon exposure to specific antigen the granules of the mast cells disappear, releasing pharmacologically active substances such as vasoactive amines, e.g. histamine. In hypersensitive patients these substances produce symptoms, as in hay fever. The release of histamine may act as a defence mechanism; its production, when microbes react with IgE at the cell surface, may amplify the inflammatory responses and facilitate rejection of the pathogen.

Antibodies are often defined operationally; that is they are named after the type of reaction which occurs between the antigen and the antibody *in vitro*. There is evidence that the nature of the interaction *in vivo* may resemble that *in vitro* and indicates their probable role in host defence. Some of these will be considered.

(a) *Neutralizing and protective antibodies*

Exotoxins of microbial pathogens (Chapter 5) are completely neutralized by specific antibodies called antitoxins. IgG immunoglobulins are most effective in neutralizing exotoxins. Antitoxins essentially block the interaction of the exotoxin with its specific substrate. This may be demonstrated both *in vitro* and *in vivo*. The specific neutralization of *Clostridium perfringens* type A lecithinase, for instance, can be demonstrated by a Nagler plate (Figure 1.3). Serum neutralization *in vivo* may also be performed. Exotoxin and antitoxin are mixed and, after allowing time

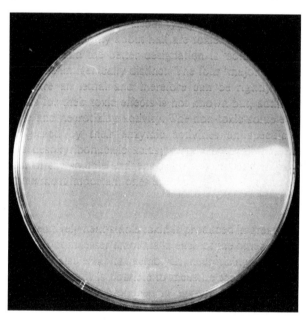

Figure 1.3. A Nagler plate. *Clostridium perfringens* is streaked across a plate of egg yolk agar. The diffuse opacity, due to insoluble fatty acids released by the lecithinase C splitting of lecithin, is inhibited on the left-hand half of the plate. This is due to the enzyme-neutralizing effect of *Cl. perfringens* type A antitoxin which was spread over this part of the plate

to interact, are injected into susceptible non-immune experimental animals. The activity of the exotoxin (e.g. the production of a skin lesion or death) is completely neutralized if sufficient specific antitoxin is used. Alternatively, the pathological effects can be minimized or obviated by first passively immunizing the animal with antitoxin prior to injection of exotoxin.

Certain antiviral antibodies neutralize the infectivity of a virus particle and thereby play a role in host defence. This is a specific reaction between virus and antibody, and is frequently used for virus identification. The reaction may be detected by neutralizing the infectivity of a virus for an experimental animal or, more usually, for appropriate tissue cultures.

Soluble antigens, such as exotoxins, and small particulate antigens, such as viruses, react with specific antibody to produce visible precipitates. Consequently these antibodies are sometimes termed precipitating antibodies. *In vitro* this reaction may be demonstrated by layering the antigen over specific antiserum whereupon a line of precipitate forms at the interface, or by allowing the two reagents to diffuse towards each other in an agar gel whereupon lines of precipitation form in the gel at the points of interaction (Figure 1.4). Interaction between

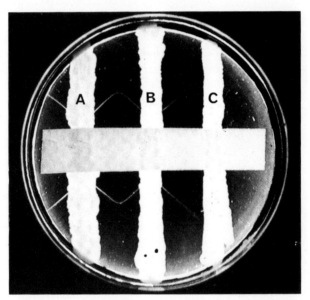

Figure 1.4. An Elek's plate to detect toxigenic strains of diphtheria bacilli. In the centre is a horizontal strip of filter paper impregnated with *Corynebacterium diphtheriae* antitoxin. Three strains of diphtheria bacilli are streaked across the plate at right angles to it. The toxins/antitoxins diffusing into the agar react to give fine lines of precipitin reaction with organism A and B but no toxin is being produced by organism C

exotoxin and antitoxin *in vivo* produces floccules or aggregates which may be removed by phagocytosis.

(b) *Opsonizing, agglutinating, and complement-fixing antibodies*

These antibodies play a role in phagocytosis. The term 'opsonization' means 'ready for the table' or 'tasty for phagocytes'. Bacteria coated with opsonizing antibodies are rapidly phagocytosed. Complement is also involved; complement-depleted animals often exhibit reduced phagocytosis. Bacteria, especially those enveloped by a capsule, may demonstrate only weak adherence to the surface of phagocytes. This can be greatly strengthened by the combined effects of antibody and complement.

(c) *The pathological role of antibodies*

Most antibodies are efficient and successful in combating the invasive properties of microbial pathogens or their products, thereby facilitating their removal from the body. Occasionally, however, some immune responses lead to harmful effects. These may take the form of an exaggerated immune response or the production of antibodies which cross-react with the patient's tissues (autoimmunity). Antigens which produce these immunologically mediated diseases may be of microbial origin but, more frequently, arise from non-microbial sources. Immunologically mediated diseases involving antibody fall into three categories based on their mechanism of action. They are summarized in Table 1.8. It must be emphasized that host damage in these mechanisms is the consequence of a hypersensitive state in the host and not of the micro-organismal antigen or toxin *per se*.

2. Cell-mediated immunity (CMI)

One aspect of CMI is the development of a state of delayed hypersensitivity in the host. Unlike antibody-mediated hypersensitivities, in which the reaction of the host to the antigen is immediate, in CMI the hypersensitive response is detected after a delay of a few days and is identified as a Type IV state of hypersensitivity (Table 1.8).

CMI arises, to a greater or lesser degree, in all bacterial infections but is more prominent in infections of a chronic nature including tuberculosis, leprosy, brucellosis, and listeriosis. In these diseases the microbes often survive and grow as intracellular parasites within the cytoplasm of macrophages after being phagocytosed. Other chronic infections producing CMI include fungal infections (such as dermatomycosis, coccidioidomycosis and histoplasmosis), and protozoal diseases (such as leishmaniasis). The skin rashes produced in some virus diseases (e.g. measles) are considered to be the result of specific CMI. Certain forms of contact dermatitis may also be attributed to CMI; others are due to immediate

Table 1.8. Mechanisms of immunologically mediated diseases (hypersensitives)

Antibody-mediated hypersensitivities	Agent involved	Target organs	Examples of clinical disease
Type I hypersensitivity Antigen combines with cell-bound antibody. Immediate hypersensitivity reactions, e.g. anaphylaxis	IgE + other immunoglobulins	Gastrointestinal tract, skin, lungs	Gastrointestinal allergy, urticaria, atopic dermatitis, rhinitis, asthma, serum shock
Type II hypersensitivity Antibody combines with antigen of host cells, 'Cytoxic antibody'	IgG + IgM	Blood cells (red, white), platelets, heart, kidney	Haemolytic anaemia of the newborn, leukopenia, thrombocytopenia, rheumatic heart disease, acute glomerulo-nephritis
Type III hypersensitivity Antigen–antibody complexes Arthus reaction (local site) Organ damage (systemic)	IgM mainly	Blood vessels, skin, joints, kidneys, lungs	Serum sickness, chronic glomerulo-nephritis, systemic lupus erythematosus
Cell-mediated hypersensitivity *Type IV*: delayed hypersenstivity	T lymphocytes	Skin, lungs, CNS, thyroid, etc.	Contact dermatitis, tuberculosis, brucellosis, mycoses, allergic encephalitis, allograft rejection

CNS = Central nervous system.

Type I hypersensitivity involving histamine release (Table 1.8). Particular substances (e.g. chromates, nickel salts) and some antibiotics (e.g. neomycin), when applied to the skin, bind to dermal constituents to form antigens which provoke states of delayed hypersensitivity.

CMI is independent of antibody. Delayed hypersensitivity can be transferred from a sensitized to a non-sensitized patient, therefore, only by transferring blood white-cell elements which are themselves sensitized. Delayed-type hypersensitivity is the basis of a number of very useful diagnostic tests. In the diagnosis of tuberculosis, tuberculin—a derivative of the tubercle bacillus—is introduced intradermally. No reaction occurs in a tuberculosis-free individual, but one with present or past infection presents a skin reaction at the site (oedema and swelling) which developes slowly, reaching a maximum after 48–72 hours.

CMI plays a role in killing intracellular micro-organisms, often present in chronic infections. It is thought that antigen-binding T lymphocytes release soluble factors (lymphokines) which confer on macrophages the power to kill phagocytosed bacteria. If unrelated organisms, e.g. *Listeria monocytogenes*, are injected together with tubercle bacilli into a patient already immune to tuberculosis, both species of bacteria are killed once phagocytosed. Thus CMI can produce a simultaneous, albeit transient, non-specific resistance against various species of micro-organisms.

(a) *The role of CMI in infectious disease*

Without doubt CMI is usually protective to the host in infectious disease. The state of hypersensitivity, which enhances the flow of macrophages to the infection site and death of micro-organisms by sensitized cells, is obviously of advantage to the host in recovery. On the other hand, delayed hypersensitivity is often an exaggerated response which may be detrimental to the host. In tuberculosis a massive reaction in a sensitized host may occur to a fresh antigen stimulation and this can awaken an old lesion in which living tubercle bacilli have remained quiescent. For this reason only minute doses of tuberculin are used in the Heaf and similar tests for the detection of tuberculin-positive humans.

FURTHER READING

Allan, D. (1980) *Outlines of Animal Immunobiology*. Baillière Tindall, London, (160 pp.).
Bellanti, J. A. (1971) *Immunology*. W. B. Saunders Co., London (584 pp.).
Keppie, J. (1964) Host and tissue specificity, In Smith, H., and Taylor, J. (eds), *Microbial Behaviour in vivo and in vitro*, pp. 44–63. 14th Symposium of the Society for General Microbiology. Cambridge University Press, London (296 pp.).
Klainer, A. S., and Beisel, W. R. (1969) Opportunist infections: a review. *Am. J. Med. Sci.*, **258**, 431–56.
Roitt, I. (1977) *Essential Immunology*. Blackwell Scientific Publications, London, 3rd edn, (324 pp.).
Wilson, G. S., and Miles, A. A. (1975) *Topley and Wilson's Principles of Bacteriology, Virology and Immunity*, 2 vols., 6th edn. Arnold, London (2693 pp.).

Chapter 2

Host–parasite interactions in health

1. General considerations

In the healthy host the internal tissues, e.g. blood, muscle, and brain, are normally free of micro-organisms. In contrast, surface tissues, e.g. the skin and mucous membranes, are being continuously contaminated. Most contaminants are incapable of establishing themselves owing to the anatomical and structural defences of these tissues or owing to their inability to compete with specifically adapted resident organisms. Those which are able to colonize particular sites are uniquely equipped both to protect themselves against the preventive defences of the host, and also to avoid inducing inflammatory or immune responses that might expel them. The mixture of micro-organisms regularly present at any site is referred to as the 'natural', 'indigenous', 'resident' or 'normal' flora. Their presence does not cause obvious signs of ill health and their relative numbers remain reasonably constant unless the balance is disturbed by disease or chemotherapy (see later).

Very little is known about the nature of the associations between the host and the normal flora but they are thought to be dynamic interactions rather than associations of mutual indifference. Indeed the association is considered to approximate to an ideal state of parasitism in which minimum disturbance is experienced by both host and parasite. Each derives benefit from the other. The normal flora derives a supply of nutrients, a stable environment, temperature, protection and transport. Benefits to the host may be less obvious but their importance is becoming increasingly recognized. In some hosts the normal gut flora synthesizes extracellular products, such as vitamins, essential for the host's nutrition, and enzymes essential for the degradation of otherwise indigestible foods. The most important general benefit is, however, the exclusion, by competition, of other micro-organisms from colonizing the site. The enteric organism *Escherichia coli*, for example, is not normally present in the oral cavity: it is unable to colonize this site owing to the presence of the highly adapted oral flora with which it must compete.

Specific biochemical interactions between surface components of bacteria and host play the principal role in selective adherence. More information is becoming available on the nature of specific adhesion for host tissues, particularly of pathogens (Chapter 4), but similar mechanisms must also apply to commensal organisms of the normal flora. The term 'bacterial stickiness' has been coined to describe this adherence but the biochemical basis is known in only a few instances. Minor differences in phenotype between strains of the same bacterial species may be sufficient to express different surface antigens to account for their unique ability to colonize particular tissues. The biochemical bases for the specific attachment of certain organisms are considered later.

Most members of the normal flora are non-pathogenic; occasionally pathogens may be present. Some usually non-pathogenic strains may assume a pathogenic role under certain conditions. *Streptococcus mutans*, and its association with dental caries, is a case in point (see later). Furthermore, mouth streptococci may be introduced into the blood stream through wounds created by dental manipulation or treatment. Where this occurs in a patient with damaged heart valves, due to congenital defect or the consequence of previous attacks of rheumatic fever, the oral streptococci may adhere to the damaged valves and initiate infection resulting in subacute bacterial endocarditis. Damage to tissue, as may occur by insertion of a catheter into the bladder, may predispose to infection by normal bowel organisms and lead to cystitis. Peritonitis may follow the accidental introduction of similar organisms into the peritoneum during operation on the bowel or by leakage of gut contents for other reasons. Diseases, such as diabetes, lower host resistance and predispose to infection by 'opportunist' organisms. Debilitated tissues, such as occur in burns, may become similarly infected. Opportunist infections have assumed greater importance recently owing to modern methods of treatment, such as prolonged and intricate surgery (e.g. heart surgery), and the use of toxic drugs. These medically inflicted conditions are the so-called 'iatrogenic diseases' (Chapter 1). These examples indicate that it is not always possible to draw a clear distinction between pathogenic and non-pathogenic organisms of the normal flora.

2. Composition of the normal flora

The flora of corresponding sites in different animal species varies widely, and is related to factors such as diet and body temperature. Some bacterial species are regularly found in particular sites on the healthy host; others are present only occasionally or at certain times of life. However, the composition of the normal flora is sufficiently constant to give general descriptions for a given animal species.

(a) *Normal flora of the human skin*

The two square metres or so of skin covering the human adult has been colourfully described as a vast empire in which contrasts of terrain and climate are as varied as those of the earth itself. Yet in spite of this great diversity, the skin is a most inhospitable habitat owing, in part, to the natural disinfectant secretions of sebaceous glands and to the regular assaults upon it by means of soaps, water, towels and lotions. Different areas of the skin support populations differing in density and composition. The high relative humidity of the groin and axilla predispose to the survival of fairly dense populations but, with few exceptions, the density of the microbial populations at most other sites is low, generally in 100s or 1000s per cm^2.

The majority of skin micro-organisms are found in the most superficial layers of the epidermis and in the upper parts of the hair follicles. Some are found in the deep follicular canals beyond the reach of ordinary disinfection procedures. They consist of a few species of bacteria including Gram-positive cocci (Micrococcaceae, mainly *Staphylococcus albus* and *Sarcina* spp.) and diphtheroids. These are generally non-pathogenic but sometimes the potentially pathogenic species, *Staphylococcus aureus*, is found on the face and hands, particularly in individuals who are nasal carriers. They persist there even after intensive washing and present cross-infection problems among nurses and surgeons in close contact with susceptible patients; they may be introduced directly into a tissue wound. Many other organisms, not generally considered to be part of the normal flora, may be present transiently in numbers largely determined by the standard of personal hygiene. This is especially so in moist crevices of the skin, particularly those surrounding the body orifices, such as the anus. In the fatty and waxy secretions of the skin, lipophilic yeasts are common: *Pityrosporum ovalis* occurring most commonly on the scalp and *P. orbiculare* on the chest and back. In the secretions of ears and genitalia, saprophytic acid-fast bacilli (e.g. *Mycobacterium smegmatis*) and diphtheroids may be present.

The tears secreted on to the conjunctiva of the eyes contain the enzyme lysozyme, which kills many bacteria by dissolving their cells walls, thus limiting the number of organisms at this site.

(b) *Normal flora of the human respiratory tract*

The anterior nares (nostrils) are always heavily colonized, predominantly with *Staphlyococcus albus* and diphtheroids, and often with *Staph. aureus*—this being the main carrier site of this important pathogen. The healthy sinuses, in contrast, are sterile. Large numbers of other species of bacteria constitute the flora of the normal upper respiratory tract which includes predominantly mixed 'viridans' streptococci (e.g. *Streptococcus salivarius*) and *Neisseria* spp. (e.g. *N. pharyngis*), and sometimes pathogens, such as pneumococci, *Haemophilus influenzae*, *Streptococcus pyogenes*, and meningococci (*Neisseria meningitidis*).

In the healthy host the trachea, bronchi, and pulmonary tissues are virtually free of micro-organisms. Air is inhaled through the tortuous, hair-lined passages which act as effective filters of suspended particles. Bacteria adhering to these particles of dust are removed. Most organisms that pass this filter are trapped on the mucous surfaces of the nasopharynx and trachea, being swept upwards by ciliated epithelial linings of the tract, to be removed subsequently by expectoration or swallowing. Thus the mucosa of trachea and bronchi are relatively sterile.

Where the respiratory tract epithelium becomes damaged, as in chronic bronchitis—which may be caused by excessive smoking—the patient is liable to attacks of descending infection by pathogens from the nasopharynx, e.g. *Haemophilus influenzae*.

(c) *Normal flora of the genital tract of the human female*

In the sexually mature female the vagina is acid, owing to the production of lactic acid from glycogen in the epithelial linings. Consequently the flora consists of mixed acid-tolerant organisms, including the predominant *Lactobacillus acidophilus* (Döderlein's bacillus), together with corynebacteria and anaerobic streptococci. Mycoplasmas also have been isolated. Infections, such as vaginal thrush due to *Candida albicans*, are particularly likely in pregnancy or in patients with diabetes. Before puberty and after menopause the vaginal secretions are mildly alkaline and contain normal skin organisms, e.g. staphylococci, streptococci, coliforms, and diphtheroids. The organs of the urinary tract, other than the lower urethra, are generally free of bacteria as a result of the flushing action of urine.

(d) *Normal flora of the human oral cavity*

The presence of particles of food, epithelial debris, and secretions makes the mouth a nutritionally favourable habitat for a great variety of bacteria. The mouth presents a succession of different ecological situations with age, and this is associated with corresponding changes in the microflora. At birth the oral cavity is bounded solely by the soft tissues of the lips, cheeks, tongue, and palate and is kept moist by the various secretions of the salivary glands. During the first year of life the primary teeth erupt and these are later superseded by the permanent dentition, usually complete by 12 years of age. During this transitional period the local conditions inevitably change as teeth are shed and new ones erupt. In addition to these changes, many other environmental factors including diet, dental treatment, and the use of artificial dentures, significantly affect the oral flora.

At birth the oral cavity is usually sterile but within the first 24 hours becomes rapidly colonized from the environment, particularly from the mother, and in the first feed. Oral streptococci, particularly *Streptococcus salivarius*, are numerically dominant, often 98 per cent of the total flora, but later fall to about 70 per cent by 1 year. Other components of the flora during the first year include species of *Staphylococcus, Veillonella, Neisseria, Fusobacterium, Candida, Lactobacillus* and *Actinomyces*. Once teeth are present, *Strep. sanguis* and *Strep. mutans* can be demonstrated and these continue so long as teeth (natural or artificial) remain: they disappear, however, from the edentulous subject. With increasing age other bacteria are common, *Bacillus melaninogenicus* and spirochaetes (e.g. *Treponema microdentium*) are especially numerous in the gingival crevices.

The yeast-like fungus *Candida albicans*, frequently present in small numbers, may multiply and cause 'oral thrush', a disease particularly of infants suffering from malnutrition and of adults with debilitating diseases, e.g. diabetes, or where the normal flora is disturbed by treatments, e.g. antibiotics (see later) or cytoxic drugs (Chapter 1).

There is strong evidence that material adhering to teeth (dental plaque) is closely linked with dental caries and periodontal disease. Plaque consists of many species of bacteria embedded in an organic matrix derived from salivary glycoproteins and microbial extracellular products. Electron microscopic observations indicate that bacteria colonize the clean tooth surface within a few hours of exposure to the oral environment and, if not removed by tooth brushing within a few hours of formation, build up into a thick, adherent film of material which cannot be easily removed by washing. In some situations the plaque becomes calcified and is then referred to a 'calculus' or 'tartar'. By far the most dominant species in dental plaque are *Strep. sanguis* and *Strep. mutans*, both of which are considered responsible for the formation of plaque.

The first stage is a weak reversible association between the bacteria and salivary glycoproteins forming a pellicle on the teeth. This is followed by a stronger attachment through two sticky glucose polymers (glucans) synthesized from dietary sucrose. The glucans are attached to the *Strep. mutans* by a glucosyl transferase enzyme; the specificity of the adhesion has been proven by the fact that the attachment can be prevented using a specific antiserum against the enzyme.

3. Normal flora of the alimentary tract of man and animals

The flora in this site, in health and disease, has been extensively investigated and a fuller treatment is presented. The alimentary tract of man and animals is heavily colonized by complex mixtures of micro-organisms. The composition of these mixtures differs between animal species and with age within the same animal species. *Escherichia coli*, for example, is present in smaller numbers in herbivores than in carnivores. In guinea pigs lactobacilli contribute approximately 80 per cent to the flora, the remainder consisting of organisms from soil, vegetation and air. Consequently, the constitution and importance of the intestinal flora must be considered in relation to each animal species. Unlike the flora of herbivores, discussed below, the flora of carnivores is probably not essential to life. This is indicated by the fact that various animals have been successfully reared on balanced and sufficient diets under completely sterile conditions without showing stunted growth (see later). In the 'conventionally' raised animal, however, administration of broad-spectrum antibiotics to animals over a long period may result in vitamin deficiency, presumably because these vitamins are normally synthesized by intestinal organisms. Human patients on broad-spectrum antibiotics may be given supplementary vitamins to offset any possible deficiency consequent to the effects of the antibiotics on the gut flora.

In contrast to carnivorous or omnivorous animals, which depend wholly upon the secretion of enzymes into their alimentary canal to digest their food, herbivorous animals rely upon the enzymic activities of micro-organisms to bring about the initial digestion. In these animals the gut flora is essential to life and a symbiotic relationship exists between the host and the microbial flora of the gut. Consequently oral antibiotics should never be administered to ruminants. The breakdown of vegetable matter requires a large chamber where the food can be retained and slow transit through the digestive tract to allow sufficient time for microbial degradation to take place. The alimentary tract of herbivorous animals is modified to include in ruminants an enlarged portion or portions of the oesophagus, including rumen and reticulum, in addition to the true stomach (Figure 2.1).

(a) *Origin and variation of the alimentary flora*

In utero, the alimentary tract of the foetus is normally sterile but during birth the mouth and anus become contaminated from the birth canal and the immediate environment. This initial mixture of organisms disappears in a few days and is replaced by successions of gastrointestinal micro-organisms typical of the animal species. These are derived by the animal sucking and licking its mother, and from the faecally contaminated environment. As the animal develops, and with the process of weaning involving a change from a milk to a solid diet, a fairly stable flora becomes established characteristic of adults within the species. In mice, for instance, the indigenous microbiota reaches a stable climax by 4 weeks. Throughout this period new bacterial species appear almost weekly.

During the first week the non-secreting gastric epithelium and the lumen throughout the gut become colonized by lactobacilli and Lancefield group N streptococci. By the end of the first week of life, anaerobic spiral organisms and facultative anaerobes (coliforms and enterococci) are found. The spiral organisms colonize the epithelium of the large intestine and reach their highest numbers by the end of the second week. Coliforms and enterococci colonize throughout the intestine, especially in association with the mucus on the epithelium of the large intestine; the numbers remain high until the middle of the third week and then decline to levels characteristic of the adult. Strict anaerobes, especially the non-sporing rods, appear by the middle of the second week; their numbers increase rapidly only to subside later.

During the fourth week, filamentous bacteria arise in association with the epithelium of the small intestine and species of *Torulopsis* are found on the secretory surfaces of the stomach. Similar sequences in the microbiota, unique to each animal species, are experienced during the early life of non-ruminant animals; the special case of ruminants is considered later.

Once the flora is established in non-ruminant animals the distribution of micro-organisms exhibits marked differences both at longitudinal levels (i.e. along the

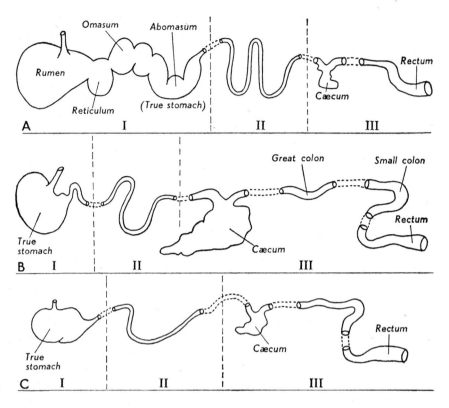

Figure 2.1. The digestive tracts of different animals showing how ruminants differ from other animals. (**A**) Herbivorous ruminant, e.g., cow; (**B**) herbivorous non-ruminant, e.g., horse; (**C**) omnivorous animal, e.g., rat: (**i**) pre-absorptive part of gut, (**ii**) small intestine, (**iii**) large intestine (after Kon, S. K. (1953) *Voeding.* **14,** 137) (Reproduced by permission of *Voeding*, Neth. J. of Nutrition)

alimentary tract) and at horizontal levels (i.e. across the alimentary tract where particular bacteria are found attached to the intestinal wall while others are found in the lumen of the gut). Upon swallowing, the majority of vegetative organisms originating in food are destroyed by hydrochloric acid and other gastric juices in the stomach which, when empty, is relatively free of living organisms. The few survivors find the acid conditions of the duodenum unsuitable for multiplication. In the small intestine the influence of bile secretions becomes gradually apparent; the highly buffered foods revert to an alkaline pH conducive to microbial multiplication.

The microbial flora of the small intestine is readily influenced by the quantity and frequency of gastric emptying but bacteria are relatively few—in man these consist chiefly of acid-tolerant lactobacilli and streptococci. Disorders which impair peristalsis of the gut lead to a greatly increased bacterial population in the small intestine. A variety of pathological conditions can do this including gastric achlorhydria or deranged immunological mechanisms such as hypogamma-globulinaemia. This indicates that the flora of this region is in part controlled by local, intestinal antibodies (IgA). The distal part of the small intestine contains a richer and more permanent flora and bacterial counts in this region may fall between 10^5 and 10^7 per ml—these include coliforms and *Bacteroides* spp., in addition to lactobacilli and streptococci.

The flora of the large intestine compares closely with that of freshly voided faeces—it contains a wider range of micro-organisms, higher levels, and is relatively stable in composition (Table 2.1), particularly between repeat samples from the same individual. The most prevalent group of organisms are the non-sporing anaerobes, especially *Bacteroides* spp. Bifidobacteria are also present in large numbers; coliform bacilli and the streptococci are constantly present, and lactobacilli are usually present. Others are isolated irregularly in small numbers, often dependent on external factors imposed on the individual.

Certain microbial types live free in the lumen of the gut, some grow in close association with the epithelial surfaces, while others colonize the crypts of Lieberkühn. The latter are often of microscopic dimensions and constitute microhabitats.

There is frequently a very close association between specific micro-organisms of the gastrointestinal ecosystem and particular gut tissues. Many microbial types are known to adhere specifically to the gastrointestinal epithelial surfaces and this has been found in all animal species studied. For instance, in chickens, lactobacilli colonize the squamous epithelium lining the crop, segmented filamentous microbes colonize the epithelium of the small intestine, and a mixture of Gram-positive cocci colonize the caecal epithelium. The mechanisms of attachment are relatively specific. Lactobacilli colonize the crop soon after the chick is hatched and persist throughout life. The biotypes of lactobacilli, regularly present, are unique to chicken; lactobacilli from mammals do not colonize the bird. The association is not affected by drastic dietary changes. Feeding penicillin, however, does eliminate them and they are replaced temporarily by coliform organisms but, after withholding the antibiotic, the same biotypes of lactobacilli once again reappear to colonize the crop epithelium. Either they persist in small numbers during treatment or the chick becomes reinfected from its environment.

It is thought that lactobacilli attach specifically to epithelia by polysaccharides or glycoproteins. Other indigenous bacteria attach by mechanisms analogous to those of pathogenic *Escherichia coli*, which exhibit filamentous structures, (e.g. K88 antigen, Chapter 4). Others change the ultrastructure of the cell membrane and become embedded in the host cells.

(b) *The stability of the gut flora*

Unless modified by infection or antibacterials (see later) the gut flora, once established, is a stable, self-regulating system. This is true both qualitatively and

Table 2.1

Occurrence of selected bacterial groups in the large intestine of normal people

Log_{10} bacteria/ml intestinal contents. Mean (Range)

Number of subjects	Entero-bacteriaceae	Bacteroides	Streptococcus	Lactobacillus	Gram-positive non-sporing anaerobes	References
11 (Appendix)	6.9 (6.7–9)	7.1(4–9.3)	7.0(5–9.7)	6.4 (4–9)	5.6 (4–8.6)	Seeliger and Werner (1962)
3 (Caecum)	6.2 (5.6–7.4)	7.9 (6.4–9.1)	2.6 (N–3.6)	D (N–2.8)	5.2 (2.3–7.6)	Gorbachh et al. (1967)

Occurrence of selected bacterial groups in faeces

Log_{10} bacteria/gram faeces. Mean (Range)

Entero-bacteriaceae	Bacteroides	Streptococcus	Lactobacillus	Gram-positive non-sporing anaerobes	References
7.6 (7–7.8)	8.4 (6.9)	7 (5–9.7)	6.4 (4–9)	6.6 (4–8.6)	Seeliger and Werner (1962)
7.1 (6.5–7.6)	10 (9.5–10.5)	4 (3.2–4.5)	4.5 (4–5)	D	Zubrzychi and Spaulding (1962)
8.4	8	7.4	7.5	8.1	Ketyi and Barna (1964)
7 (5–8)	8.8 (8–9.5)	5 (5–5)	3.6 (3–4)	6.6 (5–8)	Gorbachh et al. (1967)
6 (4–9)	10.5 (10–11.5)	5 (2–8)	4 (2–7)	10.5 (9–11)	Draser et al. (1969)
6.6 (N–10.1)	9.8 (8–11.4)	4.9 (N–9.4)	3.5 (N–10)	5.6 (N–10.8)	Finegold et al. (1970)

N = Not detected; D = Less than 100 bacteria.
(After Draser and Hill, 1974.)

quantitatively. The various bacterial species interact with each other to regulate the total numbers of each and their frequency of multiplication. For instance, as a consequence of microbial metabolism the redox potential (E_h) of the gut falls, thereby favouring an increase in the numbers of anaerobes, especially the Gram-negative non-sporing species. In contrast, the E_h is significantly higher in germ-free animals (see later) and in conventionally raised animals in which the gut flora is suppressed by antibiotics. Another consequence of microbial metabolism is the production of volatile fatty acids by *Bacteroides* spp. and some coliform bacilli, and these suppress the multiplication of other species. This is not so much a pH effect as a toxic effect of the non-ionized molecules. A direct pH effect is the production of lactic acid by lactobacilli in the stomach flora of the newborn pig. Some gut organisms liberate products lethal for other organisms. Certain coliform bacilli, for instance, produce colicins which can kill other strains either present in the indigenous flora or entering from outside. The overall consequence of these microbial interactions is to regulate the composition of the flora and the numbers of each species, but also to reduce the multiplication rate to a low order of 2 or 3 divisions per day. Local antibody also may contribute to this low rate of turnover.

Since the normal gut flora is so highly adapted it is very difficult for new strains to become established, even of species closely related to those already present. The environment, and especially food, are the main sources from which new strains may be acquired but unless these are adapted to the gut situation or have specific adhesion properties for the gut tissues, they fail to persist. This may be illustrated by attempts, in man, to displace the resident flora by lactobacilli. Based on the premise that certain bowel conditions are the consequence of 'toxic' members of the gut flora, attempts have been made in the past to cure the condition by feeding large numbers of innocuous lactobacilli. This proved generally unsuccessful; a temporary imbalance was achieved but the lactobacilli were detected only transiently and then only as long as large doses were ingested.

By virtue of its stability, the gut flora provides a unique defence against alimentary infections with gut pathogens. The indigenous flora may prevent colonization by a pathogen, the resident flora competing more successfully for attachment sites on the gut epithelium, or for essential nutrients. Products of metabolism, e.g. volatile fatty acids toxic for salmonellae or bacteriocins lethal for specific organisms, produce conditions inimical for certain pathogens. This host defence may fail under certain circumstances, as in the very young before a stable flora is established, but during these early days of life the flora is in part controlled by immunomechanisms of the gut. The stability of the climax flora, and hence this host defence, can be easily disturbed by antibiotics, especially if administered by the oral route. These suppress large numbers of the gut flora which are sensitive to the drug being used. As a result the gut defences are less effective and pathogens may colonize more readily or the gut may become overgrown by antibiotic-resistant species some of which may produce pathological effects (see later).

(c) *Variations in the gut flora within an animal species*

The stability of the normal flora within one animal species is essentially a relative term. Some fluctuations, between samples taken from the same individual and between individuals, can be detected although this is less marked within the same animal. For instance, the numbers of *Escherichia coli* per gram of faeces remains relatively constant but the range of serotypes fluctuates from day to day—some are regularly present, others only occasionally present (Figure 2.2). Those regularly present may constitute the resident flora—the others entering in the food are 'transients'.

People living in different parts of the world show marked differences in their gut flora. This may reflect exposure to different ranges of micro-organisms due to differences in location, culture, and diet. There is evidence that healthy people in India and Guatemala have higher bacterial counts in the small intestine compared with those living in temperate climates. Where the diet includes animal protein higher *Bacteroides* counts are encountered, and fewer coliforms and enterococci compared to those on a vegetable diet. Despite these differences marked changes in diet do not generally bring about major changes in the gut flora of the individual.

Sudden change of circumstances can result in specific strains of gut organism assuming a pathogenic role. Travellers' diarrhoea is a case in point. A person who remains in one geographical situation becomes adapted to the strains of gut organisms regularly present in his environment—this most probably involves immune defences. The same person travelling to foreign parts will encounter new strains of *E. coli* to which he has no immunity. Not infrequently travellers experience acute diarrhoea shortly after arrival.

(d) *The influence of antibiotics on the gut flora in non-ruminant animals*

Antibiotics and other antibacterials, such as sulphonamides, are administered orally to animals or humans for a variety of purposes including therapy, prophylaxis, preoperative surgery, and growth promotion. Whatever the purpose, the stability of the normal gut flora may be disturbed to a greater or lesser degree. This will depend on whether or not the antibacterial has a narrow or broad spectrum of activity, and the dose, duration, and route by which the drug is administered. Oral administration has a greater effect on the gut flora than the parenteral route and prolonged treatment has a greater effect than a short period of treatment. Even very low doses, if given orally over a long period, can have a marked effect.

The spectrum of activity of the drug being administered will variously suppress aerobes or anaerobes, and Gram-positive or Gram-negative species, according to their natural susceptibility or insusceptibility. For instance, *Bacteroides*—the major component of the human gut flora—are resistant to neomycin and

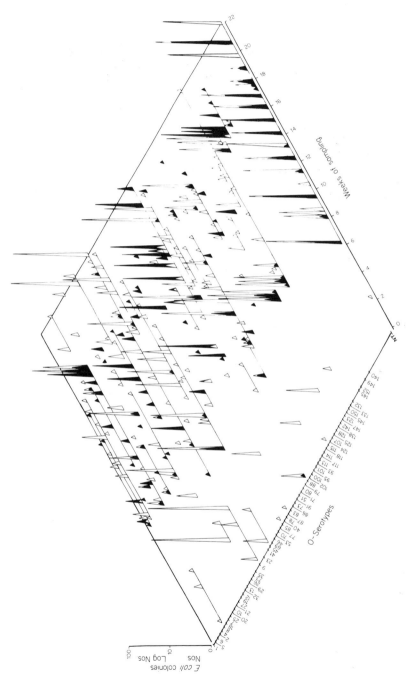

Figure 2.2. Fluctuations in the *Escherichia coli* population in a normal pig. The three-dimensional diagram is an attempt to show the relationships between the O-serotypes of *E. coli*, the numbers of each serotype within 100 colonies picked from each faecal culture on unsupplemented media, and their antibiotic sensitive/resistance properties. The pig was not receiving antibiotics in food or for therapy. The shaded areas indicate strains of *E. coli* resistant to at least one antibiotic: the unshaded areas indicate strains fully sensitive to a wide range of antibiotics. The non-numerical arrangement of the serotypes was to avoid overlapping in the figure. (Reproduced, with permission, from Linton, A. H., Handley, B., and Osborne, A. D. (1978) *J. Appl. Bact.*, **44**, 285–98)

.ut sensitive to lincomycin. The range of strains suppressed by the
. also depend on whether or not naturally susceptible ones carry
.nce determinants (R factors) to the drug (Chapter 11). Where this occurs,
.se carrying the R factors will survive while other members of the same species
will be suppressed. Thus the naturally insusceptible species and the resistant
strains of normally susceptible species will survive in the presence of the antibiotic
and increase in numbers to replace those which have been suppressed. The gut
flora, in consequence, becomes generally resistant to the drug.

The stability of the gut flora is thus modified and the balance of sensitive and
resistant strains to the antibiotic is disturbed. This continues as long as the drug is
administered but the balance is usually restored sooner or later after withdrawal
of the treatment. In man, antibiotic-resistant *E. coli*, for instance, usually continue
to be excreted for up to 10 days, and are then replaced by antibiotic-sensitive
serotypes common to the host. In many domestic animals, by virtue of continuous
exposure to antibiotics over many years (see later), the gut flora regularly includes
large numbers of resistant *E. coli*. In animals, therefore, resistant *E. coli* persist
long after the antibiotic has been withdrawn, and often indefinitely.

Disturbance of the gut flora can have marked effects on the health of the
individual. For instance, mice treated with a single dose of penicillin or
streptomycin have been shown to be susceptible to a dose of 10 salmonellae compared to untreated mice which can be infected only with doses of the order of 10^5
organisms. Normal resistance to salmonellae could be restored to treated mice by
feeding normal mouse faeces and restoring the normal flora. Suppressing the
normal flora favours the multiplication and colonization of the large intestine by
salmonellae. Rather than protect a patient, antibiotics may render the patient
more susceptible to infection. The use of preoperative antibiotics has met with
similar dangers. In the past antibiotics, often in combination, were prescribed
prior to abdominal surgery in order to render the bowel virtually sterile, thereby
avoiding septic complications. It is now considered a highly dangerous procedure.
Preoperative antibiotics have led to the development of acute staphyloenterocolitis
in some patients, the gut being overgrown by antibiotic-resistant strains of
Staphylococcus aureus. This condition has been experienced less frequently in
recent years, possibly because more gut organisms are antibiotic-resistant and
survive the therapy to persist and inhibit excessive overgrowth by resistant
staphylococci. Superinfection by other resistant organisms, such as *Candida
albicans*, may produce complications. These and similar effects indicate that
antibiotics should only be used where clinically indicated.

In addition to their use in the treatment of animal disease, antibacterials have
been used as feed supplements. The growth-promoting activities of antibiotics was
discovered fortuituously. The feeding of by-products of fermentation in the
production of antibiotics, which contained traces of antibiotics, resulted in
improved growth. Young animals fed low levels of a wide variety of agents
exhibited more rapid growth and more efficient food conversion rates. Many

attempts have been made to discover how these agents work, but the fact that agents of widely differing antibacterial mechanisms of action and spectra produce similar results makes a specific mechanism unlikely. All agents, however, are antibacterials and have no growth-promoting effects in gnotobiotic animals (see later). For these reasons it is generally agreed that their modes of action depend on the inhibition of growth-retarding bacteria or their toxins in the intestinal tract. The gut wall of conventionally raised animals is much thicker than that of gnotobiotic animals. It may be considered to be undergoing mild inflammation, due to microbial activity; this results in thickening of the wall which retards optimum absorption of nutrients.

The practice of using antibiotics as growth-promoters is not, however, without certain disadvantages. Even sub-inhibitory levels select for antibiotic-resistant strains, and consequently the gut flora in animals on these food supplements becomes dominantly antibiotic-resistant. Where this resistance is mediated by resistance plasmids (Chapter 11) the normal flora can transfer the resistant determinants to sensitive gut organisms or enteric pathogens when infection occurs. Treatment is thereby complicated. In the hope of reducing this risk the practice of feeding certain therapeutically useful antibiotics as growth promoters has been prohibited in the UK since 1971.

(e) *Microbiology of the rumen*

The food of ruminants contains carbohydrates (mainly cellulose), proteins, fats, numerous organic compounds, and minerals. Cellulose, derived from plant cell walls, is broken down by the enzyme cellulose secreted by bacteria. This energy-yielding carbohydrate is used as a nutrient for microbial multiplication and yields metabolic by-products including carbon dioxide, methane, and ammonia. Acetic, propionic, lactic, and butyric acids are also produced. The bacterial proteins and their products of metabolism, including vitamins, are food for the host.

A complex mixture of micro-organisms is involved in the breakdown of plant food. A dense population of many species of bacteria and protozoa results from the favourable conditions provided in the rumen. The rumen consists of a system not greatly different from that of a laboratory steady-state continuous culture with a regular supply of ingested food and a continuous removal of fermentation products and food residues. The rumen bacteria are adapted to live at acidities between pH 5.5 and 7.0, under anaerobic conditions, and at a temperature of 39–40°C. Only insignificant numbers of intestinal organisms common to man and other omnivorous animals have been demonstrated. The specialized flora for the breakdown of cellulose and its degradation products include micro-organisms unique to the rumen.

The diversity of bacteria in the rumen is striking. Some types are found regularly in the rumen flora; others are highly host-specific. The flora varies according to the chemical composition of the substrate; some bacteria have a

ıd, and others a narrow, range of biochemical activity. Some of the typed
cies of rumen bacteria are considered below according to the substrate
attacked.

The cellulose-digesting (cellulolytic) bacteria include Gram-negative rods (e.g.
Bacteroides succinogenes, Butyrivibrio fibrosolvens) and Gram-positive rods (e.g.
Clostridium lochheadi). Cellulolytic cocci include *Ruminococcus* spp. Succinic
acid is the main metabolic product. The starch-digesting (amylolytic) bacteria
include *Streptococcus bovis, Bacteroides amylophilus*, and *Selenomonas
ruminantium*. Their main metabolic products are formic, acetic, propionic,
butyric, and lactic acids. Most cellulolytic and amylolytic bacteria are able also to
attack water-insoluble carbohydrates, the hemicelluloses.

Yeasts of the genera *Candida* and *Trichosporon* have been found in small
numbers in the rumen of the cow, sheep, and goat. Many protozoa of varied kinds
are found in the rumen, the majority being ciliates. The part they play in rumen
digestion is not clear. It has been shown that certain species (e.g. *Entodinitum
neglectum*) fail to grow in artificial culture if cellulose is omitted from the medium.
It does not follow, however, that cellulose is digested by the protozoa since in all
cultures there is active fermentation by cellulolytic bacteria. The necessity for
cellulose in the medium could therefore be due to a requirement by the protozoa
for the metabolic products of the bacteria digesting the cellulose. It is generally
thought that protozoa play no vital role in rumen metabolism.

(i) *Synthesis of vitamins of the B complex* The amount of B vitamins in rumen
contents is found to be several times greater than that expected on the basis of the
vitamin content of the food ingested. This is the result of vitamin synthesis by
rumen bacteria but not by protozoa. The vitamin concentration in the rumen is
not influenced by the amount present in the diet; metabolic control of bacterial
synthesis maintains a steady level. This explains the fact that the concentration of
these vitamins in ruminants' milk remains steady irrespective of fluctuations in the
diet. Although the synthesis of vitamins by bacteria in the alimentary canal is not
limited to ruminants, it is probably of particular importance to these animals since
for the larger part of the year they graze on dried pastures of poor quality
deficient in vitamins.

(ii) *Protein synthesis.* The most common nitrogen source for ruminants is
provided as protein in forage. The proteins are broken down by ruminant micro-
organisms into peptides, amino acids, and ammonia, these products being used in
different ways for the synthesis of microbial protein. This in turn becomes available to the host when bacteria die and are digested by the ruminant. Protein is
therefore used by the host in the form of rumen bacteria and constitutes the major
part of the ruminant's nitrogenous food.

This knowledge has opened up a new approach to nutritional problems; urea or
ammonium salts are added to diets deficient in protein, enabling cheaper, high-
fibre feedingstuffs (e.g. straw) to be utilized economically. The microbial degrada-

tion of protein in the rumen inevitably means loss through ammonia and urea. Efforts are therefore being made to protect food proteins (e.g. by formalin) so that they pass through the rumen to the abomasum before undergoing digestion.

4. Gnotobiotic animals

It is clear that the microbial flora of the ruminant is essential for its survival but is this also true of the non-ruminant animal? By sophisticated techniques it is possible to remove the foetus aseptically from the uterus, or the unhatched chick from the egg, at full term and raise them in a germ-free environment. Many animal species have been maintained successfully for long periods of time, and rodents and guinea pigs have been bred through many generations, under germ-free conditions. These findings indicate that the normal flora, particularly the gut flora, is not essential for the life of these animals. However, certain differences have been noted between so-called 'conventionally' raised animals and their germ-free counterparts. For instance, the caecum of germ-free rodents and rabbits is greatly enlarged—this has not been satisfactorily explained but may indicate certain physiological differences. It has already been noted that marked differences are found in the structure of the gut wall. Germ-free animals possess a thinner gut wall, with optimum adsorptive properties, and a reduction in the rat of desquamation and replacement of epithelial cells. In addition, they demonstrate a reduced reticuloendothelial response due to an absence of bacterial challenge.

Gnotobiotic animals have proved very useful in research such as in investigations to determine the effect of a single bacterial species, or mixtures of known species, on the gut structure and function.

5. Specific pathogen-free (SPF) animals

In order to raise animals as free as possible from pathogens, the young may be taken aseptically from the dam, as for gnotobiotic animals, and reared under clean conditions. The object is to avoid contact with infections which the dam may carry, and subsequently to rear the animals in isolation to avoid later exposure to infections. Such stock are superior in having a better food conversion and optimum growth rates. They possess a less complex mixture of bacteria in their normal gut flora compared to conventionally reared animals. As with gnotobiotic animals, should a pathogen gain access to the herd or individual, SPF animals are more susceptible than the conventionally raised ones, doubtless because they lack specific immunity by lack of prior exposure to the pathogen.

FURTHER READING

Clarke, R. T. J., and Bauchop, T. (eds) (1977) *Microbial Ecology of the Gut.* Academic Press, London (410 pp.).

Draser, B. S., and Hill, M. J. (1974) *Human Intestinal Flora.* Academic Press, London (263 pp.).

Hungate, R. E. (1966) *The Rumen and its Microbes*. Academic Press, New York and London (533 pp.).

Klainer, A. S., and Beisel, W. R. (1969) Opportunist infection, a review. *Am. J. Med. Sci.*, **258**, 431–56.

Skinner, F. A., and Carr, J. G. (eds) (1974) *The Normal Flora*. Society for Applied Bacteriology. Symposium no. 3. Academic Press, London (264 pp.).

Chapter 3

Host–parasite interactions in disease

Infectious disease results when a parasitizing microbe damages its host; it is the consequence of the interaction between host and parasite. Both host and parasite are modified by this interaction. Under microbial attack the host tissues may be altered physiologically (e.g. a change in pH or E_h), nutritionally (e.g. damage to tissue cells may result in leakage of microbial nutrients into the interstitial spaces) and functionally (e.g. damage to the CNS may result in paralysis of vital muscles such as those controlling respiration). In the process of interaction the host responds increasingly with time to the antigenic stimulus of the parasite by producing specific immunity against the parasite or its products. On the other hand, the altered environment in the host as a consequence of disease may in turn bring about changes in the microbe. This may be in altered metabolism which may result in new products being formed, such as exotoxins, in increased virulence, or in the selection of a few genetically more competent organisms from an erstwhile heterogeneous population.

1. General considerations

(a) *Defining the terms 'pathogenicity' and 'virulence'*

In considering the subject of microbial pathogenicity, the two terms 'pathogenicity' and 'virulence' are often used. Various meanings have been attributed to these terms and frequently, in the field of animal and human pathology, they are used as synonyms. It is important therefore to define these terms. The definitions of Miles (1955) are adopted here and may be illustrated by reference to a typical usage. The following is a well-documented observation:

> Within the pathogenic species *Corynebacterium diphtheriae*, both virulent and avirulent strains are found.

In this example the term pathogenicity is applied to a species of *Corynebacterium* which regularly includes strains capable of producing the disease diphtheria in susceptible humans. In other examples the term may be applied to wider or narrower groups of micro-organisms such as a genus, a serotype or a phage type.

49

The frequency with which virulent strains are experienced within a group of micro-organisms determines whether or not we apply the term pathogenic to that group. Thus we speak of pathogenic and non-pathogenic groups of micro-organisms. For instance, *C. diphtheriae* is a pathogenic species of the genus *Corynebacterium*, whilst *C. hofmannii* is a non-pathogenic species.

Not all members of *C. diphtheriae* are necessarily virulent, i.e. are capable of producing disease, however. The isolation of an organism from a human throat with morphology, culture, biochemistry, and serology characteristic of *C. diphtheriae* does not invariably imply that this isolate is virulent. Whilst the term pathogenicity applies to the species, each isolate must be assessed for virulence by suitable tests. Virulence is thus the attribute of an individual strain and describes its capacity to produce disease.

Thus the term pathogenicity is the broader term; it describes a group of micro-organism in which individual strains are likely to be virulent. Virulence, on the other hand, is the attribute of the individual strain within a group of pathogenic micro-organisms.

(b) *The unique properties of microbial pathogens*

Compared to the large number of non-pathogenic microbes, pathogens are relatively few. To be pathogenic they must have unique properties. These properties are essential to enable them to become established in the host, to overcome the resistance mechanisms of the body, to invade the tissues, and subsequently to produce damage to the host which is recognized as disease. It is of interest therefore to know what are the properties that make a microbe pathogenic at some stage in its life-cycle. Pathogenicity in some species is associated with unique structural components of the cell, e.g. the presence of a capsule; in others, active secretion of substances either protect the organism against the host defences or actively damage the host tissues (Chapter 4).

The mechanisms by which pathogenic organisms produce disease are often obscure. With pneumococci, the principal cause of lobar pneumonia in humans, the processes which result in the pathological changes occurring in the lungs have not been elucidated. Similarly, in many other diseases the pathological interactions also remain unresolved at the present time.

The mode of action is relatively simple when pathogenicity is determined by a single bacterial product as, for instance, the production of exotoxins by *Clostridium tetani* and *Corynebacterium diphtheriae*. These exotoxins, which may be produced by growing the organisms in artificial culture and separated as bacteria-free filtrates, can mimic the pharmacological action of the natural disease when injected into experimental animals (Chapter 5). Evidence of the direct relationship of the exotoxin with the disease syndrome is shown by the fact that prior protection of susceptible animals with specific antitoxin renders the exotoxin innocuous. Further, strains which do not produce exotoxin prove to be avirulent.

Where evidence of this kind is available it is safe to identify the specific exotoxin as the virulence factor.

In the majority of infectious diseases, however, pathogenicity cannot be related to a single microbial product. At least twelve extracellular, soluble antigens have been detected in culture filtrates of *Cl. perfringens* (Chapter 5), the cause of gas gangrene in man. In situations like this attempts have been made to associate certain of the products with the disease syndrome and, in a few instances, to define the mechanism of action in biochemical terms. With other pathogens, the mechanisms of pathogenicity are even more difficult to evaluate.

2. Methods of investigating microbial pathogenicity

Historically the first essential was to relate specific pathogens to each disease. This was the approach of the early microbiologists who, during the last 25 years of the nineteenth century, discovered the majority of bacterial causes of disease in man and animals. The causal pathogens were isolated in pure culture on artificial, inanimate media and the association of each as the causal organism of a particular disease was determined by applying the principles of Koch's Postulates. These stated that:

(1) The organism should be found in all cases of the disease and its distribution in the body associated with the lesions observed.
(2) The organism should be cultivated in pure cultures for several generations outside of the body of the host.
(3) The organism so isolated should reproduce the disease when introduced into other susceptible animals.

These postulates cannot always be applied. *Mycobacterium leprae*, the cause of leprosy in man, has not at present been isolated in artificial culture (although a number of claims have been made), and therefore it has not been possible to apply the postulates to this pathogen. Nevertheless the regular presence of this organism in affected tissue, and the fact that the disease can be reproduced in mice by inoculating bacilli from human tissues, leaves no doubt that this organism is the causal pathogen. The absence of a suitable experimental animal for certain human infections, e.g. gonorrhoeae (though this has been done with human volunteers), has also limited possible investigations since it is ethically improper to experiment on man.

(a) In vitro *methods of investigation*

Knowledge of the association of a pathogen with a specific disease provides very little information on the mechanisms by which the disease process is brought about. It is possible to determine the route of entry, whether the microbe remains localized or produces generalized disease, and the specificity, if any, of the tissues affected. Early attempts to unravel the mechanisms of pathogenicity were primarily aimed at examining the activities of microbes *in vitro*, i.e. in glass or, more explicitly, under conditions of growth in artificial culture. The pathogen is grown in laboratory media and its various properties and secretions studied. These included morphological features, such as the presence of a capsule, which may be present regularly on virulent strains but absent from avirulent ones; or products of metabolism, such as enzymes, which can be shown to act on specific substrates in the laboratory; or toxic products capable of producing tissue damage or death in experimental animals.

The fact that under conditions of artificial culture a microbe produces an enzyme which will attack a specific substrate has often led to the fallacy that the same substrate may be attacked in the disease process. Haemolysis of red blood cells *in vitro*, for instance, does not necessarily indicate that the same organism will cause intravascular haemolysis in the host. Hyaluronidase is produced by many pathogenic streptococci *in vitro* but antiserum against the enzyme will not protect against infection with the living organism. This indicates that even if hyaluronidase plays a role in disease production it is not the principal virulence factor. Whilst knowledge of these properties, therefore, is useful, the range of phenomena exhibited under conditions of artificial culture may bear little relation to the natural infection.

It is well known that microbes isolated from pathological lesions often differ significantly from those present in the infected host. For instance, growth on artificial media is often accompanied by loss of certain morphological features. These may include a loss of ability to produce a capsule, a degeneration from smooth colonial properties to rough, and changes in the morphological shape and size of individual organisms. From observation with *Bacillus anthracis* some of these changes have been shown to be due to the nutritional inadequacy of many culture media. Where anthrax bacilli are grown in fluid media with a high serum content the bacilli produce capsules resembling those seen in infected tissues. Frequently virulence may be reduced, or even totally lost, by growing in artificial culture; it may, however, be restored by passage of the strain through a susceptible animal. The animal host may act as a biological filter selecting the few virulent organisms still present in a population of otherwise avirulent organisms rather than induce virulence to arise in the avirulent members.

Differences in behaviour of microbes in the host with those in artificial culture is illustrated by the fact that living, attenuated vaccines are usually more effective immunizing agents than dead ones. This may be due to the fact that the process of killing micro-organisms for vaccine production may alter delicate antigens but the more important reason is that antigens closely associated with the disease process may be exposed, or only synthesized, by living organisms in the host tissues.

It is extremely difficult, and usually quite impossible, to distinguish virulent and avirulent members of the same microbial species under artificial cultural conditions. Frequently pathogens and non-pathogens possess similar antigens. Most of the antigens detected by *in vitro* serological tests may have no direct connection with virulence. For instance, nearly 2000 antigenically different salmonellae are known but food poisoning in man, which may result from any of a large number of these, is clinically similar for each serotype causing the infection.

It is evident that static *in vitro* cultural systems cannot resemble the changing environment in the host, as occurs in the pathology of tissue inflammation, and are thus inadequate to reveal the various components of virulence of a pathogenic micro-organism.

(b) In vivo *methods of investigation*

The popularity of *in vitro* systems for investigating mechanisms of pathogenicity is a reflection of the greater ease of investigation compared with studies of disease processes in animals. Since virulence is, however, the resultant complex of interacting factors contributed by both host and parasite, the full armoury of virulence factors is more likely to be produced in the diseased animal. Despite the greater difficulties of analysis, the *in vivo* system is therefore more informative since it resembles natural infections. In fact it has now been shown that certain virulence factors are produced only, or only in significant amounts, under *in vivo* conditions. Thus virulence may be determined by the nutritional adequacy of the animal host, by selection within the host of virulent clones from a mixed population of virulent and avirulent micro-organisms, by selection of strains with small genetic differences which cannot be expressed *in vitro*, or by the presence of dynamically balanced physiological conditions which favour the production of virulence factors.

In vivo systems of investigation are based on the artificial production of disease in a selected host using inocula of pure strains of a micro-organism. Apart from the pathological changes which are observed, various fluids and tissues at specified stages of the infection may be separated and subjected to analysis. The fullest examinations are obviously made at the time the animal dies or is sacrificed. Biological and biochemical analyses are made to determine the physiological changes consequent on infection, and attempts are made to correlate these findings into a mechanism of pathogenicity. The *in vivo* characters of the micro-organism themselves are also observed after harvesting.

For ethical reasons (human infections), and on the ground of economics (large animal infections), the majority of virulence tests are performed on small susceptible laboratory animals such as mice, guinea pigs, rabbits and hamsters. The experimental animal is thus not the natural host and, in fact, may not be susceptible to natural infection by the pathogen. *In vivo* systems of this kind must not therefore be automatically related to the conditions prevailing in natural infections. How faithfully these experimental systems reflect the pathogenicity mechanisms in the natural host may be open to question. Also compared with natural infection, artificial infections are often induced by inoculating abnormally large numbers of the pathogens into the experimental animal, often by an unnatural route. For instance respiratory pathogens are often inoculated intraperitoneally or subcutaneously; some may even require the simultaneous inoculation of substances like mucin to inhibit host defences. Virulence tests on pneumococci, for instance, are usually done in mice by intraperitoneal injection of broth cultures. However pneumococci normally infect man by the respiratory route where the primary lodgement occurs but lack the virulence factors to establish a primary lodgement in the respiratory tract of mice. As few as five organisms of Type I pneumococci, if injected intraperitoneally, will multiply and kill a mouse in 24 hours. Even if the subsequent lethal process is similar to the mechanism of pathogenicity in man, which is doubtful, the factors which determine natural infectivity are unknown. For these reasons, experimental animals must be considered as convenient, self-incubating, self-adjusting, test tubes but not synonymous with the natural host. Despite these reservations, many virulence tests in experimental animals show a close correlation with the virulence of the organism for the natural host. For instance, the lesions produced in the guinea pig by the subcutaneous inoculation of cultures or filtrates of *C. diphtheriae* do not resemble diphtheria in man; however much of our knowledge of immunity to diphtheria has been built on information gained from virulence determinations in guinea pigs. Similarly death caused by experimental anthrax in guinea pigs is thought to be due to shock processes closely resembling those in cattle.

Care must be exercised in extrapolating data from one animal to another. The Vi antigen (virulence antigen) of *Salmonella typhi* was assumed for many years to play an important role in virulence for man. It had been shown that strains of *Salm. typhi*, either containing the Vi antigen or injected with Vi antigen from another strain of *Salm. typhi* or other organism, were more lethal for mice compared with those which were Vi antigen-negative. The protective action of typhoid vaccine or antisera for mice was shown to be proportional to the Vi antigen content of the vaccine or the quantity of Vi antibody in the antisera respectively. But these observations made on mice are not evidence that the Vi antigen is important in human infections with *Salm. typhi* and, in fact, its importance in man is now considered doubtful.

Tissue culture, the regular tool of virologists, whilst a useful *in vivo* system, must also be regarded as an unnatural model. Cell lines established in tissue culture are different from the parent tissues in the whole animal and, although they may be derived from the organ normally infected by virus, must be regarded as different. Consequently tissue cultures may vary in their susceptibility to certain viruses. Tissue cultures from an organ fully susceptible to a virus in the intact animal may not support the virus *in vitro*: mouse lung cultures, for instance, do not support influenzavirus. To quote Enden (1952):

> Just as repeated transfer *in vitro* may bring about alteration in the
> virulence or pathogenicity of bacteria, serial passage of viruses in
> tissue culture has frequently led to changes in their capacity to induce
> disease when inoculated into a susceptible animal. Most often under

these conditions a decrease in virulence has been noted. In some cases, however, especially when an agent is propagated in tissue derived from its natural host, an increase in pathogenic properties may follow.

In summary, it must be appreciated that experimental infections in laboratory animals and tissue cultures are essentially artificial systems. Virulence, described in these systems, must always be defined in relation to the host used and not extrapolated without qualification to the natural host. It is essential to distinguish 'laboratory virulence' from 'natural virulence'.

It must therefore be appreciated that in only a limited number of cases is it possible to be reasonably certain that the mechanisms of pathogenicity in the laboratory animal and the natural host are the same. The classic examples where these extrapolations can be made with fair certainty are those of bacterial toxins. Tetanus toxin almost certainly kills mice by the same means as it kills man and domestic animals.

Enough has been said to indicate the need for caution in the interpretation of *in vivo* results if an unnatural host is being used. In choosing an animal in which to determine virulence or mechanisms of pathogenicity it is essential therefore to select a system which approaches, as closely as possible, the conditions of the natural infection. Despite these remarks, there is no doubt that the greatest advances in recent days have been realized by *in vivo* studies. The mechanism of pathogenicity of anthrax has been worked out mainly in guinea pigs, as an artificial host. In other infectious diseases the natural host has been used, as with brucellosis in cattle. These examples, considered in greater detail elsewhere (Chapter 5), have set the stage for similar investigations in the future.

3. The measurement of virulence

A virulent micro-organism produces changes in the host and this is recognized as disease. Disease is therefore the physical evidence that an organism is pathogenic for a particular host and any estimate of virulence is inevitably a measure of the disease produced within the host range. Certain microbial pathogens are highly virulent, others are of a low order of virulence. Disease may vary from minor local damage, to more general histological changes or malfunction, even to death of the host in the most severe form. It is possible to arrange pathogenic micro-organisms in a continuous spectrum from those which are highly virulent and cause disease in healthy individuals, to those which produce relatively mild illness. This is illustrated in Table 3.1 which, in addition to frank pathogens, also includes a few commensal organisms, which only rarely cause disease. This table is incomplete and should not be taken too literally. Even recognized pathogenic micro-organisms do not always succeed in producing clinical illness; in fact the majority of infections are subclinical (Chapter 8). As already considered in Chapter 1 the success or otherwise of the microbial pathogen to initiate infection does not solely

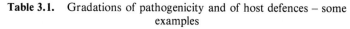

Table 3.1. Gradations of pathogenicity and of host defences – some examples

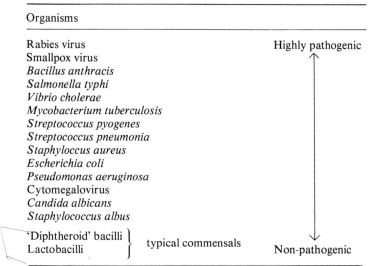

Organisms	
Rabies virus	Highly pathogenic
Smallpox virus	
Bacillus anthracis	
Salmonella typhi	
Vibrio cholerae	
Mycobacterium tuberculosis	
Streptococcus pyogenes	
Streptococcus pneumonia	
Staphyloccus aureus	
Escherichia coli	
Pseudomonas aeruginosa	
Cytomegalovirus	
Candida albicans	
Staphylococcus albus	
'Diphtheroid' bacilli ⎱ typical commensals	
Lactobacilli ⎰	Non-pathogenic

depend on the properties of the micro-organism but also depends on the body defences of the host. For these and other reasons the distinction between low-grade pathogens and 'non-pathogens' is not sharp. Consequently the measurement of virulence of low-grade pathogens is extremely difficult.

Since, even with the same microbe, the severity of disease may vary from strain to strain, the majority of virulence tests are essentially qualitative. Measurements of the degree of virulence must depend, however, on more precise tests. These usually involve comparing the virulence of different micro-organisms in similar hosts under standard conditions. Various manifestations of disease may be used for measurement. These may range from the production of a local zone of erythema at a local skin site, to death. Death of the host, or the time taken for death to occur, is the factor most commonly used, being least affected by minor differences in host resistance. Estimates of this kind are often complicated by the variability between individuals in the experimental host population. It is essential therefore to use reasonably stable and homogeneous populations of fully susceptible animals. Even then it is extremely difficult to maintain stability of the test animals since non-specific variations imposed by changes in diet, breeding, and environment may alter the susceptibility of the host to the pathogen. However, by standardizing these variables as far as possible the host variation is often ignored and differences in host response are used as a comparative measure of virulence between pathogens.

Assuming the host population to be reasonably homogeneous and stable it is possible to estimate the degree of virulence by determining the numbers of micro-organisms that produce the same measurable disease effect, or by determining the differences in the severity of the disease produced by similar doses of different strains of the same microbe.

No population of animals will be so homogeneous that at one dose level all infected animals will die. In other words, the relationship between dose and response is not regular. In similar batches of animals, as the dose is increased by constant increments, the mortality rate increases relatively slowly over the ranges 1–10 per cent and 90–99 per cent, but most rapidly around 50 per cent. For this reason the 50 per cent killing rate is the most accurate estimate of virulence; it ignores those members of the population which are unusually resistant and those which are unusually susceptible. Consequently this value (the dose which kills 50 per cent of the population, or LD_{50}) is the most useful measure of virulence.

The LD_{50} for biologically different micro-organisms varies considerably according to the nature of the microbes. For Type I pneumococci, the narrow dose range of 5–100 organisms covers the mortality range of 1–99 per cent in mice; for *Salmonella typhimurium* the same mortality range is covered by a broad dose range of 10 to 1,000,000 organisms (Table 3.2). Plotting the two series of values, i.e. mortality rate against the logarithm of the dose, two sigmoid curves—one shallow, the other steep—result (Figure 3.1). A comparison of these two curves demonstrates the difficulty of comparing the dose–responses of two dissimilar micro-organisms. By choosing the same mortality rate for both organisms (e.g. LD_{50}) it is, however, possible to compare the equivalent dose–response of each. It must be emphasized, however, that these dose–responses are only valid for the equivalent mortality rate. If multiples of the

Table 3.2. A comparison of the doses of *Streptococcus pneumoniae* and *Salmonella typhimurium* administered to mice in order to produce a range of LD effects (data from Miles, 1955)

	No. of organisms		
LD value	*Strep. pneumoniae*	*Salm. typhimurium*	Ratio
LD_1	5	10	1:2
LD_{10}	10	200	1:20
LD_{50}	20	3,400	1:170
LD_{90}	45	50,000	1:1,100
LD_{99}	100	1,000,000	1:10,000

dose increments for the LD_{50} are used the dose–responses for each organism will no longer be comparable. For instance, five increments of the LD_{50} dose of pneumococci would probably kill all mice, whilst five increments of the LD_{50} of *Salmonella typhimurium* would probably kill only 78 per cent. Multiples of LD_{50} doses can only be compared if the dose–response curves for different organisms are similar. Where the slopes are different the mortality rates for each organism will be different (Figure 3.1; Table 3.3).

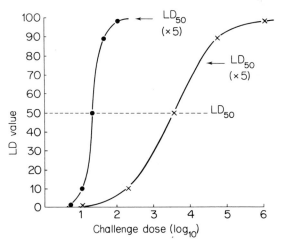

Figure 3.1. A comparison of the doses of *Streptococcus pneumoniae* and *Salmonella typhimurium* required to produce a range of LD effects in mice (data as in Table 3.2—after Miles, 1955)

Table 3.3. Comparison of the LD values for *Streptococcus pneumoniae* and *Salmonella typhimurium* for mice by increasing the LD_{50} challenge dose five-fold (data from Miles, 1955)

	No. of organisms	
	Strep. pneumoniae	*Salm. typhimurium*
LD_{50}	20	3,400
Increase dose five-fold	100	17,000
LD value corresponding to a five-fold increase in dose (as read from Figure 3.1)	LD_{100}	LD_{78}

FURTHER READING

Miles, A. A. (1955) The meaning of pathogenicity. In Howie, J. W. and O'Hea, A. J. (eds), 5th Symposium of the Society for General Microbiology, *Mechanisms of Microbial Pathogenicity*. Cambridge University Press, pp. 1–16.

Smith, H. (1964) *Microbial behaviour in natural and artificial environments*. Smith, H. and Taylor J. (eds), 14th Symposium of the Society for General Microbiology, *Microbial Behaviour*, 'in vivo' *and* 'in vitro'. Cambridge University Press, pp. 1–29.

Smith, H. (1958) The use of bacteria grown *in vivo* for studies on the basis of their pathogenicity. *Ann. Rev. Microbiol.*, **12**, 77–102.

Chapter 4

Mechanisms of microbial pathogenicity

Microbial pathogens possess unique properties by which they can overcome the host defences, multiply and produce damage to the host. In this chapter various components of pathogenicity will be considered. It is recognized that our knowledge is limited, particularly with certain pathogens, but an attempt is made to give a general account of what is known at the present time. Information derived from both *in vitro* and *in vivo* studies is considered.

Most microbial pathogens are not toxic *per se*; there are a few exceptions (see later). Disease is usually the consequence of the metabolic activities of virulent microbes within the host. These activities enable the microbe to utilize the body tissues and fluids as sources of energy and as nutrients for the synthesis of microbial cellular constituents. This often involves microbial breakdown of host substances resulting in damage to tissues. In addition many pathogens produce enzymic by-products of their metabolism some of which are highly toxic to the host: these exotoxins specifically and often irreversibly damage vital functions of the host metabolism.

1. The process of infection

Virulence factors operate at different stages of infection. Before proceeding therefore, to consider the microbial armoury which renders a pathogenic organism virulent, a brief and simplified description of the process of infection will be presented.

Pathogens vary in the extent of damage they cause. Primarily this depends on the biological nature of the microbial species and the virulence of the infecting strain. Some produce local lesions only, subsequently the micro-organisms are overcome and the lesions resolved by the host defences. There are exceptions. For example, infections with *Clostridium tetani* occur by contamination of a local wound; the organisms remain localized and multiply in the anaerobic conditions that prevail in the damaged tissue. However, the lethal neurotoxin (Chapter 5), produced locally, is transported along the neurones of the nerves to the central nervous system. Vital muscles, served by affected nerves, become paralysed and this results in death of the patient. In this disease a localized infection results in generalized symptoms.

Other infections commence as local lesions, but spread to other parts of the body by direct extension, or via the lymphatic or circulatory systems, and produce generalized infections. By these means local foci of infection are set up in organs distant from the initial site of infection. This is the consequence of pyaemia (literally pus in the blood, e.g. from a local abscess) or bacteraemia (bacteria in the blood, e.g. organisms entering the blood through the gut mucosa). In some peracute or acute infections, e.g. pasteurellosis in cattle, the organisms multiply in the blood, a condition referred to as septicaemia (i.e. sepsis in the blood). In septicaemias, the organisms rarely localize and, unless proliferation of the organism is arrested, the course of the infection terminates fatally.

Fortunately, after a period of illness, the majority of patients recover from most infections; the body defences kill and eliminate the organisms from the body, a process which may be aided by chemotherapy. Sometimes, however, the microbe gains the supremacy and the patient dies as a consequence of tissue damage or toxaemia (toxins in the blood), the toxins interfering with some vital physiological function. The progress of a hypothetical disease, followed by recovery, is illustrated in Table 4.1; the footnotes explain the sequence of events.

2. Factors associated with primary lodgement

The first stage of infection is the colonization by the pathogen of the appropriate portal of entry (Table 8.4). To achieve this it must first be able to withstand the structural or primary defences of the body. In contrast to a number of protozoal and viral diseases, few bacterial pathogens are introduced directly into the tissues

Table 4.1. The progress of a hypothetical infection in an animal host in which recovery was complete (after Lovell, 1958)

	Regional lymph node	Liver	Spleen	Other tissues	Blood stream	Antibodies
A	+	−	−	−	−	−
B	+ +	−	−	−	−	−
C	+ + +	+	+	+	±	−
D	+ + + +	+ +	+ +	+	+	−
E	+ + + +	+ + +	+ + +	+ + +	+ + +	+
F	+	+	+	+	−	+ +
G	−	−	−	−	−	+ + + +

+, + +, + + +, + + + +, = presence and number of organisms or level of antibody.
A = Invasion.
B = Multiplication.
C = Further multiplication and spread to blood stream.
D = Further multiplication especially into reticuloendothelial depots (probably corresponds to onset of symptoms).
E = Bacteraemia and initial antibody response.
F = Clinical recovery, clearing of blood stream, and increase in antibody titre.
G = Complete recovery with elimination of the infective agent.

by a vector bite. Most bacterial pathogens initially lodge on the skin or the mucous membranes of the respiratory, alimentary or urogenital tract. The differing abilities by which each microbial pathogen accomplishes colonization may explain, in part, why some infectious diseases are more communicable (e.g. brucellosis, tularaemia and tuberculosis) than others (e.g. mumps and anthrax).

(a) *The size of infecting dose*

Our knowledge of the mechanisms by which pathogenic organisms achieve the primary lodgement is very limited. Their ability to do this is the more surprising since, in contrast to artificial animal virulence tests, most natural infections are initiated by very small numbers of organisms. The tularaemia agent (*Francisella tularensis*) is notoriously infectious and it is probable that a single organism is able to initiate a fatal infection in mice. In other infections, e.g. salmonellosis in calves, an infective dose of 10^5 salmonellae is needed to establish clinical infection but repeated small doses, as occurs naturally by ingestion of food or licking other animals, must be equally effective (Chapter 14).

One mechanism, by which small inocula are enabled to colonize, has been elucidated. A partial relationship between primary lodgement of the whooping cough organism, *Bordetella pertussis*, and the production of toxin has been noted in mice infected by the intranasal route. *In vitro* studies have demonstrated that the pertussis toxin is able to inhibit the cilia action of the respiratory tract epithelium—an action which can be neutralized by specific antitoxin. It is assumed that inhibition of cilia action predisposes to initiation of infection in the natural disease and this may play a very significant role in the establishment of small numbers of the organism. The ability to produce the toxin is an essential component of pathogenicity in *Bord. pertussis* and, for this reason, the toxoid is incorporated in whooping cough vaccine.

(b) *Specific attachment of pathogens to mucous surfaces*

Many species of bacteria have a remarkable affinity for specific sites in the body. As previously noted, this is fundamental to the establishment of the normal flora in the healthy individual. Within the streptococci a wide range of adhesion mechanisms is experienced. In the human oral cavity *Streptococcus mutans* and *Strep. sanguis* adhere to the teeth, *Strep. salivarius* to the tongue, and *Strep. mitis* to the buccal cavity. Reference has been made previously to the mechanism by which *Strep. mutans* initiate colonization of the tooth surface and of its role in dental caries. Certain of the oral streptococci, particularly *Strep. sanguis*, and *Strep. mutans*, if they gain access to the blood stream—which may occur during dental manipulation—adhere to the heart valves and cause subacute bacterial endocarditis. For this to occur prior damage to the heart valves is necessary, e.g. as a sequel to rheumatic fever, and this provides the surface to which the streptococci may adhere by virtue of their lipoprotein surface structures.

Other streptococci, particularly virulent strains of *Strep. pyogenes*, adhere preferentially to the epithelium of the throat. These organisms possess an outermost layer of cell wall M-protein which facilitates adhesion. Loss of M-protein, by treatment with trypsin, or pretreatment with anti-M-protein serum, inhibits their adherence to human epithelium.

The adhesion of gonococci to epithelial cells of the urinogenital tract, by which they withstand the flushing action of urine, is considered to be associated in part with their surface pili. The association between virulence of the Kelloggs Types 1 and 2 and the presence of pili, has been noted for some years. In contrast, Types 3 and 4 are non-piliate and of lower virulence. However, other factors must also play a role since piliate and non-piliate gonococci have been shown to adhere to epithelial cells in the human Fallopian tube. It is clear that we do not yet fully understand this mechanism.

Organisms which infect by the alimentary route also must possess unique adhesive properties to withstand the continuous flow of the gut contents. Specific affinity has been noticed by particular pathogens for different sites on the same anatomical structure. Thus *Vibrio cholerae* stick to the epithelial cells at the base of the intestinal villi whilst certain strains of *Escherichia coli* stick to brush-border cells on the tips of the villi of the upper small intestine. The biochemical basis for these differences is generally unknown but extensive work has been carried out on enteropathogenic strains of *E. coli* which cause neonatal diarrhoea in piglets.

Two components of pathogenicity, in piglet enteropathogens, have been determined; the K88 antigen and an enterotoxin. Both are essential for the full expression of the disease. Since both are controlled by extrachromosomal DNA (plasmids, Chapter 11) it is possible to manipulate each factor separately and transfer them to other serotypes of *E. coli* and to *Salmonella* spp. and thereby study their pathogenic significance. The K88 antigen is essential for the *E. coli* to adhere to the mucosa of the small intestine and this can be prevented by pretreatment with K88 antiserum. The K88 antigen has been shown to be a surface structure of proteinaceous pili which act as a 'bacterial glue' causing the organisms to bind specifically to the brush-border cells of the small bowel mucosa.

The association of this antigen with specific adhesion seems to be limited to *E. coli* serotypes since transfer of the K88 plasmid to salmonellae does not produce an adhesive effect. Some piglets do not inherit specific receptors for the K88 antigen—the so-called 'non-sticky piglets'—and these are resistant to infection. Once *E. coli* attach to the gut wall, the production of enterotoxin is essential for the expression of pathogenicity (Chapter 5). In calves the K99 antigen is responsible for adhesion, and similar, but as yet uncharacterized, antigens are thought responsible for adhesion in human enteropathogenic *E. coli*.

3. Multiplication of microbial pathogens in the host

Once primary lodgement in the host is attained the process of infection can continue only if the pathogen multiplies. This depends on at least two factors, which are difficult to study separately. Firstly, the nutritional status of the host tissues must be adequate to support growth of the pathogen. Secondly, the pathogen must be able to overcome the secondary or physiological defences of the host. This may be accomplished by the production of microbial substances toxic to host tissues, or the microbe may be protected against the host defences.

(a) *The nutritional status of the host tissues*

The tissues of an animal host contain, or are bathed in fluids which contain, nutrients adequate to support the growth of most non-exacting pathogens. It is the rare event to find a bacterial pathogen which is avirulent in a particular animal host by virtue of the tissues being nutritionally deficient. There are a few well documented examples where the ability to grow, the rate of growth, or virulence to the host may be dependent on the nutritional status of the tissues.

Both naturally occurring and artificially obtained mutants of pathogenic species of bacteria are known which have become exacting for a particular nutrient. If the nutrient is absent from the tissues of the host the micro-organism cannot multiply and is thus avirulent to that host. Purine-dependent strains of *Salmonella typhi* and *Yersinia pestis* are avirulent to mice and guinea pigs which naturally lack purines in their tissues. Both organisms may become virulent in these hosts if purines are injected simultaneously with the purine-requiring strains.

Occasionally the overabundance of a tissue component may interfere with virulence but not growth. The virulence of the diphtheria bacillus, for instance, is dependent on the production of a powerful exotoxin. This toxin is produced only when the exogenous concentration of iron falls below a certain threshold level. The external concentration of iron will decide, therefore, the amount and time of toxin production (Chapter 5). The predilection for a particular organ or tissue is often associated with the presence of a growth-promoting factor in the tissue.

It is important to stress that microbial cells grown *in vivo* are markedly different from those grown *in vitro*. This has been illustrated by the frequent loss of virulence in isolates grown in artificial culture. Old laboratory strains often fail to induce immunity to infection by virulent strains of the same microbial species. The lethal exotoxin of *Bacillus anthracis*, which for so long evaded demonstration *in vitro*, was shown to be produced in the plasma of infected animals (Chapter 5). In anthrax the exotoxin is protected *in vivo* from destruction by the organism's autolytic enzymes but similar protection is absent *in vitro*, the toxin rapidly disappearing from culture supernatants after about 6 hours incubation.

(b) *The rate of growth of pathogens in the host*

The *in vivo* rates of growth of most pathogens are unknown but in the few which have been studied their rate of growth *in vivo* has been shown to be relatively slow. This is not due to nutritional insufficiency but to many other factors, including competition with resident flora and the handicap of combating the host defences. The usual method of determining growth rates, by preparing serial dilutions and setting up viable counts, if applied to the *in vivo* situation, gives a false estimate. By this approach the viable counts measure not the growth rate but the net result of microbial multiplication on the one hand and death due to bactericidal components of the tissues or removal by phagocytes on the other. To overcome these difficulties Meynell and his colleagues superinfected the pathogen with a phage which made possible an estimate of the true *in vivo* division rate distinct from the superimposed death rate. This was done by lysogenizing a strain of *Salmonella typhimurium* by phage P22. The infecting organism was also superinfected with a non-excluding mutant of phage P22. The lysogenizing phage could be detected on one test strain and the superinfecting phage on another. Since the mutant phage was usually stable and did not divide during bacterial growth, the proportion of bacteria carrying the mutant fell steadily with time in each bacterial generation (Figure 4.1). It was assumed that removal or death of the organisms *in vivo* was similar whether or not the mutant phage was carried. Comparison between the proportion of superinfected bacteria and others at the start and after a known period made possible a determination of the number of bacterial generations which had occurred in that time. Viable counts made on homogenized spleens from mice infected with the lysogenized *Salm. typhimurium* revealed that the populations of viable cells doubled every 24 hours but the mean division time was of the order of 8–10 hours compared with 30 minutes *in vivo*. These figures do not

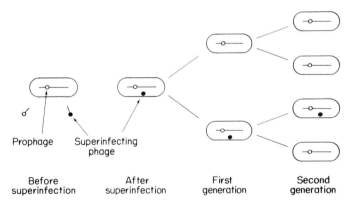

Before superinfection	After superinfection	First generation	Second generation

Figure 4.1. A method of determining growth rates of bacteria *in vivo* by the use of a superinfecting phage among dividing lysogenic bacteria. Each bacterium carries the prophage which multiplies synchronously with the bacteria. The superinfecting phage does not multiply. Hence, if each bacterium initially contains one superinfecting phage particle, the proportion of superinfected bacteria halves in each bacterial generation (Meynell, G. G. (1959) *J. Gen. Microbiol.*, **21**, 421–37. Reproduced by permission of Cambridge University Press)

necessarily apply generally to other pathogens in different hosts; it is obvious that in the terminal stages of illness, as in anthrax, when the body defences are virtually overcome, the growth rate must be very rapid. It is to be expected, therefore, that the rate of growth will vary with the stage of the illness which, in turn, will depend on whether or not the body defences have been overcome.

(c) *Products of microbial origin which aid multiplication in the host*

Substances, structural or secretory, which are non-toxic but impede the host defences and allow multiplication to occur were termed 'aggressins' by Bail at the beginning of this century. This term is widely used, and will be used here, but implies a noxious rather than blocking action on the host defences. A more functionally appropriate term 'impedin' has been adopted by some workers but has not displaced the more widely used term aggressin. This group of microbial substances includes any non-toxic factor capable of inhibiting host defences. Since defence mechanisms are varied in their mode of action it is not surprising that aggressins likewise form a similarly heterogeneous group of chemical substances.

(i) *Aggressins against host bactericides.* Many tissues of the host possess substances which are actively bactericidal. These may be found in blood serum as well as organized tissues. The virulence of pathogens is, in part, associated with their resistance to these natural bactericides and this may influence their success at both the primary lodgement and subsequently during growth and tissue invasion. Anthrax bacilli, for instance, are protected against the anthracidal components of certain host sera (e.g. β-lysins) by virtue of their non-toxic, polyglytamic acid capsules and additionally by another non-toxic, intracellular component present in a lipoprotein fraction of cell homogenates. The former is thought additionally to protect the organism against phagocytosis (see later) whilst the latter has anticomplementary activity. In a similar manner *Brucella abortus* has been shown to possess an immunogenic cell wall component which protects it against bactericidal substances in bovine serum.

(ii) *Aggressins which inhibit phagocytosis.* Phagocytic activity of the wandering and fixed cells of the reticuloendothelial system forms the main protective defence of the body once the microbial pathogen has penetrated the primary defences. This activity is non-specific but is greatly enhanced in the presence of antibody specific for the invading microbe. Since phagocytosis involves contact with the microbe followed by ingestion and finally digestion, the armoury of the successful pathogen may be directed at one or more of these successive stages of the process.

(α) *Aggressins which inhibit phagocyte migration.* Contact with phagocytes is the first essential and chemotaxis, by which phagocytes may be attracted to the infected area, is a possible target for microbial interference. Little is known of this—but fractions of tubercle bacilli are known to inhibit leucocyte migration.

(β) *Aggressins which inhibit ingestion of bacteria by phagocytes.* More is known about properties of microbes which interfere with the next stage of phagocytosis, i.e. the ingestion of the organism by phagocytes. Very few pathogens can withstand digestion once they get inside the phagocyte (see later). It is important therefore that the majority of pathogens avoid ingestion by phagocytes. Aggressins play an important role in resisting phagocytosis. They consist of structural components of virulent pathogens sited on the cell surface or outside the cell in the form of capsules. Organisms which possess these structures may resist ingestion altogether or, if ingested, may resist digestion and be rejected without damage to themselves or the phagocyte (see below). The following are examples of typical aggressins: the capsular polysaccharide of *Streptococcus pneumoniae*, the M-protein component of the cell wall and capsular hyaluronic acid of *Strep. pyogenes*, the capsular poly-D-glutamic acid of *B. anthracis*, the O-somatic antigens of some Gram-negative organisms, the Vi antigen (poly-*N*-acetyl-D-galactosaminuronic acid) of *Salm. typhi* and the protein carbohydrate envelope substance of *Yersinia pestis*.

These protective structures have no feature in common, either chemical or physiological, that would suggest a common mode of action. The surface structures may resist phagocytosis by a physical effect due to lipophilic groupings on polysaccharide side-chains of somatic antigens but this is not proven. It should also be noted that not all capsulate organisms, e.g. *Bacillus* spp. (other than *B. anthracis*), are resistant to phagocytosis; it is the chemical nature of the capsule that determines resistance. Further, the presence of a capsule on a potential pathogen (e.g. *B. anthracis*) is only one of a number of factors involved in virulence. If the other factors are lacking, the capsulate organism is avirulent.

(γ) *Aggressins which inhibit digestion by phagocytic bactericidins.* Once an organism has been ingested by a phagocyte it becomes exposed to the various bactericidins in the phagosome. Most ingested organisms are readily killed and digested but a few are able to resist the bactericidins and grow intracellularly. They include virulent strains of tubercle bacilli, gonococci, meningococci and brucellae. Their ability to survive, and even grow intracellularly, must play an essential role in their mode of pathogenicity. During the early stages of infection, before they have developed their full complement of virulence factors, an intracellular survival in the host may be a protective one. The phagocyte may also provide a protective environment for long-term survival in chronic infections or in the carrier state.

Intracellular parasites survive by virtue of their aggressins which interfere with the bactericidal mechanisms of the host cell. Many factors have been thought to act as aggressins in the survival of *Br. abortus* in phagocytes. These include the presence of growth factors in the cells, which specifically supported the growth of virulent cells, and the higher catalase content of virulent strains to counteract the hydrogen peroxide within the phagocyte. Neither of these has proved, however, to be the basis for survival. It is now known that a cell wall component of virulent

Br. abortus interferes with the intracellular bactericidal mechanisms of phagocytes. This is not present in brucellae grown in artificial culture, unless bovine placenta tissue is added, but is present regularly on virulent organisms grown *in vivo*. The cell wall material separated from virulent brucellae inhibits the intracellular destruction of avirulent strain and its effect can be specifically neutralized by antiserum against live virulent brucellae. The cell wall component is distinct from the one mentioned earlier which protects the organism against the extracellular bactercidal activity of phagocytes in serum.

Extracts of polymorphonuclear leucocytes have been shown to be bactericidal for certain pathogens, e.g. staphylococci. This is doubtless the principal means by which these phagocytes kill staphylococci. It has been shown that staphylococci grown *in vivo* (in rabbits) became more virulent for the same host and this is thought to be due to the development, by the microbe, of aggressins against the phagocytic bactericides.

The various phagocytic cells (e.g. monocytes, polymorphonuclear leucocytes, etc.) produce different intracellular bactericides. Some are active against Gram-positive organisms, others against Gram-negative organisms, and some show activity against a range of organisms. Hence, ideally, to combat all the bactericidins present in different phagocytes a bacterium would need to produce many aggressins. This ability is rarely, if ever, found. In fact, some pathogens can survive in one series of phagocytes but not in others. *Yersinia pestis*, for instance, can survive and grow in mouse and guinea pig monocytes but not in polymorphonuclear leucocytes of the same animals. These differences may play an important role in the process of infection. Organisms, initially of low virulence, may be able to survive in phagocytes which are less bactericidal for them but, during the period of intracellular existence, they may develop their full range of aggressins and subsequently be more virulent and able to resist other phagocytes to which they were previously sensitive. *Yersinia pestis*, grown in monocytes, have been shown to be subsequently resistant to ingestion by polymorphonuclear leucocytes. The intracellular state may also explain why antibiotic-sensitive organisms persist despite therapy; it may be that they survive intracellularly during therapy only to be released later and set up clinical illness.

(δ) *Products of bacteria which damage phagocytes.* In addition to protecting themselves against phagocytosis many microbial pathogens actively damage phagocytes. Many microbial pathogens, in particular the Gram-positive pyogenic organisms, secrete extracellular enzymes which kill leucocytes to form pus. These include the leucocidins of the pyogenic cocci, and fall into the general group of toxic products, the exotoxins (Chapter 5).

4. Invasive activity

The ability of bacteria to survive and multiply in the host tissues is only one aspect of the infective process. Virulent bacteria produce substances which act positively against the host by breaking down the secondary defences of the body. Even in the non-immune animal the non-specific defences have to be overcome but in the immune animal, the specific immune defences raise an additional obstacle to pathogen survival. To break down these defences bacteria must be not only present in large numbers and endowed with protective structures, but must also be capable of damaging the host defence mechanisms. The course of infection leading to clinical illness, and recovery or death, is the outcome of the interplay of these factors.

(a) Toxicity of whole cells

A microbial pathogen is not usually pathogenic by virtue of toxic chemical constituents of the organismal cell. Related virulent and avirulent microbes often possess the same chemical composition both in their cellular and extracellular structures. There are exceptions; both pathogenic and saprophytic mycobacteria, for instance, contain the toxic phthioic acid which, when injected into animal tissues, produces tubercles, a chronic type of inflammatory response.

Pathogenic mycobacteria, e.g. *Mycobacterium tuberculosis*, are able to induce progressive disease and tubercles arise in many parts of the body to which the organisms spread. Saprophytic mycobacteria, whilst not able to produce progressive disease, nevertheless upon injection into a susceptible host may give rise to extensive lesions closely resembling those of tuberculosis. Nodules and caseation are formed around the the acid-fast bacilli. Following intramuscular or subcutaneous injection the lesions remain localized. Intraperitoneal injection results in lesions widely distributed; the organisms are carried from the site of injection to various parts of the body by lymph flow or by phagocytes carrying the organisms. Intravenous injection of dead or living saprophytic strains can cause severe illness and sometimes death.

(b) Toxic products of microbial pathogens

Two groups of toxic bacterial products have been recognized, namely endotoxins and exotoxins. Endotoxins were considered to be structural components released from dead cells of Gram-negative organisms and exotoxins, soluble products of living cells. This subdivision was based chiefly on the historic mode of their discovery.

It is now considered that a rigid concept of two clearly distinct groups, based on sites outside or inside the cell, is only approximate. For instance, many exotoxins are found to accumulate within the cell and can be extracted from it. It has also been shown that endotoxins may be released from intact living cells. Thus the notions of 'exo' and 'endo' toxins are essentially ambiguous.

To overcome these difficulties a classification based on the chemical nature of the bacterial toxins is preferred. The two groups are still recognized but not on their siting in the cell:

1. *protein toxins*, which include the conventional exotoxins of Gram-positive bacteria but also a smaller number of toxins produced by a few Gram-negative bacteria; and
2. *lipopolysaccharide–protein complexes*, which include the classical endotoxins of Gram-negative bacteria.

The knowledge available on both groups of toxins is extensive and they are considered respectively in Chapters 5 and 6.

FURTHER READING

Berkeley, R. C. W., Lynch, J. M., Melling, J., Rutter, P. R. and Vincent, B. (1980) *Microbial Adhesion to Surfaces*. Ellis Horwood, Ltd., Chichester, England (559 pp.).

Dubos, R. J. and Hirsch, J. G. (Eds) (1965) *Bacterial and Mycotic Infections of Man*, 4th ed. Pitman Medical Publishers, London (1025 pp.).

Editorial (1974) Bacterial stickiness. *Lancet*, **i**, 716–17.

Glynn, A. A. (1972) Bacterial factors inhibiting host defence mechanisms. In Smith, H., and Pearce, J. H. (eds), 22nd Symposium of the Society for General Microbiology, *Microbial Pathogenicity in Man and Animals*. Cambridge University Press, pp. 75–112.

Lovell, R. (1958) *The Aetiology of Infective Disease*. Angus and Robertson, Michigan State University.

Mims, C. A. (1976) *The Pathogenesis of Infectious Diseases*. Academic Press, New York. 246 pp.

Rutter, J. M. (1975) *Escherichia coli* infections in piglets: pathogenesis, virulence and vaccination. *Vet. Rec.*, **96**, 171–5.

Rutter, J. M., Burrows, M. R., Sellwood, R., and Gibbons, R. A. (1975) A genetic basis for resistance to enteric disease caused by *E. coli*. *Nature*, **257**, 135–6.

Savage, D. C. (1972) Survival on mucosal epithelia, epithelial penetration and growth in tissues of pathogenic bacteria. In Smith, H., and Pearce, J. (eds), 22nd Symposium of the Society for General Microbiology, *Microbial Pathogenicity in Man and Animals*. Cambridge University Press, pp. 25–57.

Smith, H. (1968) Biochemical challenge of Microbial pathogenicity. *Bact. Rev.*, **32**, 164–84.

Smith, H. (1977) Microbial surfaces in relation to pathogenicity. *Bact. Rev.*, **41**, 475–500.

Standfast, A. F. B. (1964) The correlation of properties *in vitro* with host parasite relations. In Smith, H. and Taylor, J. (eds), 14th Symposium of the Society for General Microbiology, *Microbial Behaviour*. Cambridge University Press, pp. 64–84.

Chapter 5

Bacterial protein toxins—
'exotoxins'

A. GENERAL CONSIDERATIONS

The concept that bacterial pathogens produce their harmful effects on the host by means of poisonous products is almost as old as the science of bacteriology. As early as 1872 Klebs suggested that substances of microbial origin, which he called 'sepsins', were responsible for the lesions in staphylococcal infections. Similarly, Robert Koch (1884) attributed cholera to a toxicosis. Later, Loeffler (1887) observed that in fatal cases of diphtheria, characteristic lesions distributed throughout the body were invariably found to be sterile, the organism being present only in the primary lesion ('false' membrane) in the throat. He concluded that this clearly indicated that a poison produced at the site of inoculation must have circulated in the blood. A year later Roux and Yersin demonstrated the presence of a substance in sterile filtrates from cultures of *Corynebacterium diphtheriae* which—when injected into guinea pigs, mice, and pigeons—produced symptoms of disease and death similar to infections with the diphtheria bacillus. These workers coined the term 'toxin' for the bacterial poison. They also showed that in the urine, taken shortly before death from children with diphtheria, there was sufficient 'poison' to kill guinea pigs with symptoms similar to those produced by culture filtrates. This indicated that the toxin in culture filtrates was probably the same as that produced in the natural disease. These observations gave the first real evidence in support of the concept that bacterial disease could be explained in terms of a soluble toxic substance released by the bacteria. In 1889 Roux expressed the view to the Royal Society that in infectious *malades* the cause of death is poisoning and the microbe is not merely the means of spreading infection but is also the maker of the poison. Shortly afterwards Faber & Bregel, and Frankel (1889) independently discovered the toxin of tetanus and Van Ermengem (1896), the toxin of botulism. Parallel with these findings was the important discovery, by von Behring and Kitasato (1890), that immunity against diphtheria and tetanus could be artificially achieved by a process we now call immunization, and hopes were raised that specific antitoxic immunity could be developed for all infectious diseases.

Not only was this not realized, but very few other diseases could be shown to be caused by a single toxic product. It is now known that many bacteria produce a wide range of products, both toxic and non-toxic, but it is often impossible to state categorically which of these, if any, produce the full disease syndrome. Indeed, many severe diseases have yielded no demonstrable toxin either *in vitro* or *in vivo*.

1. The nature of bacterial protein toxins

As stated in the previous chapter two classes of bacterial toxins are recognized, based on their chemical composition. In this chapter the bacterial protein toxins are considered. All are soluble proteins secreted by the living micro-organism during growth and are products of cellular metabolism. Not all toxin is extracellular; frequently larger quantities than those present in the surrounding medium can be extracted from the producing cells by autolysis. In fact, in the extraction and purification of type E botulinum toxin, the organism, rather than the culture filtrate, is used as a source of the toxin.

The production of protein toxins is specific to each microbial species. Virulent strains invariably produce the toxin, or range of toxins, whilst avirulent ones do not. Most protein toxins are produced by Gram-positive bacteria. These include *Corynebacterium* spp. (e.g. *C. diphtheriae*), *Staphylococcus aureus*, *Streptococcus*

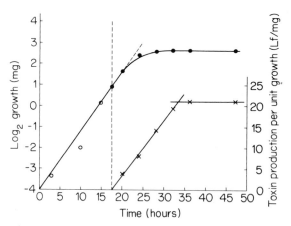

Figure 5.1. Toxin production and growth of the P.W.8 strain of *C. diphtheriae* in shake culture as a function of time. ×: toxin production in Lf per unit growth in milligrams. ●: logarithmic growth curve. (After Pappenheimer, A. M., Jr (1955) The pathogenicity of diphtheria. In Howie, J. W., and O'Hea, A. J. (eds) 5th Symposium of the Society of General Microbiology, *Mechanisms of Microbial Pathogenicity*, p. 43. Reproduced by permission of Cambridge University Press)

pyogenes, Bacillus anthracis, and pathogenic species of *Clostridium*. Gram-negative bacteria generally produce 'endotoxins' only (Chapter 6) but a few, in addition, also produce protein toxins. These include *Vibrio cholerae, Yersinia pestis, Shigella dysenteriae, Bordetella pertussis* and certain enteropathogenic *Escherichia coli.*

Not only are protein toxins distinct from other poisons in being essentially microbial in origin but they are active at much greater dilution. The dilution at which some of them are still active, compared with that of strychnine, is given in Table 5.1. This high level of activity also distinguishes them from lipo-polysaccharide toxins (Chapter 6). One reason for their great activity may reside in the generally accepted belief that all protein toxins are enzymes. As long ago as 1888, Roux and Yersin asked, 'Quelle est la nature du poison diphterique? Est ce un alcaloide ou une diastase?' It was their opinion that the diphtheria toxin was an enzyme (diastase) and not an alkaloid.

Protein toxins resemble enzymes in their protein nature, their lability, their high biological activity, and their specificity of action interfering with a specific metabolic function a component of which acts as a substrate. They are produced together with other bacterial products which, without a doubt, are enzymes (e.g. collagenases, hyaluronidase etc.)

As proteins they are strongly antigenic and, because of this property, have sometimes been referred to as 'antigenic poisons of microbial origin'. Indeed such is their specificity that antigenic differences are used to distinguish closely related exotoxins. Specific antitoxins completely neutralize the toxicity of these bacterial proteins *in vivo*. Homologous antitoxin may not, however, fully inhibit their enzymatic activity *in vitro*. This suggests that the location of the antigenic determinant structure is distinct from the enzyme determining part of the protein molecule and the degree of neutralization of each may depend on the distance between the two groups on the molecule. However, since a toxin is fully neutralized by specific antitoxin *in vivo*, but not always *in vitro*, other *in vivo* factors must play a role.

Ehrlich observed that protein toxins are not stable; in time they lose their toxic properties but retain their antigenic ones. He coined the term 'toxoid' for this product. Toxoiding of toxins can be speeded up by the use of a variety of reagents including formaldehyde, carbon disulphide, iodine, ascorbic acid, pepsin, nitrous acid, ketones, etc. Usually 0.1–0.2 per cent formalin is used. The mixture is maintained at 37°C for several weeks within the pH range 6–9. The resultant toxoids can be safely used for immunization, e.g. against tetanus, diphtheria and some of the clostridial diseases.

Since toxins are unstable it is more reliable to assay them against the more stable antitoxins. The amount of antitoxins required to neutralize a toxin preparation (determined by toxicity tests in animals) measures the potency of the preparation. This test accepts the error that any naturally produced toxoid, present in the preparation, will also take up a proportional amount of antitoxin.

Table 5.1. Properties of purified bacterial toxins. (Modified from van Heyningen, *Microbial Toxins*, ed. by Ajl, Kadis and Monte, 1970. Reproduced by permission of Academic Press Inc.)

Toxin source		Toxic dose (μg)	Host	Biological activity	Lethal toxicity compared with strychnine*
Cl. botulinum	type A	2.5×10^{-5}	Mouse	Lethal	1,000,000
Cl. botulinum	type B	2.5×10^{-5}	Mouse	Lethal	1,000,000
Cl. botulinum	type D	0.8×10^{-5}	Mouse	Lethal	3,000,000
Cl. tetani		4.0×10^{-5}	Mouse	Lethal	1,000,000
Cl. perfringens	ε-toxin	8.0×10^{-2}	Mouse	Lethal	300
C. diphtheriae		6.0×10^{-2}	Guinea pig	Lethal	2,000
Sh. dysenteriae	'neurotoxin'	2.3×10^{-3}	Rabbit	Lethal	1,000,000
Staph. aureus	α-toxin	5.0	Rabbit	Lethal	350
Staph. aureus	enterotoxin	2.5 (approx.)	Monkey	Emesis	—
Strep. pyogenes	erythrogenic toxin	5.0×10^{-6}	Man	Skin reaction	—

*Other values for comparison:

strychnine	1.0
bacterial endotoxins	0.1
snake venom	10.0

2. Biological activity of protein toxins

As enzymes attack specific substrates, so bacterial protein toxins are highly specific in the substrate utilized and their mode of action. Usually the substrate is specific for each toxin and may be a component of tissue cells, organs or body fluid. Consequently the site of damage caused by a toxin indicates the location of the substrate for that toxin. Hence terms such as neurotoxin, enterotoxin, haemolysin, leucocidin, and cytotoxin have been coined to indicate the target site of some well-defined toxins (Table 5.2). Certain toxins of staphylococci, clostridia, etc., have a fairly broad cytotoxic activity and non-specific death of tissue results (necrosis). A number of pathogenic bacteria produce phospholipases which can

Table 5.2. Biological activities of some soluble protein antigens

Source and nature of toxin		Substrate	Activity
C. diphtheriae	(Schick toxin)		
Staph. aureus	α-toxin	Skin	Cytotoxic leading to dermonecrosis
Cl. perfringens	α-toxin		
Strep. pyogenes	erythrogenic toxin (Dick toxin)	Subcutaneous tissues	Oedema and/or haemorrhage
B. anthracis	toxin complex		
Oxygen-labile haemolysins			
Cl. perfringens	toxin		
Cl. tetani	tetanolysin	Erythrocytes	Haemolysis
Strep. pneumoniae	pneumolysin		
Strep. pyogenes	streptolysin O		
Oxygen-stable haemolysins			
Staph. aureus	α-, β-toxins		
Cl. perfringens	α-toxin	Erythrocytes	Haemolysis
Cl. septicum	α-toxin		
Strep. pyogenes	streptolysin S		
Staph. aureus	leucocidin	Leucocytes	Death
Staph aureus	enterotoxin	Gut mucosa	Emesis
Cl. botulinum	toxins	Nerves	Muscle paralysis
Cl. tetanus	tetanospasmin		
Cl. perfringens	α-toxin		
Cl. oedematiens	β-, γ-toxins (phospholipases)	Cell membranes	Phospholipid cleavage
Staph. aureus	α-toxin	Variable and unknown	Lethal
Clostridia—various			
B. anthracis	toxin complex		
Strep. pyogenes	hyaluronidase	Hyaluronic acid	Polymerization
Staph. aureus	coagulase	Fibrinogen	Coverts to fibrin
Strep. pyogenes	streptokinase	Fibrin	Digest fibrin
Staph. aureus	staphylokinase		
Cl. perfringens	κ-toxin (collagenase)	Collagen	Digests collagen

cleave phospholipids in a number of different ways. Phospholipids are important components of cell membranes and cleavage of these results in destruction of the membrane and consequent death of the cell by leakage of cell contents. If this occurs with red blood cells, haemolysis results. Others, which bring about death of the host, are known as lethal toxins but although the tissue affected and the target site may be known the precise mechanism by which death occurs is not.

Not all extracellular bacterial proteins damage tissue cells and it is disputed whether or not these should be called toxins. Nevertheless their activity plays a distinct role in disease. Some attack the soluble constituents of body fluids (e.g. coagulase of pathogenic staphylococci converts fibrinogen to fibrin to form part of the abscess wall structure), others attack the interstitial cement substance, hyaluronic acid, which binds cells together in organized tissue (e.g. hyaluronidase—a depolymerase), and others breakdown the collagen between muscle fibres (e.g. collagenase). Some proteolytic enzymes (e.g. streptokinase and staphylokinase) convert the inactive plasminogin into the active enzyme plasmin which digests fibrin and prevents the clotting of blood. Bacterial enzymes which break down the integrity of tissue or interfere with the clotting mechanism of blood are sometimes referred to as 'spreading factors' since they facilitate the spread of microbial pathogens through the tissues from the primary site of infection, as in streptococcal erysipelas in man.

3. Host susceptibility to protein toxins

Not all hosts are susceptible to the same toxin. Guinea pigs are highly sensitive to diphtheria toxin whilst rats are very resistant. Between different animal species susceptible to the same toxin often considerable variation in dose is necessary to produce similar symptoms. Guinea pigs are 1000-fold more sensitive to the same preparation of dysentery neurotoxin compared with mice, and 6000-fold more sensitive to a botulinum toxin. Observations of this kind must be made on the same preparation of toxin since different preparations of the toxin often show considerable differences in their relative toxicity for different animal hosts. Differences between species are found also at the cellular level. The same phospholipase (or haemolysin) lyse red blood cells from different hosts at different rates. The α-toxin of staphylococci is 100 times more haemolytic for rabbit red blood cells than for those of human origin.

Apart from the enterotoxins, it is usual to test the potency of a toxin by parenteral administration into a susceptible animal. This has led to the erroneous conclusion that toxins are not generally active when given orally—but this is not true. Diphtheria and tetanus toxins are toxic by mouth; presumably they are absorbed from the gut; the effective oral dose is, however, many thousand-fold greater than the parenteral.

Some important examples of diseases caused by bacterial protein toxins will now be considered.

B. EXAMPLES OF DISEASES CAUSED BY BACTERIAL PROTEIN TOXINS

1. Anthrax

Most warm-blooded animals may be infected with anthrax but the degree of susceptibility varies widely. The herbivorous animals, especially cattle and sheep, are highly susceptible, as are also horses, buffalo and other wild herbivora, guinea pigs, and mice. Swine are not so susceptible—the infection running a chronic course. Dogs, cats, rats, and birds are relatively insusceptible but may be artificially infected. Man occupies an intermediate level of susceptibility; the disease in many may be fatal but more frequently occurs as a local infection and treatment with antibiotics is successful.

Two forms of anthrax are usually recognized. *Localized* or *cutaneous anthrax* occurs in a few species (man, swine, dogs, etc.). Infection may be through a wound. In man this assumes an occupational risk of dockers, butchers, etc., who carry or handle carcasses or hides imported from countries where anthrax is endemic and which may be contaminated with anthrax spores. Abrasions caused whilst carrying these contaminated materials predispose to infection. The lesion takes the form of an oedematous swelling, first hot and painful, but later cold and painless, becoming an intense, dark, bloodstained carbuncle-like lesion called a malignant pustule. In swine and dogs localized infection follows ingestion of contaminated foods, compounded animal feed containing contaminated bone meal and infected meat respectively. The organism enters the tissues of the upper part of the digestive tract, probably through the tonsil, resulting in an inflammatory oedema of the tissues of the head and neck.

Generalized or *septicaemic anthrax* may arise from cutaneous anthrax. More usually it arises from eating compounded foods contaminated with anthrax spores, as in cattle, the organism invading from the intestine, or by inhalation of spores as in woolsorters' disease in man. The spores germinate, the bacilli invade the lymph system from which they spill into the blood stream, overgrow the defence systems of the body, and produce a massive septicaemia, the organisms being found throughout the entire body at the time of death. Animals suffering with this form of anthrax usually exhibit high fever and bleeding from body openings. In the terminal stages the animal shows acute respiratory distress and shock, and death occurs within a few hours.

(a) *Mechanism of pathogenicity*

It is now known that the pathogenicity of anthrax is due to a number of microbial products which together produce the full clinical syndrome. Prior to 1954 the fact that *B. anthracis* produced a toxin was not known. Many observations on pathogenicity had been made and a number of theories of the mechanism of pathogenicity proposed. It was known that a solid immunity could only be achieved using living vaccines, indicating that the active metabolism of the organism *in vivo* was necessary to produce the protective immunogens. The sterile oedema fluid from anthrax lesions contained an immunogen which, when injected into a susceptible host, produced immunity. The association between the capsule and virulence was also established. All virulent *B. anthracis* were found to be capsulate but not all capsulate strains were virulent. This indicated that the characteristic non-toxic, polyglutamic acid-containing capsule was essential to protect the virulent organisms against anthracidal component of blood and phagocytosis by polymorphonuclear leucocytes; the capsule thereby acting as an 'impedin' or 'aggressin'. The capsule is important in the establishment of infection but probably plays no role in the terminal phase of the disease. Pathogenicity must therefore be attributed to other factors in addition to the presence of the specific capsule.

The intense bacteraemia in the blood of a septicaemic patient immediately prior to death suggested that either deoxygenation of the blood occurred or the capillaries were mechanically blocked. However, the fact that a massive bacteraemia is not always present at death in some animal species militated against these suggestions. Other workers had demonstrated a marked hyperglycaemia, an interference in calcium metabolism, and lesions in the nervous system leading to symptoms of anthrax.

Real progress in an understanding of the pathophysiology of the disease was made, when in 1954 Smith, Keppie and others demonstrated the toxigenicity of this disease. This work, together with later studies, led Sterne (1961) to write: 'perhaps we know more about the way anthrax bacillus works than we do about almost any other pathogen of similar invasiveness'. Because of this, the development of the stages leading to the present knowledge of the subject will be described.

Work prior to 1954 had concentrated on *in vitro* studies of the organism and little progress was made. Smith and his colleagues studied the behaviour of the organism *in vivo*. They attempted to produce *in vivo*-grown organisms and body fluids, adequately separated and in sufficient quantities, for chemical analysis and biological testing. The guinea pig was chosen for these studies; this animal is highly sensitive to anthrax, large numbers of animals could be used, and the same animal species could also be used for testing the biological activities of the various extracts and fractions.

Evidence that a toxin was produced *in vivo* was first established. Large numbers of guinea pigs were inoculated at the same time and, at regular intervals, groups of these animals were treated with streptomycin to terminate abruptly the bacteraemia at progressive stages of the disease. It was found that at death the number of organisms in the blood was of the order of 1×10^9 organisms per ml. If however the bacteraemia was halted before the numbers of organisms reached about 3×10^6 per ml, the guinea pigs survived; if the numbers in the circulation were greater the animals died. It was concluded that irreversible damage was caused by numbers of anthrax bacilli in the blood 1/300th of the number regularly present at death. The evidence strongly pointed to the production of a lethal toxin *in vivo*. This was confirmed by the production of extensive oedema and congestion in normal guinea pigs following infection of heparinized plasma taken from guinea pigs dying of anthrax. The tissue-damaging factors increased as the bacilli in the blood increased and eventually neat plasma was able to kill mice and guinea pigs by intravenous injection. The symptoms resembled those of natural anthrax and the lethal effect could be completely neutralized by specific antiserum.

Biological tests on the various materials separated from infected guinea pigs showed the toxin to be chiefly in the plasma and not associated with *in vivo*-grown bacilli. Traces of toxin in oedematous fluid were due to leakage of plasma into the tissues. The toxin accumulated in the plasma of infected animals in the terminal stages of the disease.

In addition to the protective polypolyglutamic acid capsule described earlier, at least three serologically active and antigenically distinct factors were demonstrated in the plasma, a protective antigen (PA), a lethal factor (LF) and an oedema factor (EF). Apart from their immunogenicity each of the three factors failed to produce significant biological activity alone but together acted as a complex toxin (Table 5.3). PA and LF combined to produce lethal activity, PA and EF combined to produce dermonecrotic activity, but EF and LF together produced no activity. The complex of all three is immunogenic, lethal, and dermonecrotic.

As a consequence of the different methods of extraction used, and the degree of purity of the products obtained, not all workers agree on the precise chemical

Table 5.3. Host responses to the plasma toxins of *Bacillus anthracis* alone and in combination (modified from Stephen and Pietrowski, 1981)

Toxins	Host response		
	Oedema in skin (rabbit)	Death (mice)	Immunity (guinea pig)
I (oedema factor)	−	−	−
II (protective antigen)	−	−	+ +
III (lethal factor)	−	−	−
I and II	+ + +	+	+ + +
I and III	−	−	+
II and III	−	+ +	+ +
I, II, and III	+ +	+ + +	+ +

− = no reaction.
+, + +, + + + = increasing degrees of reaction.

nature of the three factors. There is, however, general agreement that all three are non-dialyzable, thermolabile proteins or lipoproteins.

(b) In vitro *toxin production*

Finding a toxin *in vivo* stimulated further investigations with artificial cultures and it was soon discovered that toxins were produced *in vitro* in the early hours of bacterial growth but progressively disappeared due to autolytic enzyme destruction. Initially complex media containing serum were used but subsequently it was shown that a variety of synthetic media are suitable, the essential requirements for toxin production being the presence of bicarbonate and blood aerated with 20 per cent carbon dioxide, the carbon dioxide being necessary either for toxin production or its release from the cells.

In vitro toxin preparations are closely similar to those produced *in vivo* but important differences have been noted. For instance, injection of *in vitro*-produced toxin killed rats in 54 minutes whilst *in vivo*-produced toxin, similarly administered, killed rats in 10 minutes. These differences may be due to variation in molecular configuration or quantitative differences between the three factors, rather than fundamental differences in nature.

(c) *Assay of toxin*

The whole toxin and its three constituents may be variously assayed by their biological activity (lethality or skin oedema) and by their serological specificities. For instance, the LD_{50} of whole toxin in plasma from guinea pigs dying of anthrax can be assayed by intravenous injection of graded doses into mice or guinea pigs. The lethal components, i.e. LF and PA together, may be similarly assayed.

The production of skin oedema, following intradermal injection into guinea pigs or rabbits, may also be used. The test may be quantified either by measuring the size of the lesion or determining the dilution of the extract which produces a standard-sized lesion. The whole toxin, or a combination of EF and PA, may be assayed in this way.

At least seven animal species have been found susceptible to toxin produced *in vitro* and the pathological effect is linearly related to $\log_2$ dilutions of the concentration.

(d) *Mode of action of the toxin*

Since death may be sudden and unexpected infection may not be detected in life; the symptoms (e.g. temperature, changes in blood picture, changes in heart rate, etc.) may be virtually non-existent. Consequently different workers, often working with different animal hosts, have reached varied conclusions as to the primary cause of death. It is generally assumed that the cause of death in anthrax is the same for different species of animal hosts. Most hypotheses explain the cause of death to be due to secondary effects. Apart from the earlier suggestions that death results from blockage of the capillaries or metabolic disorders, more recent theories include cardiovascular failure, direct effect of toxin on the reticuloendothelial system, oxygen depletion, secondary shock, increased vascular permeability, and respiratory failure due to a lesion in the central nervous system controlling this function. The last four are the more probable mechanisms. The fall in blood oxygen to less than 1 per cent of normal levels (even in death due to asphyxiation the loss is about 5 per cent of normal) is the most dramatic single change, but this is not due to destruction of erythrocytes or a decreased oxygen-combining capacity of haemoglobin. Smith and his associates are persuaded that the diminished blood volume and secondary shock with impaired renal function are the immediate cause of death in guinea pigs. Others consider that capillary thrombosis is a more consistent finding. Alterations in pulmonary endothelium led to capillary thrombosis and vascular leakage with accumulation of plasma-like fluid in the tissues.

One sign common to anthrax death in all hosts, either by infection or the injection of toxin, is respiratory failure which could be responsible for the 'sudden' and unexpected terminal results frequently encountered. Some workers have suggested that this is a consequence of neurological changes which can be detected by electroencephalograms. Lesions in the CNS are considered to directly cause respiratory failure and the sequential anoxia. This interpretation, however, has not found wide acceptance since it appears that the neurological changes are secondary to the rapid effects of the toxin on the vascular system.

2. Diphtheria

This important and at one time prevalent, lethal, human disease was characterized by a local infection of the throat. It consisted principally of a membranous inflammation of the fauces, associated with an equally characteristic toxaemia, in which a powerful protein toxin exerted a specific action on certain tissues remote from its site of production. The association of *Corynebacterium diphtheriae* with the disease was first described by Loeffler (1884) and, four years later, Roux and Yersin (1888) demonstrated its toxic nature when they showed that sterile filtrates of cultures of the diphtheria bacillus were lethal to guinea pigs; the lesions produced were identical with those caused by injection of the living organism. The pathogenicity of the organism included two distinct phenomena:

(1) invasion of the local tissues of the throat—whilst this is comparatively of a low order and it is rare to find the organism elsewhere in the body it is, nevertheless, a first requirement—and subsequent bacterial proliferation; and
(2) the ability of the organism to produce the toxin.

The virulence of strains of *C. diphtheriae* cannot therefore be defined solely in terms of toxigenicity.

Three varieties of *C. diphtheriae* are recognized, namely *gravis, intermedius,* and *mitis*. These are listed in falling order of severity in the human patient but all produce the same toxin. Since the severity of the illness must in the last analysis depend on the toxin produced, the differences between the three varieties must reside in their differing abilities to produce the toxin, both in rate and quantity.

(a) *Toxigenicity of* C. diphtheriae

Of the many factors which have been investigated, two appear to have the greatest influence on toxin production; namely, a low extracellular iron concentration and the presence in the organism of a lysogenic bacteriophage.

It has been shown that toxin production in artificial cultures begins when extracellular iron is declining. Maximum toxin is formed when the iron concentration is virtually exhausted (Figure 5.1). The level of iron is extremely critical—toxin production is completely inhibited in the presence of 0.0005 mg/ml and maximally produced when the level is as low as 0.00014 mg/ml. This level falls far below growth-inhibitory levels; for instance 0.06 mg/ml does not inhibit growth. The differences in virulence between the three varieties of *C. diphtheriae* may be explained on their different growth rates in the throat. The faster-growing varieties may deplete the local iron supply more rapidly in the invaded tissues, thereby resulting in an earlier or greater production of the diphtheria toxin, and this is assumed to be characteristic of the *gravis* variety. The Halifax strain of *gravis* has a doubling time *in vitro* of 60 minutes, whereas the Park Williams *mitis* strain has a doubling time of 160 minutes. However, the PW8 strain of *C. diphtheriae* var. *mitis* (Park Williams strain) is used commercially for the production of toxin since this strain is one of the most prolific producers of the toxin of all strains tested *in vitro*—it produces 5 per cent or more of bacterial protein in the form of the toxin. It may be deduced, therefore, that *in vivo* the yield of toxin from a toxigenic strain is controlled by the physiological state of the host and this may be greatly influenced by external levels of iron.

(b) *The role of* β*-phage in toxigenicity*

Reports in the early literature (e.g. Dudley, 1919) indicated that avirulent strains of diphtheria bacilli, carried in the throats of susceptible patients in a closed community, may suddenly assume virulence and produce clinical diphtheria. No explanation of this phenomenon was available until Freeman (1951) and Hewitt (1952) demonstrated that avirulent strains of *C. diphtheriae*, lysogenically infected with a phage from a toxigenic strain, became toxigenic. It was later shown that 'curing' a toxigenic strain of its phage resulted in loss of virulence; re-lysogenization by the same phage reinstated the toxigenic property. These early observations have been amply confirmed. Only diphtheria bacilli infected with β-phage, or with certain closely related phages, are toxigenic. Many attempts to explain the relationship between phage and host cell have been proposed. Some workers considered that the prophage acts as a genome governing toxigenicity and that multiplication of phage was essential both for toxin production and its release. Others found that the degree of cell lysis, either autolytically or by phage multiplication followed by cell lysis, did not account for the amount of toxin liberated. By radioactive studies it was shown that the toxin was synthesized *de novo* from amino acids in fresh, iron-free media.

Genetic studies have clarified the position. Genetic crosses between closely related toxigenic and non-toxigenic phages of *C. diphtheriae* have revealed that the capacity of a phage to convert a strain to toxigenicity is coded by the phage genome. This *tox* character maps at a single region on the genome. By lysogenizing an avirulent strain with a mutant of β-phage (phage 45), the organism was able to synthesize a non-toxic protein serologically related to diphtheria toxin. It was released in good yield in iron-deprived media. The non-toxic protein had a shorter chain length than the toxin itself but cross-reacted with diphtheria antitoxins raised in horses and rabbits. The properties of this protein established beyond reasonable doubt that the structural gene for toxin or toxin-like protein, is carried by the respective phage genome.

It is evident that β-phage introduces into the host cell the gene coding for toxin biosynthesis. Phage multiplication is not essential for toxin production or release. The regulation of toxin yield, is however, dependent upon the metabolism of the host bacteria and it is in this way that the iron concentration exerts its marked influence. High yields of toxin are only synthesized by lysogenic bacteria with a low external iron concentration.

It is of interest to speculate on the value to the organism of synthesizing so much toxin, up to 5 per cent of the total protein synthesized. There is very little information to suggest the role the toxin plays in the life-cycle of the organism, especially since its known activity is limited to inhibiting protein synthesis in certain eukaryotic cells. Since mass immunization has been practised diphtheria has virtually disappeared and *C. diphtheriae* is no longer a component of the flora in the human throat and pharynx. It may be that the toxin played a role in the colonization of the throat in non-immune humans and, as a consequence of extensive immunization programmes, toxigenic strains have become virtually extinct.

(c) *Mode of action of diphtheria toxin*

Early observations indicated that the toxin of diphtheria produced damaging effects remote from the site of its production in the throat. The tissues affected included the pancreas, but the greatest effect was on the cardiac tissues. Electron micrographs indicated that the primary damage was in the sarcoplasmic reticulum of the myofibres. It is now known that the toxin exerts its primary effect by inhibiting protein synthesis in these susceptible tissues. The toxin specifically inactivates the translocating enzyme, the aminoacyl transferase II, an enzyme found only in certain eukaryotic animal cells, i.e. animals susceptible to the diphtheria toxin. The toxin catalyses the reaction shown in the following expression:

$$NAD^+ + \text{transferase II} \rightleftharpoons \text{ADPR-transferase II} + \text{nicotinamide} + H^+$$

Nicotine adenine dinucleotide (NAD) is converted to adenosine diphosphate ribose (ADPR), which combines with the transferase enzyme thereby inactivating it, and nicotinamide is released.

(d) *Testing isolates for virulence*

Three methods for testing virulence are in use—the subcutaneous, the intracutaneous, and an agar plate immunodiffusion method. The guinea pig is used for the biological tests. Two animals are used for the subcutaneous test; one is protected with a dose of diphtheria antitoxin (500 units) given intraperitoneally and both are inoculated with the organism under test into the chest wall. If the strain of *C. diphtheriae* is virulent, the unprotected animal dies usually in 18–96 hours. At post mortem, an extensive area of gelatinous, haemorrhagic oedema is found at the site of inoculation, the cavities (peritoneum, thoracic, and pericardium) are filled with straw-coloured fluid and the adrenal glands are haemorrhagic.

The intradermal method has the advantage that a number of strains can be tested on the same animal. To protect the animal a small dose (20–40 units) of antitoxin is given intraperitoneally at the same time as the test organisms or sterile culture filtrates are injected intradermally along the shaven abdominal surface of an albino guinea pig. Living virulent *C. diphtheriae* or toxin produce localized erythematous lesions followed by necrosis reaching its maximum effect in 48 hours. Positive and negative preparations should always be included for comparison.

The immunodiffusion test, or Elek's plate, allows antitoxin from a filter-paper strip to diffuse at right angles to the toxin from a streak inoculum of the test organism (Figure 1.4). The toxin precipitates the antitoxin where the two meet at optimal concentrations producing arrowhead lines in the agar. The test is rapid, giving results in 24 hours, and compares favourably with the *in vivo* tests.

3. Tetanus

The characteristic clinical features of tetanus and its high fatality (40–78 per cent in men) has attracted attention for centuries. Early descriptions of the disease are found in the writing of Hippocrates (460–355 BC). Tetanus occurs in man and most species of warm-blooded animals. The spores are widely distributed throughout the world. They have often been isolated from human and animal faeces and occur more frequently in soils fertilized with human and animal faeces than in barren uncultivated land. Tetanus is an infectious disease usually resulting from contamination of a natural or surgical wound by the spores of *Clostridium tetani.* The organism multiplies locally, having little power of invasion. However the symptoms often appear remote from the infection site, indicating the spread of a toxic product. Faber (1889) was the first to demonstrate production of a toxin by the organism (second in order of discovery to that of the diphtheria toxin) and it was soon shown that injection of this tetanospasmin, as the toxin was called, was able to mimic the natural disease. Another product of the tetanus bacillus—the haemolytic tetanolysin—was shown not to contribute to the disease.

(a) *The nature of the toxin*

The tetanospasmin is a simple protein without carbohydrate or lipid associations. Its molecular weight has been variously quoted as 68,000 and 148,000, the difference being due to the fact that the toxin molecules can form a dimer. The yield of the exotoxin may be equivalent to 5–10 per cent of the bacterial weight but, with no known function in the organism's physiology, the reason for its production remains a mystery.

The eleven serotypes of *Cl. tetani* differ in their ability to produce the tetanospasmin but, irrespective of the strain used, the exotoxin is antigenically and pharmacologically the same. The tetanospasmin has specific affinity for nervous tissue; it is therefore referred to as a neurotoxin. Why the toxin has a specific action on nervous tissue, to which naturally the organism has no access, is one of the many anomalies of nature. It is relatively heat-labile, being destroyed at 56°C in 5 minutes, and is also oxygen-labile. In the pure state it converts to toxoid at 0°C and even more rapidly in the presence of formalin.

The disease can be prevented by toxoid immunization. Once the disease is established, antitoxin is of little value in therapy. However, before the toxin is tissue-fixed, it can be fully neutralized by this agent. Unlike other diseases, such as diphtheria, the natural disease does not confer immunity since even a lethal dose of the toxin is insufficient to provoke an adequate antibody response.

The tetanus toxin is one of the three most poisonous substances known—the others being the toxins of botulism and diphtheria. The susceptibility of different animal hosts to the tetanus toxin differs considerably; the guinea pig is the most highly susceptible and the pigeon one of the least. Cold-blooded animals (e.g. frogs) are highly resistant at ambient temperatures (10–15°C) but their susceptibility is increased by raising the body temperature. The comparative toxicity of tetanus toxin for various hosts, published by von Behring as early as 1912, are still valid (Table 5.4).

Since no *in vitro* action of the toxin is known, the only method of assay must be carried out on the whole animal. The mouse is the usual animal of choice. The minimum lethal dose, usually the LD_{50} in 4–7 days, is used as the end-point of the *in vivo* assay. One milligram nitrogen of the pure toxin is able to kill 200 million mice; one milligram of pure total protein can kill 30 million mice.

The toxin is most potent when introduced directly into the central nervous

Table 5.4.   Relative minimum lethal dose (MLD) of tetanus toxin for various species of mammal and bird (on an equal weight basis), after von Behring (1912)

Horse	0.5	Rabbit	900.0
Guinea pig	1.0	Goose	6,000.0
Mouse	6.0	Pigeon	24,000.0
Goat	12.0	Hen	180,000.0

system; rather less by the blood stream or peripheral tissues; and least, but still effective in comparatively large doses, when given by mouth. The symptoms of tetanus have been compared with those of strychnine poisoning.

General (or descending) tetanus, follows introduction of the toxin via the intravenous route. After an initial symptomless period of ill-defined length, increasing spasticity of the jaw muscles progressively occurs. Similar changes in the skeletal muscles of the trunk and limbs follow. Finally generalized tetanic convulsions occur prior to death. This form of tetanus resembles the naturally occurring disease in animals and man.

Local (or ascending) tetanus follows injection of the toxin into a hind limb; the muscles of the affected region become first painful, then spastic. These symptoms spread in an ascending direction to muscles of the other limb, the trunk, and forelimbs. Death usually follows within 96 hours.

(b) *Mechanism of pathogenicity*

At the molecular level no known cell or body fluid is affected by the toxin, and, therefore, no recognized histological or chemical changes indicate the mode of action. Knowledge is limited to the affinity of the toxin for the nervous system. The toxin has a particular affinity for fractions of brain grey matter tissue which contains gangliosides and high levels of nerve endings (synaptosomes), especially synaptic membrane. This indicates the site of action of the neurotoxin but gives no indication of its mode of action.

Molecules of the toxin are transported to the central nervous system; they bind to the gangliosides at the motor nerve endings, are taken up into the axon by endocytosis and transported within vesicles to accumulate in the anterior horns of the respective spinal segment. The target site is essentially the neurone synapses. Having reached this site the toxin exerts an effect by stimulating the motor neurones. It acts presynaptically by inhibiting the release of neurotransmitters, such as glycine, into the synaptic cleft. The overall effect is to produce spastic paralysis of the muscles served by these nerve centres. The action of the tetanus toxin closely resembles that of strychnine except that this poison acts postsynaptically rather than pre-synaptically.

4. Botulism

The disease botulism (derived from the Latin *botulus*, meaning sausage) is an intoxication rather than an infection. The relation between eating sausage foods and a fatal form of food-poisoning was known more than 1000 years ago. In the ninth century it led Emperor Leo VI to forbid the eating of blood sausages. Since then many outbreaks have been reported. Van Ermengem (1897) isolated the causal organism, *Clostridium botulinum*, from ham, the eating of which gave rise to fatal food-poisoning in Belgium. He proved that a toxin, responsible for the disease, was present in the meat. Animal meat is not, however, the only food incriminated. Statistical analysis of more recent incidents has shown that other foods also carry the toxin. These include vegetables, fruit (especially canned), fish and marine animals for man, and fodder and bones for animals (Table 14.2). While the viable organisms and spores may be ingested in food they do not usually multiply in the body. The toxin is a product of the metabolism of the organism and ingestion of toxin-contaminated food results in botulism. It is now widely accepted that low, but toxic, levels may be present even in the spores of *Cl. botulinum*, the toxin being released by autolysis.

A number of sub-types of *Cl. botulinum* are recognized based on antigenic differences between the toxins produced by each but despite these differences all toxins have the same pharmacological activity (Table 14.2). Each sub-type produces a single toxin.

The botulinum toxin is destroyed by boiling and therefore survives only in uncooked or partially heated foods. The toxin is able to withstand the acid and digestive juices of the stomach, and be absorbed unchanged from the gut. It is a pure crystalline protein having a molecular weight of 900,000. It is the most powerful neuroparalytic poison known; 1 µg pure protein contains 200,000 MLD (minimum lethal doses) for 20 g white mice. Probably much higher doses are required to produce fatality in man. Like all protein toxins it can be readily toxoided and fully neutralized by specific antitoxin.

The morbidity rate may be 100 per cent whilst the mortality rate falls between 50 and 70 per cent. Symptoms differ from other forms of food-poisoning in that the toxin acts directly on the central nervous system. On ingestion of the toxin-containing food, the toxin is absorbed through the gastric and upper intestinal mucosa and can then be demonstrated in the blood. From the blood it becomes fixed to nervous tissue. The incubation period of the disease is usually less than 24 hours. Nausea and vomiting occur first; this is followed by paralysis of muscles affected by the action of the toxin on nerves serving these muscles, and symptoms include difficulty in breathing, in swallowing, in focusing the vision—resulting in double vision—and paralysis of the extremities. Death results from respiratory failure. In patients who recover the nervous structures affected by the toxin do not suffer permanent damage. However recovery is slow and many months may elapse before control of certain muscle movements is regained. The toxin does not exert injurious effects on the brain and spinal cord, and consequently the victim remains conscious throughout the course of the illness. Once the toxin becomes fixed to nervous tissue, antitoxin is unable to neutralize it or produce beneficial effects. Consequently by the time a diagnosis is made it is often too late to use type-specific antitoxic therapy.

The site of action of the botulinum toxin is in the peripheral nervous system. It acts at a site similar to that of acetylcholine, namely at the nerve endings (myoneural junction). The cholinergic nerve fibres are paralysed at the point of acetylcholine release. Botulinum toxin is able to affect both sites for cholinergic transmission in the autonomic system; namely, the synaptic ganglia and the parasympathetic motor end-plate peripherally located in the junctions between nerves and cell fibres.

5. Diseases caused by *Clostridium perfringens*

A wide range of clostridia produce gross tissue damaging infections in man and animals (Table 5.5). Frequently these are anaerobic wound infections and many take the form of clinical gas gangrene; this involves general and extensive tissue damage and usually death of the host. The tissue damage is caused by a range of different exotoxins produced by each species of clostridia.

Table 5.5.   Some histoxic clostridial infections

Species	Disease	Host
Cl. oedematiens		
type A	Gas gangrene	Man
	'Big head'	Rams
type B	Black disease (chronic hepatitis)	Sheep
type C	Osteomyelitis	Buffaloes
type D	Haemoglobinuria (red water disease)	Cattle
Cl. chauvoei	Black quarter	Cattle, sheep
Cl. septicum	Braxy (bradshot)	Sheep
	Malignant oedema	Cattle
	Gas gangrene	Man
Cl. perfringens		
type A	Gas gangrene	Man
type B	Lamb dysentery	Lambs
	Haemorrhagic enteritis	Sheep, goats
type C	'Struck'	Sheep
	Enteritis	Piglets
	Enteritis necroticans	Man
type D	Enterotoxaemia	Sheep
	Pulpy kidney	Sheep
type E	Enterotoxaemia	Lambs, calves

Of all the clostridia, *Cl. perfringens* has been shown to produce the widest range of toxins; twelve distinct soluble products have been identified. The toxins of *Cl. perfringens* only will be considered. A study of the various soluble products are of interest both in the role they play in pathogenicity and in their use as a basis for typing the various members of the species (Table 5.6). The four major soluble antigens (α-, β-, ε-, ι-toxins) are so called because of their prime importance in classification rather than indicating the amount of the toxin produced. The distribution of the major toxins in the various types is indicated below:

Type A	α	
Type B	β	ε
Type C	β	
Type D	ε	
Type E	ι	

From Table 5.6 it will be observed that the various types are responsible for specific diseases in different hosts. The more important of these only will be considered. Type A strains are associated with gas gangrene in man and animals in addition to being a common commensal in the gut. For gas gangrene to develop, the causal organisms must contaminate and multiply in damaged tissue, as may occur in accidental trauma and severe thermal burns. Contamination of a wound by pathogenic clostridia is very common but gas gangrene does not always follow due to the lack of favourable anaerobic conditions essential for the organisms to grow. The disease develops when contaminating organisms have multiplied and generated sufficient toxin to overcome the local defences of the tissues. Further multiplication and toxin production leads to rapid invasion of healthy tissues and the development of clinical gas gangrene. The sequence following a bullet wound is characteristic. Anaerobic bacteria are carried into the deep tissues by the bullet. The skin wound is relatively small thus limiting access of air, but extensive muscular damage and cavitation results beneath the surface. Soft tissues and blood vessels are torn, and bones may be fractured remote from the puncture wound. The damage results in anoxia of the tissues leading to death of tissues cells (necrosis) and, since collateral circulation is usually inadequate, an ideal environment is provided in which anaerobes may flourish.

A number of pathogenic clostridia may be incriminated in gas gangrene; these include *Cl. perfringens* type A, *Cl. septicum*, *Cl. oedematiens* type A, *Cl. histolyticum* and others, but by far the commonest anaerobe is *Cl. perfringens*. A sub-type of *Cl. perfringens* is characteristically heat-resistant and deficient in α-toxin, and has been associated with outbreaks of human food-poisoning in Great Britain (Chapter 14).

Lamb dysentery, caused by type B, is a lethal disease of lambs during the first 2 weeks of life and is prevalent in particular localities where the organism is present in the soil. The disease takes the form of an enteritis sometimes accompanied by extensive ulceration. *Cl. perfringens* type B are found in the dysenteric intestinal

Table 5.6. Various toxins of *Cl. perfringens* associated with different diseases (modified from Buxton and Fraser and reproduced by permission of Blackwell Scientific Publications Ltd)

Toxic and enzymic substances produced by *Cl. perfringens*

Type	Occurrence	α Alpha (Lethal necrotizing Lecithinase haemolytic)	β Beta (Lethal necrotizing, destroyed by trypsin)	γ Gamma (Lethal)	δ Delta (Lethal haemolytic)	ε Epsilon (Lethal necrotizing, activated by trypsin)	η Eta (Lethal, doubtful)	θ Theta (Lethal haemolytic, O₂-labile)	ι Iota (Lethal necrotizing, activated by trypsin)	κ Kappa (Collagenase gelatinase lethal necrotizing)	λ Lambda (Proteinase)	μ Mu (Hyaluronidase)	ν Nu (Deoxyribonuclease)
A	Gas gangrene (man) Intestinal commensal (man and animals). Soil	+++ (boxed)	−	−	−	−	±	+	−	+	−	+ (boxed)	+
A	Food poisoning (man)	+++	−	−	−	−	−	±	−	+	−	± (boxed)	++
B	Lamb dysentery	+++ (boxed)	+++ (boxed)	++	±	++	−	++	−	−	+++	+++	++
B	Haemorrhagic enteritis of sheep and goats (Iran)	+++	+++ (boxed)	−	−	+++	−	+++	−	+++	−	−	+
C	'Struck' in sheep	+++	+++ (boxed)	++	+++ (boxed)	−	−	+++ (boxed)	−	+++ (boxed)	−	−	+
C	Enteritis in lambs, calves, and piglets	+++	+++ (boxed)	−	+++ (boxed)	−	−	+++	−	+++	−	±	+
C	Enteritis necroticans (man)	+++	+	++	−	−	−	−	−	−	−	−	+++
D	Enterotoxaemia of sheep and pulpy kidney disease	+++	−	−	−	+++	−	++	−	++	++	+	+
E	Enterotoxaemia in lambs and calves	+++	−	−	−	−	−	++	+++	++	++	±	+

+++ = produced by all strains; ++ = produced by most strains; + = produced by some strains; ± = produced by very few strains.
☐ (boxed) = production of large amounts of toxin. Alpha, beta, epsilon, and iota are the major lethal toxins.

discharges. Little tissue invasion occurs but the lethal toxins, generated in the intestine, are absorbed and affected lambs die within a few hours.

Cl. perfringens type C causes a variety of enterotoxaemic diseases in animals and man. 'Struck' is a local name for a disease of adult sheep. Animals struck down by this disease die suddenly, and the only symptoms may be convulsions at time of death. The lesions, immediately after death, include severe enteritis and peritonitis—if the post mortem is postponed the muscles present the appearance of gas gangrene. Other strains of type C cause enterotoxaemia in calves, lambs, and piglets, and necrotizing jejunitis in man.

Enterotoxaemia of adult sheep and calves is caused by *Cl. perfringens* type D. There is no enteritis, probably because the organism fails to produce α-, β-, or γ-toxins. Type D is also responsible for pulpy kidney disease of sheep in which there is marked decomposition of the kidneys, often the only post mortem change. The toxins are absorbed from the intestine and the animal dies of acute entero-toxaemia.

(a) *Nature of the toxins*

Of the 12 soluble products of the various types *Cl. perfringens*, each identified by a Greek letter (Table 5.6), only about half are toxic. The term toxin is not applic-able therefore to all, and the better designation is 'soluble product' or 'soluble antigen' since each is antigenically distinct. The four 'major' soluble antigens used in identification are all lethal and therefore can be rightly called toxins. The biochemical basis for their toxic effects is not known but, additionally, they may have haemolytic and necrotizing activity. The non-toxic soluble antigens, in con-trast, are recognized by their enzymic activities on specific substrates, e.g. hyaluronic and desoxyribonucleic acids; some also exhibit haemolytic activity. No doubt most of the soluble antigens contribute at least in part to the disease syndrome but the most important ones are the major toxins and only these will be described.

(i) α-*toxin*. This relatively heat-stable toxin is produced in greatest amounts by *Cl. perfringens* type A but in lesser amounts in each of the other types (B–E). Besides being a lethal toxin it also exhibits haemolytic, necrotizing and lecithinase activity. Unlike other lethal toxins it is possible to measure the amount of α-toxin by the degree of lecithinase activity produced in egg-yolk medium, by its haemolytic activity against mouse red blood cells (these cells are resistant to other haemolytic soluble antigens such as α- and δ-toxins, Table 5.6) or by its lethal or necrotizing activity. The LD_{50} is usually determined by intravenous inoculation of laboratory animals whilst the necrotizing activity is determined by intradermal inoculation. Each of these reactions can be specifically neutralized by *Cl. perfringens* anti-α-toxin.

The α-toxin is a lecithinase C (Figure 1.3), the type most commonly found in micro-organisms. It splits lecithin into phosphorylcholine and stearyl oleyl-glyceride and is immunologically related to the lecithinases produced by *Cl. bifermentans* and *Cl. sordellii* but is far more toxic. It is immunologically distinct from those produced by *Cl. oedematiens, B. cereus, B. mycoides*, and *B. anthracis*. The haemolytic activity of α-toxin is probably the result of the lecithinase acting on phospholipids in the red blood cell membrane.

(ii) β-*toxin*. This necrotizing and lethal, but non-haemolytic, toxin is produced by *Cl. perfringens* types B and C both of which are responsible for enteritis in sheep, cattle, goats, and pigs. It is detected by intradermal inoculation into depilated white guinea pigs in which purplish, necrotic areas are produced. β-Toxin from *Cl. perfringens* type B produces irregular necrotic areas due to the presence of μ-toxin (hyaluronidase) whilst the β-toxin from type C, being free of μ-toxin, produces regular, smaller areas. It is antigenically distinct from all other clostridial toxins, is destroyed by trypsin, is readily toxoided by formalin or β-propiolactone, and is produced in highest concentration in young cultures (5–8 hours old), the concentration falling off with time, none being detected after 2 days.

β-toxin plays a significant role in the causation of lamb dysentery in young lambs (*Cl. perfringens* type B) and struck in older animals (*Cl. perfringens* type C). The occurrence of characteristic lesions in these diseases, namely haemorrhagic mucosal ulceration and superficial mucosal necrosis of the small intestine, are attributed to it. The differences between the lesions in the two diseases are thought to be due to the presence of ε-toxin (which increases intestinal permeability) and μ-toxin (which, being a hyaluronidase, facilitates spread of the β-toxin through the intestinal tissues), both of which are produced by *Cl. perfringens* type B.

(iii) ε-*toxin*. This toxin, together with a few other clostridial toxins (namely the β-toxin, ι-toxin, and the toxins of *Cl. botulinum*) shares the relatively unique property of being absorbed unchanged from the intestine. It is produced by *Cl. perfringens* types B and D but more regularly by type D which causes enterotoxaemia in sheep. It is initially produced by the organism as a non-toxic, heat-stable prototoxin which requires proteolytic digestion for activation—this is brought about naturally by κ- and λ-toxins also produced by the organism, or it may be artificially activated by 5 per cent trypsin. The highly toxic ε-toxin is heat-labile.

ε-toxin is necrotizing and lethal but non-haemolytic. It is detected by intradermal injection into guinea pigs. The toxin affects the permeability of the intestine which allows more toxin to be absorbed. At post mortem, no lesions are seen in the intestine but haemorrhages are present in the subendocardium and there is necrosis of the kidney parenchyma.

(iv) ι-*toxin.* Like ε-toxin this, too, is secreted as a prototoxin which can be activated by λ-toxin or trypsin. ι-Toxin is produced by *Cl. perfringens* type E, a non-pathogenic type, but the toxin is capable of producing necrosis and lethal effects in experimental animals.

6. Infections by pyogenic cocci

Historically staphylococcal and streptococcal infections have been responsible for most pus-forming lesions in man and many in animals. Many streptococcal infections have yielded to penicillin therapy and, in consequence, are not as difficult to control as once they were. In contrast, after an earlier reduction in the incidence of staphylococcal infections, over recent years they have become increasingly prevalent mainly due to the ability of many strains to produce β-lactamase by which they resist penicillin therapy. Both groups of organisms produce wide ranges of biologically active cellular and extracellular products which may be important in their pathogenicity (Table 5.7). The principal products of *Staphylococcus aureus* will be considered.

(a) *Staphylococcus aureus*

Infections in man are characterized by local pus formation, generally within walled abscesses in the skin and subcutaneous tissues. In more severe forms *Staph. aureus* can cause metastatic abscesses localizing in kidney, brain, and myocardium, and acute osteomyelitis. Staphylococcal infections of animals include mastitis in cows, goats, and bitches; skin lesions in lambs, horses, dogs, and poultry; chronic abscess formation in horses (called botriomycosis), and osteoarthritis in the joints of poultry.

It is clear that *Staph. aureus* is a most successful parasite; it can survive on a variety of body surfaces and has marked ability to invade local tissue. Apart from its broad armoury of pathogenic components it also has a remarkable ability to adapt to changing situations. A few genetic variants, better adapted to a new environment, may be selected and replace the parent population. For instance, treatment of an infected patient with penicillin can induce certain staphylococci to produce ever-increasing amounts of β-lactamase by which they can resist the antibacterial effects of the therapy.

Staph. aureus produces a wide range of cellular and extracellular products *in vitro* (Table 5.7). Some act as impedins whilst others possess tissue-damaging factors. Their precise role in pathogenicity is often difficult to define. A few of the cellular products will be considered.

(i) *Staphylocoagulase.* This is a non-toxic, thrombin-like substance, causing the clotting of blood plasma. Both cell-bound and cell-free coagulase are produced.

Table 5.7. Biological properties of some products of staphylococci. (Reproduced with permission from Jelaszewicz and Wadström, *Bacterial Toxins and Cell Membranes*, Copyright 1978 by Academic Press Inc. (London) Ltd)

Product	Main biological activities
Extracellular	
α-Haemolysin	Haemolysis, lethality, cytotoxicity
β-Haemolysin	Haemolysis (sphingomyelinase C), cytotoxicity
γ-Haemolysin	Haemolysis, cytotoxicity
δ-Haemolysin	Surfactant-like activity, including haemolytic and cytotoxic activities
Epidermolytic toxin	Epidermolysis *in vivo* (cleavage of granular layer)
Leucocidin	Leucotoxicity, lack of toxicity to other cells
Succinic oxidase factor	Inhibition of mitochondrial respiration
Staphylocoagulase	Prothrombin activation and fibrinogen clotting without participation of other blood clotting factors
Staphylokinase	Fibrinolysis (plasminogen activation)
Enterotoxins	Toxicity for sympathetic nerves and smooth muscles
Bacteriocins and micrococcins	Bactericidal and bacteriostatic *in vitro*
Lysozyme	Hydrolysis of peptidoglycan
Nuclease	Hydrolysis of RNA and DNA
Mitogens	Non-specific lymphocyte activation
Hyaluronidase	Hydrolysis of hyaluronic acid
Cellular	
Cell wall (peptidoglycan)	Pyrogenicity, arthrogenicity, toxicity, adjuvanticity
Capsular substance	Antiphagocytic properties
Clumping factor	Paracoagulation of fibrinogen
Protein A	Reaction with Fc fragment of immunoglobulins

The cell-bound coagulase plays a vital role in the initiation of infection. It causes fibrinogen to be absorbed to the bacteria which results in their aggregation at the site of infection or in the vascular bed; this restricts effective phagocytosis. The cocci multiply and as they do so produce a battery of enzymes including free coagulase. In the resultant abscess, the wall consists of fibroblasts embedded in much fibrin, a direct effect of coagulase on fibrinogen, typical of staphylococcal lesions.

(ii) *Staphylococcal haemolysins*. Among the extracellular proteins produced by *Staph. aureus* some are cytotoxic by damaging the cell membrane. These include the haemolysins. At least four haemolysins have been identified (α, β, δ, and γ). Other cells beside erythrocytes may also be damaged, hence they would better be called cytolytic toxins. The mechanism of action of each is clearly different.

(α) *α-Haemolysin*. Of all the soluble products of staphylococci the α-toxin is considered to play the most important role in pathogenicity. It is haemolytic, cytotoxic, dermonecrotic, and lethal (Table 5.8). The nature of the substrate attacked by the toxin is unknown but its greatest activity is associated with damage to cellular and lysosomal membranes resulting in osmotic disturbances within the cell. It lyses erythrocytes from rabbits with little activity against those from other species. The haemolytic activity is fully neutralized by specific antitoxin. Introduced parenterally it is lethal. The primary site of action is the central nervous system, especially the hypothalamus and the visual sensory region of the cerebral cortex.

(β) *β-Haemolysin.* The range of haemolytic activity of the β-haemolysin differs markedly from that of the α-toxin; it is active against sheep, ox, and human erythrocytes. *In vitro* it exhibits a hot–cold phenomenon. Exposure of sensitive erythrocytes at body temperature (37°C) shows little or no haemolysis; subsequent cooling to room temperature results in full haemolysis developing. The membrane substrate attacked by the β-haemolysin is the sphingomyelin—this is not its only activity since pig and human erythrocytes also contain this substrate but are not lysed. The β-haemolysin is strongly cytotoxic to established cell lines causing progressive death and decrease in sphingomyelin content. Sensitive cells include isolated leucocytes and macrophages. Platelets and lysosomes are also damaged and the latter may be responsible for the cytotoxic effects. Little is known, and there is confusion, about its action *in vivo.* Some workers have found no lethal action—others have suggested a cardiovascular effect. Further work is required on highly purified preparations of this toxin.

Table 5.8. Summary of the biological properties of staphylococcal membrane-damaging toxins (modified from McCartney and Arbuthnott in *Bacterial Toxins and Cell Membranes*, ed. by Jeljaszewicz and Wadström. Copyright 1978 by Academic Press Inc. (London) Ltd)

α-Toxin	Lethal, dermonecrotic and haemolytic; 1×10^3 LD_{50}/mg toxin (mouse); 2×10^4 HU/mg toxin (rabbit RBC). Causes spastic paralysis of smooth muscle. *Mode of action:* unknown; perturbation of hydrophobic region in membranes of sensitive cells? Specific receptors on sensitive cells?
β-Toxins	Lethal? Hot–cold haemolysin $> 10^6$ HU/mg toxin (sheep RBC). Requires Mg^{++} ions. *Mode of action:* enzymatic; sphingomyelinase C. Used as a probe for sphingomyelin in membranes.
γ-Toxin	Lethal, haemolytic, 4×10^3 HM/mg toxin (human RBC). *Mode of action:* unknown.
δ-Toxin	Lethal? Haemolytic for several species of mammalian erythrocytes. 10^2 HU/mg toxin (human RBC). *Mode of action:* detergent-like action on cell. Inhibited by certain phospholipids.
Leucocidin	Toxic to PMN leucocytes and macrophages of man and rabbits. *Mode of action:* creation of an ion channel due to specific perturbation of membrane lipids. This results in efflux of potassium ions and other secondary changes.

(γ) *γ-Haemolysin*. This haemolysin has been isolated in a highly purified state but little is known about its biological activities. It causes the release of lysosomal enzymes from isolated rabbit peritoneal granulocytes. *In vivo* it is lethal for guinea pigs after intracardiac injection.

(δ) *δ-Haemolysin*. Unlike α-, β-, and γ-haemolysins, the δ-haemolysin haemolyses erythrocytes from all animal species by a non-enzymic mechanism. It is relatively heat-stable and possesses strong surface and hydrophobic activity. It can lyse many cell lines *in vitro* including macrophages, granulocytes, and leucocytes in addition to erythrocytes. These effects can be neutralized by blood serum components (α- and β-globulins) which suggests it may be inactive *in vivo*.

(iii) *Staphylococcal leucocidin*. The separate identity of this product was not established until it could be purified. For a long time it was confused with the leucocidin activity of α- and β-haemolysins. It is now established that a non-haemolytic leucocidin is produced by staphylococci both *in vitro* and *in vivo* which specifically acts on leucocytes but not on lymphocytes or other tissue cell. The leucocidin consists of two fractions, the S (slow) and the F (fast). Separately each is inactive but together they act synergistically in a highly specific way. Damage occurs to the granulocyte cellular membrane giving increased permeability and impairing the sodium pump. The presence of antibodies to the leucocidin in the sera of patients with staphylococcal infections suggests a role in pathogenicity.

7. Enterotoxins formed by intestinal Gram-negative pathogens

A number of bacterial gut infections cause acute diarrhoeal diseases of man and animals, involving considerable fluid losses. Symptoms follow the rapid multiplication of the organisms in the gut with little or no invasion of the mucosa or other tissues. The effects are due to the production of enterotoxins in the gut lumen which act locally on the gut epithelium. The group includes a number of Gram-negative pathogens including *Vibrio cholerae, Vibrio parahaemolyticus, Escherichia coli*, and *Shigella dysenteriae*, and a few Gram-positive pathogens, including *Clostridium perfringens*. The term 'enterotoxic enteropathies' has been coined to cover diseases caused by these pathogens.

(a) *Cholera and cholera-like diseases*

Cholera is an acute, infectious, intestinal disease of man. Profuse, watery diarrhoea is the usual clinical symptom and vibrios are discharged in enormous numbers in the 'rice-water' faeces. Patients become rapidly dehydrated and, unless the fluid balance is quickly replaced by administering fluids intravenously, and later by mouth, death follows (Chapter 16). Typical symptoms of cholera appear 1–2 days after infection. Most infections are contracted by the oral route and are chiefly waterborne. The causative organism was originally *Vibrio cholerae* but, in recent years, *Vibrio El Tor* has been incriminated increasingly in outbreaks.

At one time it was thought that in cholera the intestinal epithelium was completely destroyed and fluid poured out from the raw surface of the gut. This has now been shown to be not so since, at post mortem, the epithelium is apparently normal and no inflammatory reaction is observed. It is now known that all the symptoms of cholera can be produced, not only by the vibrio organisms, but also by bacteria-free culture filtrates, indicating the involvement of an enterotoxin. The discovery of a toxin was not made until suitable experimental animal models were available. The first of these was the ligated segment of the small intestine in the living rabbit. Inoculation of cell-free filtrates of cholera vibrios causes the lumen of the gut to fill with fluid, indicating the presence of an enterotoxin. Dogs also proved to be susceptible both to infection and to intralumenal administration of the toxin. The symptoms produced in the dog are essentially the same as in man.

It is widely thought that the delay in discovering the cholera toxin was due to the absence of an experimental animal. It is more probable, however, that no-one expected a Gram-negative organism to produce a real toxin. Indeed, parenteral injection of cholera enterotoxin, the usual route for testing an exotoxin, does not cause diarrhoea, the chief symptom of cholera. The enterotoxin essentially acts upon the intestinal mucosa to produce the symptoms present in the natural disease. The diarrhoea factor (DF), as it is sometimes called, activates an enzyme, adenylate cyclase, which is present in the cell membrane of gut epithelium. This leads to a superabundance of adenosine-$3',5'$-cyclic monophosphate (cAMP) which, in turn, affects electrolyte transport. The outpouring of isotonic fluid is the consequence. This occurs in the upper part of the small intestine and in such quantities as to outweigh completely the absorption process in the lower part of the gut. Moreover, the effect is entirely localized to the gut epithelium. Thus the 'toxin' does not have direct toxic effect but exaggerates a specific physiological process to a highly pathological degree. For this reason some workers have reservations about calling it a true toxin.

The toxin has been shown to have a physiological effect on other tissues. For instance, it activates an intracellular lipase in isolated fat cells of the rat epididymis causing release of lipids. Such effects, however, play no role in the pathogenesis of cholera since the toxin is not absorbed into the blood stream. It has been observed that, in cholera, the endothelium of capillaries in the intestinal villi become rarefied and might be more permeable than normal. A permeability factor therefore may be involved but, if so, only to a very minor extent in the disease process.

It has been established that gut scrapings and brain, but not other tissues, inactivate the cholera toxin. The inactivating component has been shown to be ganglioside GM_1, and this is believed to act as the receptor for the absorption of toxin to the cell membrane of the gut epithelium.

The cholera toxin (choleragen) is a complex molecule of three polypeptides.

These are organized in two sub-units A and B into which the molecule readily dissociates. The intact molecule has a molecular weight of 32,000 daltons; sub-unit A has a molecular weight of 27,000 daltons (consisting of α-polypeptide, about 22,000 daltons and the γ-polypeptide, about 5000 daltons). Sub-unit B comprises four polypeptides, each of about 14,000 daltons. The structure of the whole molecule is therefore AB_4.

Sub-unit A is the component which activates adenylate cyclase in the epithelium. It can do this alone in intact cells but to a lesser degree than when associated with sub-unit B. It has low immunogenicity and is not inactivated by ganglioside.

Sub-unit B plays no role in activating the enzyme but binds the whole toxin molecule to the cell membrane. It is strongly immunogenic (often therefore termed a toxoid or choleragenoid) and combines with ganglioside.

The whole molecule therefore binds tightly to the outside of the gut epithelium through the sub-unit B–ganglioside bond. Thus the outside of the cell rapidly becomes saturated with toxin molecules and sub-unit A interacts directly or indirectly with the adenylate cyclase. The consequence of this interaction with the gut lining is the enormous flow of fluid into the small intestine.

(b) Escherichia coli *enterotoxins*

The natural habitat of *E. coli* is the alimentary tract of man and most warm-blooded animals, and carnivorous and omnivorous birds. The genus and species are defined on biochemical grounds but, within the species, many different serotypes are recognized. The serotyping scheme is based on the identification of somatic 'O', surface 'K', and flagella 'H' antigens of which 164 'O' and 56 'H' groups are currently recognized.

(i) *Infections outside of the gut.* The majority of *E. coli* are harmless as long as they remain in the gut, but some strains are able to cause infections in other sites. The most common coliform infection in the human (especially in women) is in the urinary tract. The source is usually the patient's own faecal flora, the organisms entering the bladder by an ascending infection, often being introduced mechanically by the insertion of a catheter. In addition to urinary tract infections, hospitalized patients may suffer post-operative wound infections with *E. coli* and sometimes develop septicaemia. It was once thought that all these infections were opportunist infections, the causal organism being any one of the normal flora carried by the patient and accidentally introduced mechanically into the various sites or at times when the tissues were unusually susceptible. While this is true in certain aspects it is now becoming increasingly recognized that many of the infections are caused by strains having unique properties which are not carried by the majority of strains in the faecal flora. These properties include the ability to produce haemolysins and colicin V, properties now recognized as conferring increased virulence. Strains from human, bovine, ovine, and avian sources carrying colicin V have been shown to be more invasive than others and this character is plasmid-mediated (Chapter 11).

E. coli can also cause infections outside of the gut in many animals. In the older animal, infections of the mammary gland (mastitis) has increased in prevalence over recent years, particularly in cows; other infections are found in the reproductive tract and urinary tract of various animals. Unique to poultry are infections in the air sacs, ovary, and heart sac (pericarditis).

(ii) Escherichia coli *infections within the gut.* The majority of *E. coli* exhibit no adverse effects upon the gut physiology but a limited number of serotypes are recognized as enteric pathogens both in man (Table 5.9) and animals. Not all exert their pathogenic activity in the same way; three groups are recognized dependent upon their interaction with the gut mucosa. One group, the enteropathogenic *E. coli* (EPEC) chiefly cause diarrhoea in babies and newborn animals. Like salmonellae, these *E. coli* penetrate the epithelial lining of the gut but with only minimal mucosa destruction. Clinical gastroenteritis and, occasionally, bacteraemia are the main symptoms. No enterotoxins have been detected so far and the mechanism of pathogenicity is unknown but colonization of the jejunum and upper ileum by EPEC strains is a prerequisite. At present evidence implicating these specific serotypes is based on detailed studies of the clinical history and the epidemic situation in which the same serotypes regularly occur. A second group, like *Shigella*, are enteroinvasive (EIEC); only a small number of serotypes are recognized (Table 5.9) and these are distinct from the EPEC strains. They invade the intestinal mucosa and subsequently cause extensive inflammation, necrosis, and ulceration. The organisms are located chiefly in the colon where intense inflammation of the mucosa occurs. This leads to fever,

Table 5.9. *Escherichia coli* and acute diarrhoea in man (after Rowe, 1977)

Type	Enteropathogenic (EPEC)	Enteroinvasive (EIEC)	Enterotoxigenic (ETEC)
Pathogenic mechanism	Unknown	Epithelial cell invasion	Enterotoxins (LT and/or ST)
Age group affected	Infants (rarely adults)	Adults and infants	Infants and adults
Epidemiology	Sporadic cases and outbreaks	Sporadic cases and outbreaks	Sporadic cases and outbreaks
O-groups commonly associated	26, 55, 86, 111, 114, 119, 125, 126, 127, 128, 142	28ac, 112ac, 124, 136, 143, 144, 152, 164	6, 8, 15, 25, 27, 78, 148, 159

and the stools contain blood and mucus, typical of dysentery. Infections occur in older infants and adults. The third group, also of distinct serotypes, include strains of enterotoxigenic *E. coli* (ETEC) (Table 5.9). Infections occur in the small bowel but they do not invade the intestinal mucosa. The symptoms are due to the release of enterotoxins by the *E. coli*, causing secretion of fluids and electrolytes leading to watery diarrhoea.

(α) *Enterotoxigenic* E. coli *(ETEC)*. The enterotoxigenic strains have received most attention in recent years. The rabbit ileal loop test, used to demonstrate the enterotoxin of *V. cholerae*, was first used to examine strains of enteropathogenic *E. coli* in 1956 in Calcutta where the test was being widely used in cholera work. The strains examined were from adult patients suffering from a cholera-like disease, and produced a positive loop test in rabbits. It would seem that the significance of this work went unnoticed. The role of enterotoxins in the pathogenesis of certain *E. coli* gut infections was clearly demonstrated in 1967 by veterinary workers using strains isolated from young pigs and calves suffering with scours, i.e. acute diarrhoeal disease, often called colibacillosis. Both live bacterial suspensions and cell-free filtrates of the specific serotypes produced fluid accumulation, when injected into ligated segments of the small intestine of the corresponding animal species, and diarrhoea, when administered intragastrically to young animals also of the corresponding species. Enterotoxigenic strains of *E. coli* are implicated in babies and lambs, as well as piglets and calves. Generally the young are affected more seriously than adults. Enterotoxin production is not of itself sufficient to enable strains of *E. coli* to cause diarrhoeal disease; other factors are also involved in pathogenicity. These include a colonization factor which is highly specific between the *E. coli* serotype and the susceptible animal or baby host. With porcine strains, in addition, most produce a haemolysin and have the ability to ferment raffinose. All these properties are plasmid-mediated (Chapter 11). Enterotoxin production is encoded on the 'Ent' plasmid; the colonization factor (CF) is coded on a different plasmid. In pig strains the ability to ferment raffinose is linked with the colonization factor on the same plasmid but the haemolysin (Hly) is coded on a separate plasmid.

(β) *Colonization factors*. Bacterial colonization is relatively unusual in the small intestine due to the strong bowel movements by which food and ingested bacteria are continuously propelled along the gut. In ETEC infections, however, the small intestine becomes colonized by these strains. This is made possible by the possession of colonization factors which enable the strains to adhere specifically to the gut lining. Colonization factors (CFs) are surface antigens, the nature of which is highly specific for each susceptible animal species. In the pig the K88 antigen is implicated. In calves the factor is a K99 antigen, whilst in babies, although serologically identified, the factor has not been given an antigenic designation other than CF. The colonization factor in humans has also been described as a pilus-like surface antigen. Since colonization factors are plasmid-mediated, it is possible to transfer them to suitable recipients, such as *E. coli* K 12, but naturally they are limited to a few serotypes. The colonization factor in human strains has been reported in three serotypes only (O63:H12; O78:H12; O153:H12).

(γ) *The enterotoxins.* The ability to produce both enterotoxin and colonization factor act together to produce the full clinical picture. Once the organism has adhered to the small intestine, toxin production commences. Strains which are toxigenic but which lack the plasmid coding for adherence are not able to produce diarrhoeal diseases.

The *E. coli* enterotoxins have been divided into two classes. The heat-labile class (LT) and the heat-stable class (ST). The LT class is an immunogenic toxin, consisting of at least two proteins of different molecular weight (11,500 and 25,000 daltons respectively). Crude LT preparations have an effect on bowel epithelium similar to that of the toxin of *Vibrio cholerae*, and it is generally thought to activate adenylate cyclase by a similar mechanism. One of the LT proteins is immunologically related to the B sub-unit of cholera toxin, while the activity of the crude LT toxin resembles that of the A sub-unit of cholera toxin. Unlike cholera toxin, ganglioside does not inactivate the LT enterotoxin of *E. coli*. This suggests that the binding sites in gut epithelium are different for the two toxins.

The ST class is non-antigenic and of a relatively low molecular weight (about 1000 daltons). The response of the bowel to ST is different from that of LT. Fluid immediately accumulates in the bowel but the duration of its action is short-lived and does not involve adenylate cyclase. Most enterotoxigenic strains produce both toxins. This is not invariably so; a minority of strains produce one or the other.

Although LT and ST are distinct toxins, one plasmid usually codes for the production of both. However sometimes one plasmid codes for one and a different one for the other. Since both enterotoxins are plasmid-mediated it might be expected that promiscuous dissemination among bowel *E. coli* might be experienced. In fact this is not so; relatively few serotypes naturally carry these plasmids.

(iii) *Assay of enterotoxins.* The demonstration that ligated segments of the small intestine in the living rabbit would fill with fluid following intralumenal administration of cholera toxin has proved a most useful test for demonstrating and assaying the toxin. Strains of enterotoxigenic *E. coli*, able to produce the heat-stable toxin (ST), also exhibit the same response. The test is tedious to perform and expensive if large numbers of samples are to be tested. A test yielding parallel results with the ST toxin, and proving more economical in time and cost, has been developed using suckling mice. Injection of the toxin into the stomach of suckling mice causes fluid to accumulate. This increase can be measured by comparing the weights of the intestines from the test mice with those from untreated controls. No fluid accumulates with the heat-labile toxin nor with cholera toxin, which is also heat-labile.

Both the LT toxin of *E. coli* and cholera toxin may be assayed by determining their ability to activate adenylate cyclase in a number of tissue cell lines. For instance, cultures of ovary cells of the Chinese hamster (CHO), show morphological responses due to cAMP, the consequence of adenylate cyclase activity. This activation, due to either toxin, can be specifically neutralized either by choleragenoid or anti-cholera antitoxin.

FURTHER READING

Ajl, S. J., Ciegler, A., Kadis, S., Montie, T. C., and Weinbaum, G. (eds) (1970–71) *Microbial Toxins—Bacterial Protein Toxins*, vols I–III. Academic Press, New York.

Bizzini, B. (1979) Tetanus toxin. *Microbiol. Rev.*, **43**, 224–40.

Buxton, A., and Fraser, G. (1977) *Animal Microbiology*, vol. 1. 'Clostridium', chapter 20, pp. 205–28. Blackwell Scientific Publications, Oxford and London.

Craig, J. P. (1972) The enterotoxic enteropathies. In Smith, H., and Pearce, J. H. (eds), 22nd Symposium of the Society for General Microbiology, *Microbial Pathogenicity in Man and Animals*. Cambridge University Press, pp. 129–55.

Craig, J. P., and Pierce, N. F. (1976) The structure and function of enterotoxins. *J. Infect. Dis.*, Suppl. **133**, S1–S156.

Jeljaszewicz, J., and Wadström, T. (eds) (1978) *Bacterial Toxins and Cell Membranes*. Academic Press, New York (432 pp.).

Smith, H., and Keppie, J. (1954). Studies on the chemical basis of the pathogenicity of *Bacillus anthracis* using organisms grown *in vivo*. In Howie, J. W., and O'Hea, A. J. (eds), 5th Symposium of the Society for General Microbiology, *Mechanisms of Microbial Pathogenicity*, Cambridge University Press, pp. 126–51.

So, M., and Falkow, S. (1977) Molecular cloning as a tool in the study of pathogenic *E. coli*. In Beers, R. F., Jr, and Bassett, E. G. (eds), *Science and Society*. Raven Press, New York.

Stephen, J., and Pietrowski, R. A. (1981) *Bacterial Toxins*. Nelson & Sons Ltd., Surrey, England (104 pp.).

Uchida, T., Gill, M. D., and Pappenheimer, A. M., Jr (1971) Mutation in the structural gene for diphtheria toxin carried by temperate phage β. *Nature New Biol.*, **233**, 8–11.

Willis, A. T. (1969) *Clostridia of wound infection*. Butterworths, London (470 pp.).

Wright, G. G. (1975) Anthrax toxin. *Microbiology*, 292–5.

Chapter 6

Bacterial endotoxins

1. General considerations

Endotoxins are a group of bacterial toxins chemically distinct from the bacterial protein toxins (exotoxins) considered in the previous chapter. They are invariably associated with Gram-negative bacteria although some Gram-negative bacteria evoke both endo- and exotoxins; certain enteropathogenic *Escherichia coli*, for instance, produce a typical enterotoxin (exotoxin) in addition to endotoxins characteristic of Gram-negative organisms in general. Killed suspensions of pathogenic and non-pathogenic species of Gram-negative organism are frequently toxic on injection into animals and man. Substances having similar patho-physiological effects on the host, can also be separated from filtrates of old cultures of Gram-negative organisms, by boiling suspensions of the bacteria or by other simple extraction procedures using hypertonic saline, sodium citrate, or aqueous phenol. For these and other reasons, endotoxins at one time were con-sidered to be structural components of bacterial cells released upon autolysis of the cell or upon lysis by disruption or digestion, and hence the descriptive term 'endotoxins' was applied to them.

Pfeiffer (1892) was the first to suggest that part of the bacterial structure could be poisonous to the host when released by disintegration of the cell and it was he who coined the term endotoxin. In many ways this is still a useful term, and no acceptable alternative has been found, but too narrow an interpretation has led some workers to exclude substances present in supernatants of young cultures in which no cell breakdown was apparent. Since the release of endotoxin was deemed to depend on autolysis of the cells, early workers harvested endotoxin after prolonged periods of culture incubation. It is now established that the O-antigens of Gram-negative bacteria are part of the same molecular complex which exhibits endotoxic activity. Since the O-antigens are sited at the cell surface the complex must be part of the outer cell envelope and liberally exposed. Separation of cell wall material from cytoplasmic membranes has shown the bulk of the endotoxin to be present in this structure. The release of endotoxin is not, however, limited to autolysis of cells. Endotoxins can be released from intact cells without noticeable structural change to the cell envelope as viewed under the light and electron microscopes. This release may occur in substantial amounts from intact young, growing cells.

2. The chemical nature of endotoxins

Chemically endotoxins consist of protein-free, lipopolysaccharide complexes, but frequently include non-essential complex polysaccharides, lipoid components, polypeptides, and metallic ions; the presence or absence of these additional components depends on the methods of culturing the organisms and their extraction. Recent work on the chemical composition of the envelope of Gram-negative organisms has indicated the possible relationship between lipopolysaccharide structure and function, but a complete association remains to be elucidated.

Most work has been done with species of *Salmonella*. The O-specific antigens on cells from smooth colonies are located at the cell surface and consist of macromolecular complexes of polysaccharide and lipid which, in the cell, are bound to the cell wall mucopeptide. These lipopolysaccharides are long-chain phosphate-containing heteropolymers composed of three distinct regions. Presented schematically they are linked through region I to II to III (Figure 6.1).

Region I consists of repeating units of oligosaccharides. Particular sugars in this structure, especially the terminal ones, confer immunological specificity on the respective O-antigens and the property of 'smoothness' to the strain. Loss of this region leads to the strain becoming 'rough'.

Region II is the core structure which, with minor variations, is common to all salmonella lipopolysaccharides but is structurally different in other genera of Gram-negative bacteria. In rough strains the core occupies the terminal position on the lipopolysaccharide. This structure varies in complexity between strains, from the most complex (the least rough) to the simplest (the most rough). At least five different chemotypes of decreasing complexity have been recognized with sequential loss of *N*-acetylglucosamine, glucose, galactose, and glycine.

Region III is the lipid component. This links the lipopolysaccharide macromolecule to the cell-bound protein. Endotoxic properties are associated with this region. Two lipid components can be separated. Lipid B is easily detached and biologically inactive; lipid A is strongly bound to polysaccharide and is responsible for endotoxicity. It appears to be a polymer of phosphorylated β-1,6-linked glucosamine disaccharide.

The properties of lipid A and its relationship to toxicity have been elucidated by a study of a rough, heptose-less mutant of *Salm. minnesota*. In this organism the lipopolysaccharide included lipid A, but not lipid B, and the only sugar present was 2-keto-3-dioxyoclonic acid (KDO). Most biological properties of endotoxins were expressed by this structure. Removal of KDO by mild hydrolysis resulted in free lipid A which exhibited very low endotoxic activity; this suggested that the sugar was necessary for the expression of toxicity. However, if the free lipid A was linked with a protein of non-bacterial origin, e.g. bovine serum albumin, a water-soluble complex was formed having potent endotoxin properties. The evidence therefore suggests that the polysaccharide or the protein parts of these complexes were not essential for toxicity *per se* but acted in an ancillary way for its expression. Polysaccharides in lipopolysaccharides, and KDO in glycolipid,

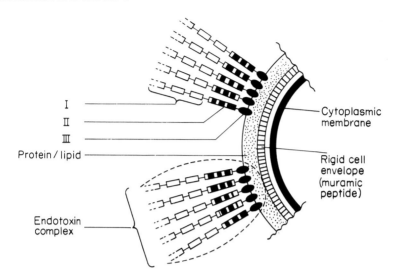

Figure 6.1.   Diagrammatic representation of part of the wall structure of a typical Gram-negative organism. Region I—repeating O-specific units of oligosaccarides. Region II—the core structure common to all members of a single genus. Region III—lipid component closely associated with the cell-bound protein

act as water-solubilizing carriers for the otherwise insoluble lipid A, in an analogous way to the bovine serum albumin in the lipid A complex. The structure of the polysaccharides greatly influences water-binding capacity at the cell surface; highly branched polysaccharides bind large amounts of water. Rough strains, which lack polysaccharides, have low water-binding capacity and this may account for the frequently observed tendency to spontaneous agglutination.

Virulence and the property of 'smoothness', associated with region I, are regularly present in many Gram-negative bacterial infections. On the other hand lipid A, the biologically active structure of endotoxins, is present in both rough and smooth strains. Virulence does not depend therefore on the presence of lipid A alone, but requires also the presence of region I. The smooth antigens probably aid the organism either in adherence to gut tissues or in resisting phagocytosis to which rough strains readily succumb.

3. Physical properties

Endotoxins are extremely heat-stable, a physical property which distinguishes them from most other microbial poisons. For instance, endotoxin derived from pseudomonads are inactivated by autoclaving for 7 hours at 126°C but not by heating for 30 minutes at 145°C. Raising or lowering pH reduces destruction by moist heat. Oxidizing agents, including permanganate, dichromate, hypochlorite, and hydrogen peroxide, are effective in degrading endotoxins. However, such methods and reagents cannot be used to destroy endotoxins which may be present in fluids, such as glucose drips used for intravenous injections. It is therefore essential in the pharmaceutical preparation of these fluids that bacterial contamination, which may release endotoxins, be rigorously excluded. Adequate safeguards, therefore, must depend on rigorous testing of adequate samples of each medicinal product to ensure, beyond reasonable doubt, freedom from endotoxins. One approach of limited application is to remove endotoxins either by ultrafiltration or absorption techniques.

4. Biological properties

In comparison with bacterial protein toxins, endotoxins have a much lower order of potency. They are good antigens but seldom produce immune responses which give full protection to the animal against a challenge by the endotoxin. They cannot be toxoided. Irrespective of the bacterial source all endotoxins produce the same range of biological effects in the animal host. The most obvious are diarrhoea, prostration and antemortem convulsions which occur within a few hours of injecting the endotoxin. The signs often resemble those of secondary shock resulting from peripheral circulatory failure. At post mortem the main findings are congestion, oedema and haemorrhage in the gut, abdominal viscera, and lungs.

Most of our knowledge of the biological activities of endotoxins derives not from a study of natural disease but by challenge of experimental animals. The most enhanced effects are produced by injecting the endotoxin into the host. There is little evidence to suggest that endotoxins are absorbed from the gut following ingestion. A few reports have indicated, however, that the ingestion of large numbers of *Aerobacter aerogenes* causes diarrhoea and nausea, and this raises an important question since the feeding of bacterial protein as food is a possible means of supplementing the deficient world protein source. Endotoxins are more likely to be absorbed from the respiratory tract following inhalation; aerosols of endotoxins have been shown to cause interstitial pneumonitis.

Some of the principal host responses in experimental animals will be considered.

(a) *Lethality*

Administering a large dose of endotoxin by a suitable parenteral route results in death in most mammals. The sequence of events follows a regular pattern. Signs of distress occur after a latent period; these include diarrhoea and prostration leading to death. How soon death occurs depends partly on the dose of endotoxin given but more significantly on the species of animal; rabbits, for instance, are much more sensitive to endotoxins than mice (approximately a 1000-fold difference).

Other factors which influence an animal's sensitivity to endotoxins include genetic differences between strains of animals and differences in age. In some species mature animals are more sensitive than young, e.g. rabbits; in others, e.g. guinea pigs and fowls, the reverse is true. In some species of animal sensitivity appears to be dependent on exposure to Gram-negative bacteria in the intestinal flora. Germ-free animals are more resistant than conventionally raised ones and animals with a gut flora including large numbers of Gram-negative bacteria are more sensitive to endotoxins than those with few. These observations suggest that prior exposure to Gram-negative bacteria increases sensitivity to endotoxins. It is sometimes possible to obtain the opposite effect; greater resistance or tolerance in normally susceptible animals can be induced by deliberate exposure to endotoxins (see later).

In the same animal species lethality to an endotoxin varies with the route of administration. In rabbits, for instance, the order of susceptibility increases in the following sequence: intradermal, intraperitoneal, and intravenous routes. In mice, age and route of administration have little effect.

(b) *Pyrogenicity*

The word 'pyrogen' was used as early as 1875 to describe a substance capable of inducing fever but, as a specific term for filterable, heat-stable substances of bacterial origin, was not used until 1923. The temperature response in animals to endotoxin is the most characteristic and consistent biological effect produced by these substances, and for this reason has been most thoroughly studied and used as a biological test for endotoxins.

Different animal species show greater or lesser sensitivity to pyrogens. In the mouse and rat a decrease in body temperature results; in rabbits, dogs, horses, and man a temperature rise is experienced. Man is the most sensitive. The temperature elevation and its persistence following intravenous injection is dose-related. Very small amounts of endotoxin (of the order of 0.001 mg per kg body weight) are pyrogenic in man and rabbits. In rabbits no reaction is observed for 10–20 minutes; this is followed by a temperature rise reaching a peak a little over 1 hour from the time of injection. A small dose produces a relatively low fever response which returns to normal in a few hours. Larger doses produce a high initial temperature response after 1 hour; the temperature then falls throughout the following hour, after which a second rise occurs, often higher than the first, reaching a maximum in 2–4 hours from the time of injection. The temperature usually returns to normal within 6 hours. This biphasic fever, which occurs in man, dog, and horse as well as in rabbits, is one of the most typical host responses to bacterial endotoxins (Figure 6.2).

The fever may be mediated either by an immunologically based type of

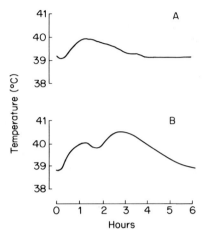

Figure 6.2. Fever responses in rabbits to different doses of endotoxin inoculated intravenously at time 0. **A:** A low dose—after a latent period of 10–20 minutes the temperature rises to a peak by a little over 1 hour. **B:** A larger dose—this is followed by a shorter latent period, the temperature then rises to reach two peaks, the second reaching a higher maximum after about 3 hours

hypersensitivity or the release into the circulation of an endogenous protein pyrogen by direct endotoxic tissue damage which acts directly on the hypothalamic thermoregulatory centres. Subsequent to the pyrogenic stimulation the animal becomes tolerant to the endotoxin and this tolerance last for several weeks.

(c) *Local and general host reactions to repeated injections of endotoxin*

If two injections of endotoxin are given to an animal at correctly spaced intervals of time, one of two responses may be elicited: a local, dermal reaction, or a generalized reaction. *The local phenomenon* is known as the dermal Shwartzman reaction after its discoverer in 1937. It may be reproduced in rabbits. An endotoxin is injected into the skin; 24 hours later, endotoxin is injected intravenously. Within the following 6 hours haemorrhagic necrosis occurs at the skin site of the first injection. The material used for the intravenous injection may be endotoxin from the same or a different bacterial source or even unrelated non-endotoxic substances—e.g. extracts of Gram-positive bacteria or tissue (e.g. liver glycogen). These non-endotoxic substances, whilst able to provoke the reaction in a sensitized animal, cannot induce sensitivity in the skin.

The *generalized endotoxic reaction* was first discovered by Sanarelli (1924). He inoculated rabbits intravenously with sub-lethal doses of *Vibrio cholerae* and 24 hours later gave bacterial endotoxins intravenously in the form of culture filtrates of *Escherichia coli* and *Proteus* spp. Most of the rabbits died and at post mortem lesions of haemorrhagic necrosis were found in the mesentery and kidneys; viable *V. cholerae* could be recovered from the mesenteric lesions. This reaction is often called the generalized Shwartzman reaction.

The local and generalized reactions were thought to be fundamentally the same. However certain differences have been demonstrated. In the local reaction, after the second injection, the small venules and capillaries of the skin lesion become occluded with leucocytes and platelets. In the generalized reaction the glomerular capillaries become occluded by homogeneous, amorphous, fibrin-like material.

(d) *Effects on the vascular system*

The vascular system is very susceptible to endotoxins. For instance, small doses of endotoxin render blood vessels more susceptible to adrenalin. This is similar to the state of hyperreactivity which occurs in the Shwartzman–Sanarelli reactions described above. Following intravenous injection, initial functional disturbances of the small blood vessels lead to more extensive effects on the circulation resulting in a fall in blood pressure which, in severely poisoned animals, culminates in shock and death. These effects are apparently dependent on the release of endogenous vasoactive substances like histamine, serotonin, noradrenalin, and plasma kinins.

Endotoxic shock has been compared with shock resulting from trauma and haemorrhage. In dogs, for instance, endotoxins cause an immediate fall in systemic arterial blood pressure, an initial slowing of heart beat (bradycardia) followed by an accelerated pulse (tachycardia). The magnitude of the response is dose-related.

(e) *Haemorrhagic necrosis of tumours*

Towards the end of the last century it was observed that mixtures of bacteria damaged malignant tumours by selectively producing extensive haemorrhages in the tumour mass. This led to attempts to use various preparations, such as 'Coley's mixed toxins', in the therapy of cancer. Little attempt was made to determine the active ingredients and it was not until 1943 that it was shown that endotoxin prepared from *Serratia marcescens* was 1300 times more active than the crude mixture. The toxic somatic antigens of Gram-negative bacteria were shown to be the damaging factor.

Endotoxins selectively damage tumour tissue without affecting adjacent normal tissue. They do not, however, damage tumour cells growing in tissue culture and it is thought that the damage to tumours *in vivo* is due to the effect of the endotoxin on the vasomotor (constriction of blood vessels) and metabolic responses of the whole animal through which damage to the abnormal tissue is mediated.

(f) *Endotoxin as adjuvant*

The immune response of an animal to an antigenic stimulus can be accentuated by the simultaneous injection of an endotoxin. The administration of endotoxin to an animal receiving an antigenic stimulus provokes an accentuated response. Antibody can be detected earlier, the level increases more rapidly and a higher titre is attained. This effect was discovered when Gram-negative bacteria were present in polyvalent vaccines—the lipopolysaccharide was shown to be the enhancing ingredient. It is thought to act by causing cellular damage, thereby altering the permeability of cells at the antibody-producing sites and also by releasing nucleic acid fragments which are known to act as stimulators of other cells involved in antibody formation.

5. The role of endotoxin in disease

The biological activities so far considered are based almost entirely on artificial inoculation of endotoxins into experimental animals. There is no doubt that similar effects may be encountered in patients receiving large volumes of fluids containing bacterial contaminants and their endotoxins intravenously. Patients on drips are usually debilitated through illness, trauma, or age; and endotoxic shock, followed by death, may be encountered. Nevertheless the introduction of the endotoxin in these medicinal fluids is essentially an artificial exposure.

The role of endotoxin in natural disease is not easy to elucidate. As was pointed out earlier, endotoxins are not limited to pathogenic Gram-negative bacteria and therefore are not the primary determinant of pathogenicity. It is assumed, however, that they are released from the infecting organism in the host body. There is little evidence that endotoxins are absorbed from the gut; to be able to affect the host the endotoxins have to be released into the tissues of the host. Once an infection is fully established, particularly in the terminal stages of many diseases caused by Gram-negative bacteria, the signs often resemble endotoxic poisoning. The number of organisms present at the time of death is relatively constant also, indicating that a certain concentration of endotoxin must be generated *in vivo* to be lethal. Fever is a frequent sign of infections by Gram-negative bacteria, as in typhoid and paratyphoid fever, and infections with *Brucella* spp. It has often been observed that treatment of undulant fever with broad-spectrum antibiotics is followed by a rise in temperature, doubtless due to release of endotoxins consequent upon death of the micro-organisms.

Great uncertainty is experienced regarding the mode of action of endotoxins in disease and their possible role in the hypersensitive immune state. In some ways endotoxic effects resemble allergic phenomena. Thus animals may become tolerant to endotoxins by exposure to sub-lethal doses. Most observations on this phenomenon have been made on tolerance to the pyrogenic effect. A first dose of endotoxin renders the animal partially tolerant whilst repeated doses increase its resistance. This is indicated by the elimination of the second peak of the biphasic fever curve and an incomplete loss of the first. It is not, however, specific; tolerance induced by one endotoxin may also show tolerance to other serologically unrelated endotoxins. Also substantial resistance develops within 18 hours and, under continuous infusion, within a few hours, both of which indicate that this is not classical immunity.

In contrast to tolerance, an animal treated with endotoxin may become hypersensitive or hyperreactive to subsequent doses of the same or unrelated non-toxic antigens. This is particularly evident in the Shwartzman–Sanarelli reactions and comparison has been drawn between these and the Arthus and tuberculin reactions which are distinctly immune hypersensitivities.

Hence the situation remains unclear. It would appear that the immediate effects of endotoxins are due to primary toxicity which may be manifested in many ways. Many of these expressions of primary toxicity are not linked to endotoxins but would result by damage of the same susceptible cells by other agents. The less immediate effects, especially those produced by repeated doses of endotoxin (e.g. Shwartzman–Sanarelli reactions and induction of tolerance), may prove to have an allergic element but the evidence for complete parallelism with classical immune responses is lacking. At the present time we have little understanding of the precise role of endotoxin in natural infections.

6. The assay of endotoxins

In considering the various pathophysiological effects produced by endotoxins, it was stated that a number of these effects are dose-related. Most bioassays of endotoxins are based on toxicity for the intact animal and are time-consuming. They include the ability to produce fever, abortion, synovial inflammation, dermal necrosis, or death. Toxicity assays can detect relatively low levels (e.g. 0.0003 μg) but are not accurate quantitative estimates since the effect may be reduced or potentiated by the presence of metabolites released from injured tissue. Other approaches which are more specific and quantitative include antigen assays, such as fluorescent antibody to detect endotoxic antigen, inhibition of the agglutination of endotoxin-sensitized erythrocytes, or measuring the amount of antibody stimulated by endotoxin.

More recently an *in vitro* test has been devised, called the Limulus lysate coagulation test. The sensitivity of the test has been greatly increased to measure as little as 10 picograms endotoxin. The test involves a lysate of amoebocytes from the blood of the adult horseshoe crab, *Limulus polyphemus*. Blood is taken by cardiac puncture, the cells are harvested, washed in buffer and homogenized in the cold. After storing for 48 hours at 4 °C the suspension is centrifuged and the clear supernatant diluted with buffer until a solid gel can still be obtained with a crude endotoxin of *E. coli* O26:B6, at a concentration of 10^{-6} μg/ml, in less than

3 hours. The final lysate is stored at −60°C. The test is performed by incubating a mixture of 0.1 ml lysate with 0.1 ml of endotoxin containing material at 37°C and observing the formation of a viscous gel, usually within 4 hours, which denotes the presence of endotoxin. Controls of the endotoxin preparation alone, and buffer and lysate alone, are incubated at the same time.

FURTHER READING

Ajl, S. J., Cieglar, A., Kadis, S., Montie, T. C., and Weinbaum, G. (eds) (1971) *Microbial Toxins, Bacterial Endotoxins*, vol. 5. Academic Press, New York.

Landy, M., and Braun, W. (1964) *Bacterial Endotoxins*. Institute of Microbiology, The State University, New Brunswick, NJ.

Levin, J., Tomasulo, P. A., and Oser, R. S. (1970). Detection of endotoxin in human blood and demonstration of an inhibitor. *J. Lab. Clin. Med.*, **75**, 903–11.

Yin, E. T., Galanos, C., Kinsky, S., Bradshaw, R. A., Wessler, S., Luderitz, O., and Sarmiento, M. E. (1972). Picogram-sensitive assay for endotoxin. *Biochem. Biophys. Acta*, **261**, 284–9.

Chapter 7

Viral pathogenicity

1. General considerations

Many aspects of pathogenicity, such as host responses, are common to bacterial and viral infections. Some, however, are unique to viruses and these will be emphasized in this chapter.

(a) The nature of the virus

Viruses are less evolved micro-organisms than bacteria or fungi and are therefore more 'primitive', i.e. less complex in structure and function. As a result viruses cannot exist as free-living agents but replicate intracellularly where they utilize existing biosynthetic pathways. In this respect viruses are totally parasitic upon their host cells, whether these are bacterial, plant, or animal cells.

The individual virus particle, the virion, consists of a minimum essential quantity of either DNA or RNA (never both) in an intimate and stable association with a variety of structural and non-structural (enzymic) proteins. The minimum essential quantity of nucleic acid is the least amount necessary for the virion to code for its replication. When it undertakes this it utilizes the biosynthetic enzymes and pathways of the cell it parasitizes. In general terms, the smallest virion is least complex in structure and composition, contains least nucleic acid, and has greatest parasitic dependence upon existing cellular enzymes and biosynthetic pathways to undertake its replication. The largest virion contains most nucleic acid and, in consequence of its capacity to produce a greater number of virus enzymes, is least dependent upon existing cellular enzymes; it is nevertheless still dependent to some degree on the cellular biosynthetic pathways. The parasitic utilization by the virus of the cellular biosynthetic pathways produces the cellular lesions and malfunctions which are collectively termed cytopathic effects. Cytopathic effects may be observed both in the individual virus-infected cell and also in the tissue of which that cell is a component. The cytopathic effects, together with the defence reactions elicited by other cells and tissues, constitute the pathogenesis of a virus infection in the animal body.

All viruses are cytopathogenic to some degree upon their host cells. Not all viruses, however, are necessarily virulent. The virological term 'virulence' describes the relative infectiousness of a virus, i.e. the propensity of a virus to produce rapid and profound malfunctioning of its host cells. The terms cyto-pathogenicity and virulence are therefore interdependent but not synony-mous.

(b) *Virus strategy*

The prime strategy of a virus is perpetuation of its genome and this requires that the genome is replicated over successive generations. In order to expedite this, viral mRNA is required for the production of two essential proteins, a nucleoprotein which is intimately associated with and protects the genome, and a replicase to mediate replication of the genome.

The greater the virulence of a virus the greater the possibility that it will deplete the numbers of natural hosts available for its replication. Depletion may occur either by death or by naturally enhanced immune resistance of the infected hosts. Since either situation threatens its chance of survival, a virus may exhibit an alter-native strategy of lowering its virulence and fostering a continuing and stable virus–host relationship by adaptive regulation of its replication process.

The most successful viruses are those which multiply readily and rapidly in a diversity of hosts while producing only a minimum of damage. A good example is vaccinia virus, a man-made hybrid of cowpox and alastrim viruses, developed for the control of smallpox by vaccination (Chapter 10).

The diversity of virus species which exist today indicates a variety of modes of intracellular replication, each representing specific adaptations between a virus and a particular host cell. As may be expected, the cytopathic effects produced by different viruses are equally diverse.

Many aspects of virus replication still require elucidation. For instance, we do not know how nucleic acid survives in the hostile environment of the cell cytoplasm after its release from the infecting virion nor why the cell allows it to undertake subsequent transcription and translation.

Transcription is the production of new polynucleotide chains off a polynucleotide template and is the method by which mRNA and new viral genomes are duplicated. Translation is the reading of the mRNA codons at the ribosome for the production of polypeptide chains (see Figure 7.1).

(c) *Studies of viral pathogenicity*

The intracellular activity of a virus produces specific cytopathic effects in infected tissues of the host. An understanding of these effects, and new insights into the pathology of virus diseases, have been obtained since the use of cell culture tech-niques by virologists.

Viruses are usually studied in cultures of cells derived from normal animal hosts. However, while cell culture offers the advantage of an easily manipulated system free of any intrinsic immune and hormonal regulations it also suffers from

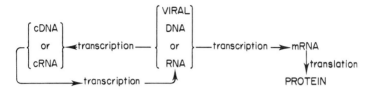

Figure 7.1. Transcription and translation. mRNA = messenger RNA; cDNA and cRNA = complementary DNA and RNA having the mirror-image configuration of the template upon which they are formed

some disadvantages, such as variations in cell susceptibility to infection, the presence of endogenous viruses and mycoplasma derived from the parent animal, low levels of spontaneous mutation, and the production of interferon (see later).

The alternative to cell culture is the experimental laboratory animal. Animals are not ideal experimental systems since they vary in susceptibility to infection, may carry latent viruses (which can be a source of virological confusion), and present handling problems. In addition, extrapolation of the results from the experimental animal to the disease problems in the natural host species is always open to question. On balance, observations based on cell culture experiments have proved as useful a means of elucidating the mechanisms of viral pathogenicity as have experimental animals—both have made significant contributions to knowledge and both are considered below.

2. The virus and the host cell

(a) *Cellular infection*

Infection of a cell may be considered to occupy the first two phases of the viral replication cycle, namely *adsorption* and *penetration* (Table 7.1).

The chance of an inert particle (a virion) being adsorbed to the dynamic cellular plasma membrane is greatly increased the closer they are brought together. This is aided on the one hand by Brownian movement of the virion and on the other by the active undulations and microvillous formations of the plasma membrane. Initial *adsorption* involves the electrostatic forces existing on virion and membrane surfaces, and some viruses (e.g. polioviruses, poxviruses) have considerably lowered adsorption rates in the absence of Ca^{++} and Mg^{++} ions in the extracellular fluids. Other viruses adsorb to specific receptor substances on the plasma membrane, as does influenzavirus to N-acetylneuraminic acid. This receptor substance serves also for adenoviruses, polyomaviruses and encepalomyocarditis viruses. Under optimal conditions total adsorption may occur in 30 seconds.

Penetration of the plasma membrane by the adsorbed virion may occur in one

Table 7.1.   A virus replication cycle

Phase	Time scale	Consequence
1. Adsorption ⎱ 2. Penetration ⎰	30 minutes ⎱	Infection ⎱ A prelude to CPE* and evidence of
3. Eclipse	2 hours	viral pathogenicity
4. Synthesis	5 hours	⎰ Replication ⎰
5. Maturation	30 minutes	
6. Release	6–8 hours +	

While the cycle may be arbitrarily divided up, as shown, in reality the six phases follow each other without discernible pause.

*CPE = cytopathic effects.

of two possible ways: either by viropexis (viral pinocytosis) initiated by the cell or by fusion of the virion envelope (if present) with the plasma membrane. Both processes result in the virion entering the cytoplasm when infection is achieved.

(b) *Intracellular replication*

The cell normally digests foreign particles which enter the cytoplasm. Hence the next phase in virus replication is fusion of a lysosome with a virion, or virion-containing vesicle, to form a phagolysosome. Preliminary digestion by lysosomal enzymes results in loss of virion integrity and this constitutes the *eclipse* phase.

The phase of *synthesis* commences with the escape of the nucleic acid genome of the virion from within the phagolysosome and the initiation of a 'factory area' within the cell. Nucleoproteins complexed with the viral genome serve to protect it from destructive lysosomal nucleases. Once the nucleic acid genome has escaped digestion it becomes productive; should escape not occur the infection is abortive and non-productive. The 'factory area' may be within the cytoplasm or the nucleus or both, depending upon the virus. Subsequent cycles of viral synthesis provide the structural and non-structural proteins and nucleic acid which are needed for the assembly of progeny virions during the succeeding phase of *maturation*. The structural proteins have the propensity for self-assembly during this phase; this is demonstrated during infection with some cubical viruses (e.g. hepatitis B virus) in which enormous numbers of 'empty' virions may be generated.

Depending upon the virus, assembly of progeny virions occurs either at internal or external membranes by a process of budding-out or within the cell by a process of condensation. Those which bud-out, e.g. enveloped viruses, are generally less pathogenic since they do not inevitably require cell lysis for their release; those which undergo a condensation process intracellularly, e.g. cubical viruses, are generally more pathogenic since cell lysis is necessary for their release.

(c) *Cytopathic effects*

Replication of viruses intracellularly results in a number of recognizably different cytopathic effects (CPEs) in host cells. The same virus may produce more than one CPE in sequence and all usually culminate in *necrosis* as the virus overwhelms the cell's capacity to continue metabolism. The various types of CPEs are considered below.

(i) *Inclusion formation.* Some, but not all, viruses produce discrete and dense 'factory areas' during maturation which stain specifically to reveal cellular inclusion bodies. They may be intracytoplasmic or intranuclear in location; inclusions may be so characteristic of the virus infection that they are diagnostic. In the early days of virology inclusions were named after the worker who first described them, e.g. Negri bodies (intracytoplasmic, in rabies infection) and Lipschutz bodies (intranuclear, in herpesvirus infections).

(ii) *Syncytium formation.* Syncytia are the fusion products of infected cells rather than the result of a single cell undergoing nuclear division (karyokinesis) without associated cytoplasmic division (cytokinesis). They are erroneously termed 'giant cells'. Some viruses (e.g. morbilliviruses, paramyxoviruses) readily induce syncytia in cell culture due to their insertion of viral antigens (haemolysins) into the plasma membrane which initiate membrane fusion.

(iii) *Vacuolation.* A few viruses induce vacuolation of the nucleus or of the cytoplasm, but rarely both together. Ectromelia virus induces nuclear vacuolation, measles cytoplasmic vacuolation, and endogenous simian viruses (SV and VA series) massive cytoplasmic vacuolation. The mechanism of vacuolation is not known.

(iv) *Hyperplasia (cellular proliferation).* This may be a transient feature of some infections (e.g. herpesviruses, poxviruses) or a more prominent feature of other (e.g. papovaviruses, adenoviruses). Where hyperplasia occurs in the absence of any terminal necrosis, and where there is also a simultaneous interaction of viral genome with host cell genome, transformation may occur (see below).

(v) *Latency or other phenomena of persistency.* A latent infection is a concealed and insidious infection, usually chronic in nature, in which a degree of equilibrium is established between the virus and its host cell resulting in only sparse necrosis. If a sufficiently strong stimulus is given to the infected cells they may enter a productive phase of infection. This is exemplified by herpes labialis in man, in which the ensuing latent infection of the trigeminal ganglion may be stimulated by a superimposed common cold to produce 'cold sores' on the buccal mucosa. In cell cultures it is often easy to obtain 'carrier', or chronically persistent, infections in which all cells bear virus antigen but few produce progeny virions, (e.g. with measles virus). Occult viruses are not directly observable in the infected cells and may require special techniques for their detection (e.g. rabbit papillomavirus). The terms latent, subclinical, inapparent, and occult are sometimes used inter-changeably in the literature, which leads to some degree of confusion. Persistent infections of any sort owe their existence to a failure of the immune system to eradicate them.

(vi) *Transformation.* This describes the synchronous co-existence of virus and host cell in which there occurs a stable integration of part or whole of the virus genome into the genome of the cell leading to profound transformational changes in the cell. Transformed cells may demonstrate new enzymes or antigens, faster multiplication rates, loss of growth inhibition, and altered morphology, all of which are now considered to be characters indicative of oncogenesis, i.e. tumour formation. Transformational changes are especially observed in cell cultures infected with RNA-containing retroviruses and DNA-containing papovaviruses and adenoviruses.

Retroviruses are unique in being able to employ their endogenous RNA-directed DNA polymerase ('reverse transcriptase') to transcribe a complementary DNA (cDNA) copy off their RNA genome as template. It is now recognized that the interaction between this cDNA and the chromosomal DNA of the host cell results in the acquisition of nucleotide sequences from each other. In any subse-quent replication, the viral RNA genome will contain sequences transcribed off chromosomal DNA of the cell. Thus, while the cell will be transformed by acquisi-tion of the virus, the virus too may be provided with additional co-oncogenic characters. Recent studies of the mechanism by which retroviruses are able to induce oncogenic transformations in normal tissues have revealed that many cells contain silent chromosomal gene sequences which closely mimic retroviral oncogenic gene sequences (or 'oncogenes'). These silent sequences may become activated under the appropriate conditions.

In an infected cell culture the first sign of transformation taking place is usually 'micro-tumour formation' or 'cell piling', that is, hyperplasia.

(vii) *Necrosis.* Necrosis is cell death and is the terminal CPE of all virus infection! In cell cultures it is seen as a rounding-off or balling-up of the infected cells which makes them appear more refractile in bright field microscopy. The necrosing cell stains more intensely with histological stains because of fluid loss and concentra-tion of constitutive proteins. In the final stages of necrosis autolysis sets in.

(d) *Haemagglutination and haemolysis*

Viral receptor substances on host cells may also be found on the plasma mem-brane of red cells (e.g. receptors for influenzaviruses, poxviruses, group B arboviruses). Sufficient concentrations of virus cause red cells to agglutinate *in vitro* and this phenomenon of *haemagglutination* is used in the laboratory to detect either the presence of haemagglutinating viruses or specific antibodies which inhibit the haemagglutination (haemagglutination-inhibition). *In vivo* viraemias do not yield sufficiently high concentrations of virus to produce intravascular haemagglutination.

A number of viruses (e.g. morbilliviruses, parainfluenzaviruses, pneumoviruses) which possess surface antigens capable of causing syncytium formation amongst infected host cells (see above) can also make red cell membranes 'leaky'. This latter phenomenon of *haemolysis* has been used for the laboratory detection of viral haemolysins. As with haemagglutination, in viraemias there are insufficiently high concentrations of circulating virus to produce haemolysis.

(e) *Other associations with host cells*

(i) *Viroids*. These are infectious agents of several diseases of higher plants. They consist of naked lengths of low molecular weight species of RNA and exhibit the properties of resistance to heat, to ionizing and u.v. radiation, and to the various chemicals which inactivate orthodox viruses. The infectious agents of the spongiform encephalopathies in man (Creutzfeld–Jakob disease, kuru) and in animals (scrapie, mink encephalopathy) share these properties and there is a growing opinion that these are animal viroids, though whether they are RNA or DNA in nature is unknown.

Plant viroids are unique infectious agents because of the very small amount of genomic RNA they possess, sufficient only to code for a single protein of around 120 amino acids in length. This may be an RNA transcriptase. In this respect they resemble an RNA copy of a self-replicating rogue gene. It is obvious that they rely totally upon their host cell for completion of replication and must have an intimate association with them. If the animal diseases mentioned above are truly caused by viroids then a search for these agents will be extended to a number of other insidious diseases for which no orthodox agents have yet been found.

(ii) *Provirus*. Our understanding of the intracellular replication of oncogenic DNA (e.g. SV40) and RNA (e.g. Rous sarcoma) viruses, in which the virus becomes integrated into the host cell genome in the virus replication cycle, has been extended by new findings.

It has been demonstrated that the SV40 genome becomes stably integrated and covalently bonded with chromosomal DNA as a provirus. The SV40 provirus has been shown to survive over many hundreds of generations in the cells and still remain capable of mRNA transcription. Not all the viral genes held within the provirus are expressed, indicating that the provirus is under cellular control like the cellular chromosal DNA. Infected cells revert to normal behaviour upon loss of the provirus through mutation or excision.

Rous sarcoma virus (RSV), which also produces a similar transformation of infected host cells, has a virion-associated RNA-directed DNA polymerase ('reverse transcriptase'). This enables it to have a proviral stage in its replication cycle and RSV provirus similarly survives over hundreds of cell generations.

3. The virus and the host animal

(a) *The virus*

As stated earlier the prime strategy of the virus is perpetuation of its genome. To facilitate this the virus must meet and infect a susceptible host; three aspects of host infection will be considered; namely, the routes of infection, the infecting dose and tissue affinities.

(i) *Routes of infection.* There are three principal routes—namely respiratory, alimentary, and traumatic. Of these the respiratory route affords the easiest, most direct, means of entry to susceptible tissue and, under conditions of close crowding, very rapid rates of spread can be obtained. The alimentary route requires either intimate host contact or ingestion of contaminated food or water. The traumatic route affords direct entry into subcutaneous tissues via the bites of insects or animals or other trauma. Virulent viruses which spread rapidly in a population are usually associated with the respiratory (e.g. influenzavirus) or alimentary (e.g. poliovirus) routes of infection.

(ii) *Infecting dose.* The more virulent the virus the fewer the number of virions needed to infect, and for some the actual number of virions needed to infect may be less than 10. During the foot-and-mouth disease outbreak in England in 1967–68 it was shown from epidemiological data that rapid downwind spread was obtained despite the considerable dilution and dispersal suffered by the virus in the winds; this outbreak displayed the highly virulent nature of the virus.

(iii) *Tissue tropisms.* Early attempts at virus classification were based upon their tissue tropisms (Table 7.2), but this classification scheme was short-lived since

Table 7.2. Tissue tropisms and representative viral genera

Tropism	Representative genera
Dermotropism	All poxvirus genera
Pneumotropism	Influenzavirus
Viscerotropism	Enterovirus (poliovirus spp.)
Neurotropism	Lyssavirus, poliovirus spp.
Pantropism	Morbillivirus, vaccinia virus

some viruses (e.g. poliovirus) possessed more than one tropism. Many viruses have a restricted range of host tissues in which they can lodge, replicate, and produce their principal lesions.

(b) *The host*

Acute infection of the host is usually associated with rapid virus multiplication and numerous lesions, whereas chronic or persistent infections are associated with slow multiplication and few lesions. The most successful virus is one which occupies a position between the two extremes; i.e. rapid multiplication to ensure good host-to-host spread while producing minimal disease in the host, as exemplified by influenzavirus.

When a virulent virus is introduced into a susceptible population one possible outcome is death of all hosts with an inevitable total loss of virus. This extreme circumstance is rarely encountered in real life, indicating that there must be many adaptations, both of virus and host, which serve to promote the survival of each.

(i) *Host factors which influence the course of disease.* As with bacterial infections, there are a number of host factors, other than cellular and humoral immune mechanisms (see later), which affect resistance of the host to infection; these include age, sex, passive antibody status, genetic constitution, and nutrition. Aspects unique to viruses are considered below.

Some viruses cause disease in the female animal only, especially in the mammary tissues. Both the mammary carcinoma virus of mice (the Bittner agent) and the mammillitis virus of the ox produce frank disease in susceptible females at the time of lactation. The males of these species are refractory to the respective viruses though in the case of the Bittner agent males, infected when suckling, may have disease induced by later oestrogen treatment.

Often within a single host species marked variation in susceptibility to virus infection has been recorded due to the genetic constitution of the species. In children, for example, there is considerable variation in resistance against childhood virus diseases, such as chickenpox and measles, and much of this variability is considered to be due to the individual's genetic constitution. Because of the experimental difficulties encountered in demonstrating resistance loci in man, early studies were conducted in laboratory animals. The first major advance was the discovery of resistance genes in mice which were linked with T-cell and macrophage function. In mice the H-2 histocompatibility gene loci constitute a major regulator of antigenic specificity, with numerous non-H-2 linked gene loci acting as regulators for various other single antigens. The situation is therefore complex in the mouse and may be similarly complex in other species. Subsequently in man HLA (human lymphocyte antigen) histocompatibility loci associated with virus diseases, were recognized. It is clear that the individual host's genetic constitution plays a crucial role in innate resistance.

For each virus species there is a minimum size of host population which will support its continual replication in the community. For measles virus, which is shed from an infected host over a short time period, the population size is 500,000; for chickenpox, which is shed intermittently, it is 100; for lymphocytic choriomeningitis or for leukaemia viruses of rodents, which are shed over prolonged periods, it is an order of magnitude lower. The most difficult populations to infect with virus diseases are therefore those of sparse numbers scattered over large areas such as may exist in rural agricultural provinces; the easiest are those in which considerable numbers are concentrated into small areas, such as exist in densely populated towns and cities. Communal living conditions offer ideal conditions for cross-infection.

(ii) *Non-specific mechanisms of host defence.* In the early phases of infection an inflammatory reaction occurs at the primary site due to local tissue damage. This results in the release of cell constituents which act chemotactically and attract leucocytes. In a first infection with a virus no antibody will be present to combat the infection and so the host must rely upon the immediate defences afforded by cellular and extracellular products.

(α) *Macrophages.* As in bacterial infections, the macrophage plays an important role in the early defence of the host. Macrophages are located in the circulating blood and in the tissues. The free alveolar and peritoneal macrophages, together with the fixed hepatic macrophages (Kupffer cells), have a common origin with the circulating macrophages. All arise from precursor stem cells of the bone marrow.

The macrophage actively destroys phagocytosed virions, a property termed intrinsic resistance. This is distinct from extrinsic resistance by which macrophages affect virus replication in other host cells by secretion of macrophage factors (e.g. interferon, see later). They also cooperate in the immunological responses of tissues to viruses.

In the adult animal macrophages provide an early local cellular barrier to infecting viruses through their intrinsic resistance. By this means they inhibit spread of the virus from the primary multiplication sites to secondary sites in vital tissues such as the liver and brain. Where infection has persisted sufficiently long for an antibody response to have occurred there will be an enhancement of macrophage activity due to naturally occurring Fc receptors on the cell surface. The Fc receptors permit more rapid attachment of viruses/antibody complexes and of their phagocytosis; removal of the Fc receptors with trypsin has been shown to destroy this antibody-dependent enhancement. It follows that viruses which stimulate the least antibody responses, or which are sufficiently virulent to initiate peracute infections in advance of significant antibody levels, are best able to avoid this enhanced macrophage activity. The more virulent a virus, the greater its ability to avert digestion within the macrophage; the most virulent may initiate replication within the macrophage. When this occurs, and because macrophages are mobile cells, the macrophage will transport the virus to other tissues of the body. In the very young animal the ability of macrophages to digest intracellular viruses is less well developed and relatively avirulent viruses may replicate readily in these cells.

Macrophages play a significant role in the genetic resistance of animals to certain virus infections and this has been demonstrated in mouse Kupffer cells. For instance, the resistance of the C3H Swiss strain of mice to mouse hepatitis virus and the susceptibility of the PR1 strain of mice to the same virus is largely due to the respective high and low levels of intrinsic resistance of their Kupffer cells.

(β) *Polymorphonuclear neutrophils (PMNs)*. The major antiviral roles are played by macrophages and T-cells (see later), but the role of the PMN is more enigmatic. Generated by myeloblasts of the bone marrow, they are the most numerous of the circulating granular leucocytes but while these cells have a clearly defined role in clearing bacteria from the tissues no clear antiviral role has yet been assigned to them.

(γ) *Cellular products*. Many acute viral infections last only a few days and, since the primary immune responses to infection develop slowly, they cannot play a fully effective role in the early phase of infection. Host defence must therefore rely on other antiviral mechanisms to give early protection until the specific immune responses have developed; interferon provides this protection.

Interferons are potent antiviral glycoproteins. Three types have been recognized: β interferon produced by fibroepithelial cells, and α and γ by leucocytes. Interferon production is initiated by the presence of viral (i.e. foreign) nucleic acids in the infected cell. They may be detected as early as 6 hours after infection, suggesting that some interferon may be present preformed in producer cells.

Interferons do not interfere with adsorption of the virus to host cells and therefore do not interfere with the initiation of infection. They can, however, totally block the synthesis of structural proteins and enzymes which are essential for virus replication. Absorption of interferon by a virus-infected cell induces the formation of cellular mRNA which codes for an inhibitor protein. This protein disrupts polysomes assembled around viral mRNA, thereby inhibiting the synthesis of viral proteins in the cell cytoplasm. The individual virus-infected host cell does not usually survive, since the interferon acts at a mid-stage in virus replication, but the interferon induced in the infected cell will subsequently interfere with virus synthesis in other cells, and the overall effect is to lower virus synthesis in the tissue as a whole and aid the process of recovery.

Current knowledge about the mechanism of interferon indicates that recovery from a virus infection is multifactorial and in the early stages of infection factors other than interferon are involved. One of these may be complement but the evidence is not as clear as for interferon.

Specific immune responses in host defence

Viruses, like bacteria, are potent immunogens and consequent upon infection stimulate both humoral and cell-mediated immune responses. The cooperative activity of both plays a decisive role in recovery from virus infection.

The role of humoral antibody

The addition of antiserum to a virus suspension results in attachment of immunoglobulins to the virions, the effect of which is neutralization of infectivity. While the physical size of a virion dictates the number of immunoglobulin molecules able to attach to it, kinetic studies have shown that neutralization of a virion may be achieved by a single antibody molecule.

As with other antigen/antibody complexes, the association is reversible, at least in the initial stage of attachment, and the antibody can be dissociated from the virus by appropriate treatment, e.g. by dilution, when the infectivity of the virus is partly regained. The power to dissociate diminishes rapidly with time, indicating a subsequent firm association which, *in vivo*, aids the body in eliminating the infection.

Antibody neutralization of virus can be detected by loss of infectivity for susceptible host cells, e.g. in cell culture. It follows that antibody complexed with virus must block its adsorption, penetration, or intracellular replication. Studies with phage M13 and poliovirus indicate that antibody interferes with the initial stages of infection of these viruses, whereas other studies, with macrophages equipped with receptors for specific antigen/antibody complexes, indicate that intracellular replication is the site of inhibition.

In the mammalian body circulating immunoglobulin can attach both to free virions released from infected cells and to those still attached to cell surfaces. The virus/antibody complex absorbs complement and, where the virus is of the enveloped variety or is attached to a cell plasma membrane, lysis of the envelope or cell membrane results. Attachment of antibody to a non-enveloped virus renders it more amenable to phagocytosis and subsequent digestion. These events occur extracellularly and limit the spread and multiplication of virus in the body, thereby reducing the clinical intensity of the infection.

The mucosal surfaces of the respiratory and gastrointestinal tract provide the main portals of entry for the majority of viruses. These surfaces secrete IgA and IgE antibodies, though IgM and IgG also function there. For respiratory viruses the primary humoral response consists of IgM, IgG, and IgA, and secondary responses tend to be anamnestic IgG responses.

There are a number of other considerations which may modify immunoglobulin activity. The density of immunoglobulin in the tissues must be at a critical level before immune recognition commences and lysis, phagocytosis, or cytotoxicity takes place. With low densities of immunoglobulins, such as are obtained early on in the humoral response, modulation of infected cells occurs before lysis. The sparse immunoglobulin interaction with viral antigens at the surfaces of infected cells initiates altered modes of viral replication and favours the selection of persistently infected cells. Clinical reports are published periodically which demonstrate the remarkable persistence of viruses in the partly compromised host. There are specific tissue considerations superimposed on the above. With measles virus persistence is best seen in the presence of existing levels of immunoglobulins; it is considered that the persistence of the virus in brain tissue, which gives rise to subacute sclerosing panencephalitis, is favoured by the absence of complement in this tissue and by its lower levels of immunoglobulins compared with serum levels. There are also iatrogenic considerations. Cortisone treatment of measles infection, for instance, will predispose towards giant cell pneumonia by virtue of the immunodepressive effects of the drug. A number of viruses are very capable of replicating in the B-cells, thereby inducing a state of immunodepression. Often such infections generate defective strains of the virus which are able to initiate persistent infections, such as are obtained with measles, lymphocytic choriomeningitis, and bovine leukosis viruses. In the case of measles virus the best replication occurs in mitogen-transformed lymphocytes, and only 1–2 per cent of the lymphocytes release infectious virions, indicating an increasing predominance of defective interfering virions as the replication proceeds.

Humoral immunity is not always of benefit to the host. Viruses may initiate immune complex diseases and autoimmune diseases. Mice vertically infected with lymphocytic choriomeningitis virus suffer from immune complex disease in contrast to those which are horizontally infected in later life. Similarly, mice infected with lactic dehydrogenase virus have circulating virus–antibody complexes and, since such complexes are preferentially filtered out in capillary beds, frequently suffer from life-threatening nephritis as a complication of the virus disease. In other subacute or chronic virus disease there may be generated anti-immunoglobulin antibodies of various specificities, indicating the initiation of an autoimmune disease condition. Measles and influenza are good examples of virus diseases in which autoimmunity develops. The replication of measles virus in lymphocytes leads to the generation of antilymphocyte antibodies. Similarly, infection of lymphocytes with retroviruses may lead to massive lymphoproliferation, generating frank lymphomas with autoimmune potential due to aberrant expression of viral antigens, as with murine MuLV-M oncornavirus.

To escape attack by humoral antibodies viruses and virus-like agents may resort to 'withdrawal' of immunogens. This withdrawal may be initiated by the host cell itself when it enters a resting phase under infection conditions in which the pace of production of viral antigens is in step with the pace of metabolism in the cell, as is observed with measles, SV40 and murine oncornaviruses, or it may be initiated by antigenic modulation of the virus, as is also observed with measles virus. Failure to produce adequate quantities of antigen points to a consequential failure to produce high levels of homologous antibodies resulting in persistence of infection.

In some diseases, no homologous antibodies have been detected despite observed multiplication of the agent. This is so with the virus-like agents of scrapie in sheep, mink encephalopathy, and kuru and Creutzfeldt–Jakob diseases in man. Thus, while the infectious and transmissible natures of these agents are not in doubt the diseases they generate cannot be controlled by vaccination and, in the absence of detectable immune responses, the virulence of each agent will alone determine the survival of the infected host.

The role of cell-mediated immunity (CMI)

T-lymphocytes, involved in CMI, are generated in the thymus. Their major antiviral role is one of cytotoxicity.

For this antiviral cytotoxicity to develop T-cells must first be primed by encounters with virus antigens; this is aided by released macrophage lymphokines. Priming may also result in interferon release and more macrophage recruitment. In order to initiate priming the viral antigens need to be properly presented to the cells. Undegraded virus is known to initiate the best priming whereas formalin and detergent treatment of virus effectively prevent priming.

Once primed, T-cells locate virus-infected cells in the tissues, aggregate about them and mount a cytotoxic attack which results in their lysis. Effective location of infected cells is most effective with viruses which insert antigens into host cell plasma membranes (e.g. orthomyxo- and paramyxoviruses, morbilliviruses, pneumoviruses and herpesviruses). Antigenic variation, as exemplified by orthomyxoviruses and polioviruses, constitutes one form of virological defence in hosts already primed by a first encounter with the virus.

Cytotoxic attack has its greatest effect when mounted early in the virus replication cycle, before virus maturation, cell lysis, and extracellular liberation of virus occur.

While arbovirus and enterovirus infections may be successfully controlled by early macrophage activity and later antibody formation, herpesvirus, poxvirus and morbillivirus infections require the additional and vital help of primed T-cells. T-cells are of especial benefit in these infections (in which cell-to-cell spread occurs) because of their association with interferon release and because antibodies are least effective in preventing this spread.

Since T-cells exert their antiviral cytotoxicity against cells expressing viral antigens at the surface it follows that there are a number of strategies by which viruses may escape their effects and survive for long periods in their host. For example, by multiplying in T- and B-cells themselves, and thereby interfering with the cellular immune response at cellular levels, as do a number of viruses, or by means of incomplete cycles of replication in tissue cells which do not insert viral antigens into the plasma membrane; these may also fail to stimulate neutralizing antibody formation. Some adopt host cell antigens as external viral capsid antigens while others release large amounts of viral antigen which will provide sufficient circulating immune complexes to block antibody-dependent cell-mediated cytotoxicity.

In the host experimental T-cell depletion is obtained with thymectomy or with antilymphocyte serum treatment. Both treatments provide increased susceptibility to a variety of herpesviruses. On the other hand, in nude (athymic) mice some common murine viruses, such as lymphocytic choriomeningitis, Sendai, and Theiler's disease viruses, express their pathogenicity only in the presence of T-cells, thereby indicating their ability to multiply in those cells. Naturally occurring human immunosuppressive diseases, such as ataxia–telangiectasia, in which T-cell formation is suppressed, similarly increase susceptibility to herpesviruses (simplex, zoster and cytomegalovirus), measles and vaccinia viruses, sometimes with a fatal outcome.

It is now recognized that certain virus, or virus-like, diseases exhibit the characters of prolonged persistence of the micro-organism in the tissues, and that the immune responses are ineffective in eliminating them though it may control them to a degree. With scrapie, mink encephalopathy, kuru and Creutzfeldt–Jakob diseases there is no antibody formation, and whatever immunological response may be initiated it neither controls nor complicates these fatal diseases. On the other hand, with Aleutian mink, equine infectious anaemia, lactic dehydrogenase, LCM and murine leukaemia diseases there is a persistent viraemia in the face of an adequately high antibody response, the outcome of which is the formation of immune complexes which, when removed by macrophages, tend to initiate macrophage sites of virus replication. Antibodies do not play any supporting role to either macrophages or T-cells under these conditions. By contrast, such common viruses as herpes simplex and measles are able to become latent within the body tissues (the former within dorsal root ganglia), and undergo subsequent reactivation to give 'cold sores' and subacute sclerosing panencephalitis respectively. Here the CMI response serves to control the *spread* of the viruses in the face of existing levels of antibody but does not eliminate the infection from all of the infected tissues. Measles antigens are known to suppress T-cell formation, while herpesviruses are known to depress the immune response via initial suppression of T-cell activity as well. In both of these diseases virus may be isolated from the lymphoid cells of the host which act as unexpected virus reservoirs.

FURTHER READING

Fraenkel-Conrat, H., and Wagner, R. R. (eds) (1979) Virus–host interactions. In *Comprehensive Virology*, vol. 15. Plenum Publishing Corp., New York.

Neighbour, P. A., and Bloom, B. R. (1980) Natural resistance to virus infections. In Weinstein, L., and Fields, B. N. (eds), *Seminars in Infectious Disease*, vol. 3, pp. 272–93. Thieme–Stratton Inc., New York.

Notkins, A. L., and Koprowski, H. (1973) How the immune response to a virus can cause disease. *Sci. Am.* **228** (1), 22–31.
Proffitt, M. R. (ed.) (1979) *Virus–Lymphocyte Interactions*. Elsevier, North-Holland, New York.
Stewart, W. E. (1979) *The Interferon System*. Springer-Verlag, Vienna.

PART II

COMMUNITY ASPECTS OF THE HOST–PARASITE INTERACTIONS

Chapter 8

General concepts of epidemiology

Interactions between the individual host and particular micro-organisms have been considered in previous chapters. A further important level of interaction must now be studied—that of infectious disease within the group or community of which the individual is a member. Individuals within a community exhibit wide variations in their response to an infectious agent, while between communities the severity of an outbreak by the same infectious agent may vary from time to time and from place to place. The factors which decide the level of interaction and the ultimate pattern of the disease episode are complex and numerous. In this chapter an attempt is made to analyse and summarize some of the more important ones.

1. Defining terms

The epidemiology of infectious disease straddles many disciplines including public health, community medicine, medical statistics, veterinary state medicine, and animal husbandry. Frequently terms are used by workers in these fields which mean different things to different people and it is necessary to define the more common ones as they will be used here.

Outbreaks of infectious disease may be described according to their duration and the geographical extent of their spread (Table 8.1). Within a particular community some diseases are regularly present at a low incidence of infection; these are called ENDEMIC. The types of endemic disease occurring in one place will depend largely on whether the local social circumstances provide effective barriers to their spread. In civilized communities with good sanitary facilities infectious diseases are limited almost entirely to those which can pass readily from person to person by droplet infection (e.g. influenza, the common cold, and certain childhood infections as measles and chickenpox) or, by direct transfer of infectious material, as occurs in venereal diseases. In less developed countries, where water supplies and sewage disposal are inadequate, alimentary diseases (e.g. typhoid and cholera) may be endemic. Similarly, malaria occurs in countries where insect vectors have not been controlled. The continuous presence of low levels of infection in confined localities indicates that endemic diseases are limited geographically in *space*. The parallel term, ENZOOTIC, is used for similar diseases of animals.

Other diseases, usually absent from one locality, may be introduced from

Table 8.1. Terms used to describe outbreaks of
infectious disease in man and animals

Man	Animal	Limited in
Endemic	Enzootic	Space*
Epidemic	Epizootic	Time and space*
Pandemic	Panzootic	Time

*Geographical location.

another producing a sudden outbreak in the community. The term SPORADIC is used of individual cases of infectious disease which are unconnected with one another. Provided conditions for transmission are right sporadic cases may lead to small OUTBREAKS which may escalate into an EPIDEMIC. After a period of time the disease may die out, or be stamped out by a procedure of eradication, and disappear in that locality, at least for a period of time. Epidemics and their animal counterpart—EPIZOOTICS—are limited in *time* and *space*. A good example of an epizootic in the United Kingdom is that of foot and mouth disease. Generally the UK is free of this disease but, occasionally, the infection has been brought into the country and, once established, spreads rapidly. Stringent legislative measures to contain the infection by slaughter, disinfection, and stopping all movement of animals and transport in affected areas, has succeeded in containing and ultimately stamping out the infection. Other countries in Europe and elsewhere, which do not adopt these stringent measures, experience foot and mouth disease as an enzootic disease; they rely on mass and repeated vaccination programmes to keep outbreaks under control.

From time to time epidemic diseases, normally confined to a particular location, cross geographical barriers and spread from one continent to another. The spread of new serotypes of the influenzavirus throughout the world is an example of this. The terms PANDEMIC and PANZOOTIC are used of these diseases in man and animals respectively. They usually die out with time and therefore are limited in *time* but not *space*.

The progress of an infectious disease is usually monitored by statistics of mortality or morbidity. Of these the most useful and least equivocal is that of mortality since the fact of death is incontrovertible. Mortality is the number of deaths due to a particular cause in relation to the whole population over a certain time in a certain area. Emphasis must be placed on the whole population since some workers quote the number of deaths among clinical cases caused by the same infectious agent but this should be called 'case mortality' or 'case fatality'. The difficulty with the clinical approach is in deciding the population to which these statistics relate. For instance, in poliomyelitis should clinical cases be recognized as those presenting with paralysis or include those with mild symptoms which often go undiagnosed? Case fatality measures essentially the virulence of the infectious agent, but mortality statistics are the more reliable measure of the spread of the disease with time.

Morbidity is the incidence of disease; that is the number of people falling ill irrespective of whether or not the disease proves fatal. This is much more difficult to assess. Apart from the subjective assessment of clinical illness, which may range from severe illness to trivial symptoms, it is difficult to define the population being measured. Does it include the individuals who have acquired the infectious agent but have not fallen ill (i.e. subclinical infection) or does it include those exposed to infection but which have not become infected? Extensive laboratory investigations are necessary to establish these facts.

Reference to morbidity has introduced the term subclinical infection. This term indicates that 'infection' is not synonymous with 'disease'. Within a community, individuals will respond variously to an infectious agent. A proportion, although exposed, will fail to become infected; those which are infected may show no clinical symptoms, or there may be a clinical response varying from mild or trivial symptoms to severe symptoms, even leading to death. Subclinical infection can only be determined by laboratory evidence, either by isolating the causal organism or demonstrating an antibody response to infection. Reasons for the differences in individual response to infection and their influence on the pattern of an outbreak in a community will be considered later in this chapter. Anticipating these comments, it is true to say that in a well-adjusted host–parasite relationship 'subclinical infection is the rule, disease the exception and death a rarity' (Burnet).

2. Availability of a pathogen within the community

Very few known pathogenic micro-organisms have become extinct despite worldwide efforts to eradicate them. This fact alone indicates their ability to perpetuate themselves and survive. A disease may disappear for a time, only to reappear later. Survival of a particular pathogen between episodes of clinical disease must occur as hidden or latent infections in apparently healthy hosts or in other hosts acting as reservoirs.

The perpetuation of a pathogen depends on a number of factors which include their ability to escape from infected hosts in life or after death, to survive outside the host, to be transmitted to susceptible hosts, and to infect them. These will be considered *serriatim*.

(a) *Escape of pathogens from infected hosts.*

Of the very large numbers of pathogens which are produced by propagation within the infected host only a very small proportion survive to perpetuate the species. To do this some survivors must be able to leave the infected host. The nature of the pathogen, the disease it causes, and the distribution of the pathogen in the body, will decide the particular route of exit. Examples are given in Table 8.2.

Table 8.2. Examples of routes by which human and animal pathogens are shed from infected hosts

Site of exit	Contaminated tissue or fluid	Pathogen	Disease	Host
Body surface	Desquamated skin	*Trichophyton mentagrophytes*	Athlete's foot	Man
	Hair	*Microsporum canis*	Ringworm	Man, dogs, etc.
	Lesion crusts	Poxviruses, e.g. Varicella	Chickenpox	Man
	Exudates (e.g. pus)	*Staphylococcus aureus*	Abscesses	Man, sheep, etc.
Nose	Secretions	Adenovirus	Common cold	Man
	Exudate (blood stained)	*Bacillus anthracis*	Anthrax	Cattle
Mouth	Secretions	Measles virus	Measles	Man
	Saliva	Rhabdovirus	Rabies	Dog
	Sputum	*Mycobacterium tuberculosis*	Tuberculosis (pulmonary)	Man, cattle, etc.
	Tonsil	*Erysipelothrix rhusiopathiae*	Erysipelas	Swine
Mammary gland	Milk	*Streptococcus agalactiae*	Mastitis	Cow
Anus	Faeces	*Shigella dysenteriae*	Dysentery	Man
		Mycobacterium johnei	Johne's disease	Cattle, sheep
		Poliovirus	Poliomyelitis	Man
Urogenital tract	Urine	*Leptospira canicola*	Leptospirosis	Dog
	Semen	*Brucella abortus*	Brucellosis	Bull
	Eggs	*Salmonella pullorum*	Pullorum disease	Poultry
Eyes	Tears	*Haemophilus influenzae*	Pink eye (New Forest disease)	Cattle
Wound vector syringe	Blood	*Rickettsia prowazekii*	Typhus fever	Man
	Blood	Virus B hepatitis	Serum hepatitis	Man

The dynamics of pathogen shedding varies from one disease to another. Virus diseases of the respiratory tract have been judged to be responsible for half of all episodes of human illness. This must indicate that shedding of these viruses is a highly successful procedure. Pathogens localizing in the upper respiratory tract are ejected from the respiratory tract in droplets of moisture. During talking, coughing, and sneezing air is forced, under considerable pressure, through the nose and the constricted apertures between the teeth. This results in the ejection of enormous numbers of fluid droplets. Sneezing is the most vigorous of these mechanisms and a single sneeze may generate as many as one million droplets less than 0.1 µm in diameter, and thousands of larger droplets. The large ones fall out of suspension but the smaller ones rapidly evaporate to non-volatile particles many of which will remain airborne and carry viable pathogens. Dust laden with micro-organisms arises from clothing, bedding, handkerchiefs, etc., which in turn derive contamination from desquamated skin scales, hairs, pus, and other body secretions. Alimentary infections are invariably shed in faeces, urine, and sometimes saliva. Other examples of shedding are given in Table 8.2.

(i) *The duration of shedding.* This period (i.e. the time during which a patient is infectious) is generally shorter for virus infections than for bacterial infections. However, large numbers of virus particles are usually liberated during the shedding period. A measles patient is infectious for up to 1 week. Bacterial infections of the oral/pharyngeal region, e.g. diphtheria, tonsillitis, are carried in the throat for several weeks, whilst patients with pulmonary tuberculosis remain infectious for months or years. In general, the period of shedding corresponds to the duration of clinical illness which may be extended to include part of the incubation period (Figure 8.1). In many viral diseases, for example mumps, measles, and smallpox, the infectious period begins 1 or 2 days prior to the appearance of recognizable symptoms. Where this occurs, an apparently healthy patient may shed the pathogen before infection is diagnosed and the patient isolated, and this constitutes a very important factor in the spread of infection in a community.

(ii) *The length of the incubation period.* This period, i.e. the time from infection to the first signs of illness, differs from one infectious disease to another. This can have an additional and profound effect on the pattern of an epidemic. The longer the incubation period the further an infected patient may travel before shedding the pathogen. This is particularly relevant with modern speed of travel and patients who are apparently healthy but infected in one country, may travel to the opposite side of the world before symptoms appear. Smallpox has been introduced into the UK in the past by immigrants arriving from countries where this disease was endemic. Even in patients not travelling abroad, the 'mobile shedder', i.e. the person who continues to leave home and mix with others, is far more likely to spread infection in the community than the individual who remains at home.

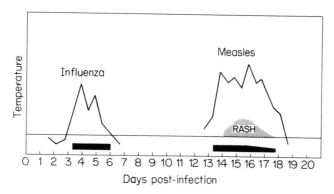

Figure 8.1. The time relationship between the incubation period and period of communicability in influenza compared with measles. Symptoms and fever are indicated by the temperature curves; the approximate period of virus liberation is shown as a solid bar (after Burnet and White, 1972, *Natural History of Infectious Diseases*, p. 125. Reproduced by permission of Cambridge University Press)

(b) *Survival of pathogens outside the host*

Once pathogens are shed their perpetuation depends, at least in part, on their ability to survive outside the host before they reach a susceptible one. Pathogens vary greatly in their ability to resist inimical conditions, such as exposure to ultraviolet light and particularly to drying. The presence of body exudates aids them in this. The gonococcus is a very delicate organism and does not survive for any length of time outside the human host; in consequence transfer of infection (from one host to another) can only occur by venereal or equally close contact. Many viruses causing respiratory infections are highly sensitive to drying and survive for only a brief time in droplet spray—close contact indoors is usually essential for their transmission. Many enteropathogens can survive much longer. Sewage containing *Salmonella typhi* and poliovirus remain a potential source of infection for a long time, and *Salm. dublin* has been shown to survive in dried excreta on the walls of animal houses for months. Sporing pathogens, such as *Bacillus anthracis*, survive for years in soil, on contaminated straw or hay and in contaminated bone meal which may be added as a mineral supplement to compounded animal foods. Their virulence usually is unimpaired and sporadic cases of anthrax occur in the UK from time to time following ingestion of contaminated food.

(c) *Transmission of microbial pathogens between hosts*

Pathogens eventually die unless they reach a susceptible host. The means by which they reach and enter a new host are very diverse and the principal

Table 8.3. Examples of vehicles by which pathogens are transmitted from one host to another

Vehicle	Nature of contamination	Examples of disease
Food	From infected animal, e.g. meat, milk, eggs	Salmonellosis
	Contaminated during processing	Staphylococcal food-poisoning
Water	Sewage contamination	Typhoid, cholera
Air	Dust contaminated with infected body exudates	Q-fever, poxviruses
	Droplet spray exhaled	Upper respiratory tract viral diseases, e.g. common cold
Vector	Mechanical carriage, e.g. flyborne	Salmonellosis
	Infected arthropods	Rickettsial diseases, malaria
	Non-arthropod vector, e.g. rabied dog, bat, etc.	Rabies
Inanimate objects	Toys	Dysentery
	Clothing	Q-fever
	Floors	Athlete's foot
	Wool	Anthrax (woolsorters' disease)

pathways are summarized in Table 8.3. A knowledge of how pathogens spread is essential to the design of control measures. Whilst the routes of transmission are diverse the routes of entry into a new host are limited to ingestion, inhalation, contact, and wounding, and the occasional vertical transmission from mother to offspring (Table 8.4). Typical examples of these are considered in later chapters but will be briefly summarized here.

(i) *Ingestion.* Alimentary infections invariably enter the host via the mouth (Chapter 14). After swallowing they must survive the acid and digestive juices of the stomach or, in the ruminants, the extensive competing flora of the rumen, and eventually reach the small and large intestine. For these reasons alone usually large numbers of organisms are required to form an infective dose and are likely to be more successful if ingested in a large bulk of food or large volume of fluid. Smaller numbers of pathogens may infect via the tonsil. Alimentary pathogens are usually present in food, water, or milk, or may be introduced into the mouth on contaminated objects or fingers, or, in the case of animals, by licking the faecally contaminated coats of other animals contaminated by faeces.

(ii) *Inhalation.* In advanced countries, where many routes of infection are controlled, by far the greatest number of human infections enter by the respiratory route (Chapter 17). Many are fairly trivial like the common cold, measles, mumps,

Table 8.4.   Portals by which pathogens enter the host

Site of entry	Mode of entry	Examples of diseases and hosts
Mouth	Ingestion	
	Food	Food-poisoning (man)
	Water	Typhoid fever (man)
	Sucking—communal nursettes in calf pens	Salmonellosis (calves)
	Licking—animals' coats, environment,	Salmonellosis (calves)
	contaminated fingers, toys	Poliomyelitis (man)
Respiratory tract	Inhalation—dust, droplets	Shipping fever (cattle)
Eye	In-ocula–rubbing, swishing of animal's tail	Conjunctivitis (cattle)
Body surface	Contact	
	Venereal	Vibriosis (cattle)
	Kissing	Glandular fever (man)
	Animals rubbing together	Ringworm (cattle)
	Wounding	
	Surgical	Tetanus (horses)
	Accidental abrasions	Anthrax (man)
	Insect bite or abrasions	Plague (man)
	Animal bite	Rabies (man, dog)
Urogenital tract	Ascending—urethra, to bladder, via ureters to kidney	Pyleonephritis (cow, man)
	Transplacental—*in utero* infection	German measles (man)
	Transovarian—infected eggs	Pullorum disease (poultry)
	Parturition—birth canal (vagina)	Gonorrhoea (man)

and chickenpox. Different pathogens tend to lodge in and initiate infection in different parts of the respiratory tract. Some have a predilection for the epithelial lining of the nose or larynx, e.g. the common cold (adenoviruses), others for the bronchial passages, e.g. influenzavirus and the diphtheria bacillus, and *Streptococcus pyogenes* for the tonsils. The psittacosis agent must be carried deep down into the lung to initiate infection. Whether or not these agents reach these predilection sites depends at least in part on the size of the particles on which they are carried; the further they must penetrate to initiate infection the finer must be the particle of dust or droplet which carries them.

(iii) *Contact*. Close body contact is necessary for the transmission of venereal diseases since the causal agents readily die outside the body (Chapter 18). Certain diseases, e.g. herpes simplex, cold sores, and infectious mononucleosis (glandular fever), may be transmitted by kissing.

(iv) *Wound*. Some pathogens enter the body through the skin, usually by way of a scratch or more obvious wound. Contact with urine-contaminated water containing leptospirae may be followed by leptospirosis. This is often an occupational hazard of sewer workers and cavers who contact water contaminated with infected rat urine, of dairyman in contact with cattle urine, or swineherds from pig's urine. Other infections may enter through breaks in the integrity of the skin resulting from animal or vector bites (see rickettsial diseases, Chapter 19) or by surgical or accidental wounds (as in tetanus).

(v) *Vertical transmission*. Direct infection from one generation to another can occur by contamination of an egg in the ovary of a laying hen (as in pullorum disease, Chapter 14) or via the placenta to the foetus *in utero* (as in rubella in humans, see later).

(d) *The infective dose*

Many susceptible individuals may be exposed to infection but only a proportion in fact become infected. A number of factors account for this but one is the number and virulence of organisms which challenge the individual. Hard facts on what constitutes an infective dose under natural conditions are difficult to obtain and must depend chiefly on observation and experimental infection of animals.

From records of laboratory-derived infections it is recognized that some pathogens more readily infect than others under conditions where similar aseptic techniques are being practised. The most common laboratory infections are tuberculosis, tularaemia, brucellosis, and Q-fever; dysentery, typhoid fever, poliomyelitis, and influenza rarely occur as laboratory infections. This would suggest that much smaller doses of the former are needed to initiate infection.

Experimental infection of laboratory animals also indicates marked differences in the infective dose. It is possible to initiate anthrax in the mouse by introducing a very small number of anthrax bacilli on the tip of a straight wire into a minor incision, e.g. at the base of the tail. In contrast, salmonella infection of mice and calves requires as many as 10^5 salmonellae to initiate infection. Whilst the broad differences in infective dose are clear, under natural conditions other factors obviously play their part. For instance, repeated intake by mouth of small numbers of *Salm. dublin*, as occurs naturally in calves may be equally as effective as one large dose. The nature and volume of food passing through the stomach may facilitate passage of a smaller number of pathogens into the small intestine. Even less definable, but possibly of greatest importance, are simultaneous stress factors which often lower the resistance, or alter the physiology, of the host and allow infection to occur more readily. Cross-infection by *Salmonella* spp. between pigs or calves, for instance, has been shown to occur at increased frequency following the rigours of transport between the producing farms and the abattoir.

3. The nature of the community

Having considered factors which affect the transmission (the sowers) of the microbial pathogens (the seed) we must next consider the various ways in which the community (the soil) responds to infection. The community as a whole, and the individuals within it, exhibit differences in response to the same pathogen. These differences are due mainly to differences in immune status at the time of exposure. The response of the community will be in proportion to the numbers and the age distribution of susceptible individuals within it. Dependent on these factors different communities respond to infection in different ways.

(a) *Virgin population*

Isolated communities, such as remote island communities, may elude exposure to a particular infection for many years. The longer this occurs the greater the proportion of susceptible individuals within it. The proportion alters mainly by death of the older members which, by prior exposure, may have been immune, and their replacement by birth of new, fully susceptible members. Subsequent exposure to infection after many years may find a completely susceptible population. The history of measles on the Faroe Islands is a classical example. An epidemic in 1781 met a fully susceptible community and attacked all ages with marked severity. A further attack 65 years later again affected the whole population with the exception of a small number of older members who had experienced in infancy the earlier outbreak which had conferred a lifelong immunity. A further outbreak in 1875 affected only persons under 30 years of age; these included those which had been born since the 1846 outbreak.

A population totally made up of non-immune individuals may be referred to as a virgin population. Exposure to a new infection will affect all ages, the most severe clinical illness being experienced in the most susceptible age groups, namely the elderly and the very young (Figure 8.2a). This epidemic pattern is experienced in modern times when a new serotype of influenzavirus appears, as in the pandemic of Asian 'flu in 1957 (see Chapter 17).

Outbreak of mild infectious diseases, such as the common cold, frequently affect communities of geographically isolated people either when visited from outside—the so-called 'stranger's cold'—or when individuals from these communities visit other countries. Refugees from the South American islands of Tristan da Cunha suffered many mild respiratory infections, to which they had not been exposed recently, when they sought refuge in the UK during days of volcanic eruption in their islands. Similarly members of the Antarctic expedition went down with 'flu-like colds when visited by relief ships and, later, on their return to their home countries.

Visitors to these virgin populations, although carrying the virus, are usually healthy. By frequent exposure to the virus in an endemic situation they build up an

(a) 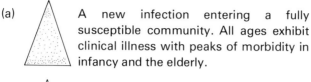 A new infection entering a fully susceptible community. All ages exhibit clinical illness with peaks of morbidity in infancy and the elderly.

(b) 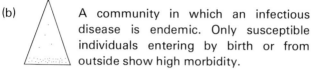 A community in which an infectious disease is endemic. Only susceptible individuals entering by birth or from outside show high morbidity.

(c) 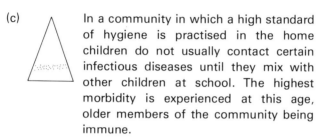 In a community in which a high standard of hygiene is practised in the home children do not usually contact certain infectious diseases until they mix with other children at school. The highest morbidity is experienced at this age, older members of the community being immune.

Figure 8.2. The distribution of clinical illness in communities with different immune status. The triangles illustrate the age distribution in a community—the bases represent the newborn members and the peaks the diminishing numbers of increasing age

immunity which is reinforced from time to time, and hence they are able to carry the virus with impunity. By this means the virus is maintained and, as contact is made with susceptible hosts, cases of clinical illness occur among them.

(b) *Populations exposed to endemic infection*

Where infection persists in a community, the individuals are regularly exposed to the antigenic stimulus which confers on them an immunity to the particular pathogen; this often persists throughout life. Protection is dependent on repeated stimulation and hence on the individual remaining in the community. Susceptible individuals in the community will be those that enter from outside or who are born into it. The highest morbidity will therefore be in the very young. This is illustrated in Figure 8.2b and is typical of the majority of childhood infections, such as measles, chickenpox, whooping cough and mumps. Sometimes, dependent on factors such as hygiene in the home and the degree of mixing with others, infection may be delayed until contact is made with other children in nursery or school

(Figure 8.2c). Sonne dysentery, for instance, which is often contracted by handling toys or school equipment contaminated by the hands of other children, frequently occurs among children attending day nurseries.

(c) *Variations in community response*

Poliomyelitis has shown a progressive shift in age incidence in different communities from time to time. It is a disease of developed countries, the incidence of which, prior to extensive immunization, failed to respond to improvements in medicine and standards of living. Infection is by the oral route. The virus multiplies in the alimentary tract and is excreted in faeces for 1 or 2 weeks. During this phase immunity is stimulated and in most individuals the episode is over without symptoms of illness. Occasionally, however, the virus enters the blood stream and reaches the central nervous system. Symptoms ranging from headache and fever (the non-paralytic form) to paralysis which may be fatal, are experienced.

To understand the shift in age incidence of paralytic poliomyelitis it is necessary to appreciate the situation which arises when a virgin population first encounters the infection. In a virgin population most cases of paralysis occur in the age range 15–25 with only low incidence in infants. In a community where poliomyelitis is endemic and in which crowding gives ample opportunity for contact infection to occur, most members of the community build up their own immunity and the majority of paralytic clinical cases arise in susceptible infants born into the community—hence the term 'infantile paralysis'. In communities with higher standards of living, and correspondingly higher standards of hygiene, many infants do not come into contact with infection in the early years of life; their first encounter is at the time they mix with other children at school. Under these conditions the peak of paralytic cases occurs during the early years of schooling. It must be emphasized that these shifts in age prevalence of paralytic poliomyelitis can only be detected in social groups in which artificial immunization has not been carried out. By mass immunization programme, first with the Salk vaccine (a formalin-treated suspension of poliovirus) and later the Sabin vaccine (an attenuated, living suspension) against types I, II, and III, the incidence of paralytic poliomyelitis has virtually disappeared in communities where this is practised.

(d) *Contact probability*

The impression that every member of a population contracts infection during an epidemic is quite fallacious. When an epidemic strikes a community, only a proportion of members will demonstrate clinical illness. Some may be immune prior to the outbreak; others may be subclinically infected during the outbreak; but a proportion of fully susceptible individuals will elude infection. It has been observed on many occasions that an epidemic 'burns itself out' long before all susceptibles make contact with the infection. This is because, as the epidemic runs its course, the probability that an individual will be exposed to infection decreases with time. Early in the epidemic the contact probability is high; that is, an infected individual is likely to come into contact with many susceptible individuals and pass on the infection. At the peak of the epidemic the ratio of infected to susceptibles levels out, and only about half the population may remain susceptible. Thereafter, the opportunity for an infected patient to contact susceptible persons becomes progressively less. Consequently, an epidemic of infectious diseases, in which the duration of shedding lasts but a few days (e.g. measles, influenza), dies out once the proportion of suceptibles to non-suceptibles reaches a certain level.

Subsequent to an epidemic, the community may remain free of further episodes by the same pathogen for years. During that time the composition of the community will change by the birth of new susceptible members, by movement of susceptible persons into the community and, in the elderly, by the natural loss of immunity in persons who previously experienced the infection. The pattern of an epidemic following a second exposure to infection will depend on the proportion of susceptible to immune individuals at the time.

(e) *Miscellaneous factors which influence outbreaks*

Epidemiological studies reveal that epidemics are often precipitated or influenced by particular factors. Some are seasonal: infections of the human respiratory tract, for instance, occur more commonly in winter when people tend to be confined indoors in badly ventilated buildings. Similar conditions often prevail in animals reared under intensive husbandry conditions although this is not limited to particular seasons of the year. Outbreaks of human plague in Bombay have in the past shown seasonal prevalence from year to year. This has been shown to follow the rise and fall of the rat populations and the associated survival and multiplication of the rat flea which transmits plague to man. This occurs most readily whilst the rat spends more of its time under moist conditions. Hence the annual peak of plague in man falls between January and April prior to the host season.

Waterborne infections, such as cholera (Chapter 16), still persist in densely populated parts of India, particularly in flat areas with extensive surface waters where sanitation is poor.

War conditions have influenced many epidemics for a variety of reasons. The movement of susceptible personnel into foreign countries where foci of infection or vectors occur, malnutrition among peoples in devastated areas often accompanied by infestation with lice and other parasites which can transmit infection, and the grouping of people from many areas. The discovery of congenital damage caused by rubella infections was based on the last of these. Young women, brought up in relative isolation in Australia, were grouped together when called into military service. Many had not experienced rubella infection before, but contracted the infection after entering the forces. This coincided with a number becoming pregnant and, in retrospect, the birth of children with congenital abnormalities was linked with rubella infection during pregnancy.

FURTHER READING

Anderson, G. W. (1965) The principles of Epidemiology as applied to infectious disease. In Dubos, R. J., and Hirsch, J. G. (eds), *Bacterial and Mycotic Infections of Man*, 4th edn., pp. 886–912. Pitman Medical Publications, London.

Andrews, C. H. (1967) *Natural History of Virus Diseases*. Weidenfeld & Nicolson (237 pp.).

Burnet, F. M., and White, D. O. (1972) *Natural History of Infectious Disease*, 4th edn. Cambridge University Press (278 pp.).

Gale, A. H. (1959) *Epidemic Diseases*. Pelican, No. A456 (159 pp.).

Wilson, G. S., and Miles, A. A. (1964) *Principles of Bacteriology and Immunity*, 5th edn., vols. I and II. Arnold Ltd, London (2563 pp.).

Chapter 9

The control of infectious disease— I: Methods to avoid infection reaching susceptible hosts

Infectious diseases are responsible for an immense amount of suffering; their control is therefore a first priority. This is attempted at different levels. One most effective means is by eradicating or reducing the sources of infection before it reaches susceptible hosts. A second is to interfere with the normal epidemiological cycle of the infective agent, thereby avoiding the infection reaching susceptibles. A third, considered in the next chapter, is to increase the resistance of susceptible individuals by increasing their immune status. A fourth means is to attack the pathogens by using specific chemotherapy (Chapter 11). These control measures will be considered individually.

1. Control of sources of infection

One of the more important means of disease control is achieved by removing or controlling sources of infection, thereby preventing pathogenic micro-organisms reaching susceptible members of a population.

Prevention, control, and eradication of sources of infection are the concern of many countries. The spread of infection from man to man, from animals to man, and to a lesser extent from man to animals, constitutes the more important pathways of cross-infection. The term 'zoonoses' has been coined to include microbial diseases which are naturally transmitted between vertebrate animals and man; more than 100 zoonoses are recognized (Chapter 13).

(a) *Eradication of infected animals*

Eradication programmes obviously cannot be practised in the control of specifically human diseases although under certain circumstances medical examination, treatment, and isolation are compulsory. Eradication is, however, proving a most successful approach to the control of a number of diseases of domesticated animals. Great Britain, by virtue of the fact that it is separated from the Continent by a sea barrier, has been able to pursue eradication schemes independent of other European countries. Already eradication of a number of serious diseases has been achieved, including tuberculosis and foot-and-mouth disease in domestic animals, and a programme is, at present, under way to eliminate brucellosis (Chapter 15). Initially a defined area is cleared of infection. This is done by testing all stock, followed by culling infected animals. The area is then designated disease-free. This area is extended year by year until the whole country is free of infection. Reliable tests to detect both clinically and subclinically infected animals are essential to success. Testing must continue after completion of the eradication to ensure that freedom from infection is maintained. For instance, in order to maintain the national herd free of tuberculosis, farm animals are regularly tested by the tuberculin test and any positive reactors immediately destroyed. Over recent years a recurrence of bovine tuberculosis in dairy cattle in the south-west of England was traced to badgers as a source of infection. This necessitated the local destruction of badgers by gassing to avoid further cross-infection to dairy herds.

The carcasses of animals which die of highly infectious diseases, or are slaughtered during an eradication programme, must be adequately disposed of. Anthrax is controlled, in part, in Britain by legally enforced disposal of anthrax carcasses of the few cases which occur sporadically. *Bacillus anthracis* readily produces highly resistant spores when exposed to air. Hence, if the death of an animal from anthrax is suspected, the carcass must not be opened for post-mortem examination; this avoids exposing the organism to the air and contamination of the surrounding ground or pasture by spores which would thereby be produced. Instead the cause of death can be confirmed by microscopical examination of blood smears taken from a peripheral vein and, if anthrax is diagnosed, the carcass is incinerated *in situ* or buried in quicklime well below ground level. In foot-and-mouth disease stringent measures of eradication are adopted in every outbreak in Britain and many other countries. The cost of compensating farmers for loss of livestock, even in such a serious outbreak as that of 1967–8 in Britain, was less than the widespread economic loss in milk yields and in the value of carcass meat which would have resulted if the disease had been allowed to become endemic in the animal population. Added to this would be the expensive vaccination programme which would have to be continuously implemented to keep the infection under control.

(b) *Isolation of potential sources of infection*

The spread of infection may be prevented by the segregation of infected animals from susceptible ones. Quarantine measures are widely practised today in the control of animal and human diseases, but are now based on precise knowledge of the length of the incubation time and the time the patient remains infectious. Importation into Britain of many animal pets is controlled at the ports by quarantine measures to avoid introducing infections not normally present in the country, such as rabies in dogs and psittacosis in birds. Many countries require veterinary certification that dogs are free of leptospiral infection before they are imported. Other measures to control animal diseases include legal notification of specific diseases under the Statutory Zoonoses Order. This makes possible a continuing appreciation of the distribution and incidence of infection. Other approaches are by the improvement of the husbandry of animals with its corresponding improvement in health of animals, therapy to reduce the length of excretion of pathogens, and the safe disposal of infected excreta. The disposal of large volumes of infected slurry from intensive animal units is presenting a real problem at the present time (Chapter 14).

On the human side, patients with highly infectious disease, like smallpox, Lassa fever, and infectious hepatitis, are isolated and treated in special isolation units in hospitals. Immigrants are medically examined on entry to ensure freedom from infection and immediately treated if infected. Food handlers, especially any returning from abroad, should be medically checked to ensure freedom from enteric infections, and known carriers are legally required not to engage in the food industry. The proper and efficient disposal of sewage also safeguards against enteric infections. Under special conditions, as during cholera epidemics, restriction on the movements of pilgrims may be imposed.

2. Control of pathways of cross-infection

Most microbial diseases of animals and man arise from exogenous sources and reach susceptible hosts by characteristic pathways. One approach to the prevention of infectious disease is to break the chain of events which results in the infective agent reaching the susceptible host from sources of infection. These measures of control are based on accurate knowledge of the natural history of disease. A number of these control measures will be considered.

(a) *Provision of safe water supplies*

Some pathogenic organisms are waterborne and when present in drinking water may infect susceptible hosts (Chapter 16). Waterborne pathogenic organisms usually gain access to water from animal and human excreta and sewage. Outbreaks of disease, such as typhoid fever and cholera in man, were at one time widespread but the control of water supplies and improved hygiene in food handling have virtually eradicated these diseases from the technically more advanced countries. Water may also be polluted by poisons of non-microbial origin from factory wastes. Constant control is therefore necessary to ensure that drinking water conforms to both chemical and bacteriological standards of purity. Drinking water may be artificially purified by storage, filtration, and chlorination.

(i) *Storage*. By storage of water in lakes and reservoirs the amount of suspended organic material normally tends to decrease, both by sedimentation and as a consequence of microbial activity. This in turn leads to a reduction in the number of micro-organisms and in particular of pathogenic bacteria which are less able to survive in competition with saprophytic species.

(ii) *Filtration*. Even if no additional pollution occurs during the process, storage of water is insufficient to render it safe for drinking; hence artificial methods of purification must be employed. The water is first piped to a filter plant which may be a *slow sand filter* suitable for reasonably pure waters and for use where adequate space is available for the large filters needed, or a *rapid sand filter*, suitable for turbid waters or where insufficient land is available.

(iii) *Chlorination of drinking water*. Neither natural purification nor filtration renders water absolutely safe for drinking; this may be achieved by the use of chlorine. Chlorination destroys all waterborne pathogenic bacteria and also pathogenic intestinal protozoa, such as *Entamoeba histolytica*, the cause of amoebic dysentery. Some chlorine combines with reduced organic material in the water and is no longer effective. This 'chlorine demand' varies considerably with different waters according to the amount of organic matter present, and consequently the amount of chlorine that must be added to give a safe residual amount of free chlorine (0.2–0.4 part per million) also varies. The small amount of free chlorine necessary to render water safe is harmless and usually cannot be detected, but waters with an abnormally large 'chlorine demand' may have unpleasant flavours due to the presence of chlorine compounds, such as chlorophenols.

(b) *The control of sewage*

Enteric pathogenic micro-organisms are excreted in faeces and this necessitates the adequate and safe disposal of sewage to avoid cross-infection. In isolated communities the use of septic tanks or the simple expedient of burial is adequate. The latter was advocated in the Mosaic law over 3000 years ago (Deuteronomy 23, 12–14).

In highly industrialized countries such as Britain, where large communities have grown up, the disposal of sewage presents an enormous problem. The simplest means of sewage disposal, still practised by coastal and estuarine towns, is to run the raw sewage into the sea or nearby river. The larger inland communities, by lack of convenient dumping grounds, have been forced to develop sewage purification systems whose end-products are mainly harmless water and some solids.

Sewage consists mainly of water containing dissolved and suspended organic and inorganic substances along with many micro-organisms. Sewage is first screened to remove solid matter. Grit and stones are allowed to settle out from the slow-flowing stream of sewage. There is a regular removal of putrefactive sludge to prevent the onset of anaerobic digestion. The remaining material, which is removed after varying periods of sedimentation (usually 15 hours in Britain), consists of liquid with suspended flocs of organic matter. This may be treated by the activated sludge process or by slow filtration, both being microbiological processes which reduce the number of micro-organisms present and render it safe.

(c) *Farm wastes*

Traditional methods of farming involve the collection of animal excreta, together with relatively large amounts of bedding, into dung heaps sited near animal houses. In these heaps aerobic decomposition (composting) occurs—this produces a rise in temperature which virtually pasteurizes the dung and destroys pathogens. After about 4 weeks the manure is safe to spread on farm land for agricultural purposes.

Modern methods of intensive animal husbandry have created enormous problems due to the vast quantities of animal effluents which have to be disposed of. These 'slurry effluents' consist of faeces and urine, diluted by large volumes of water used for hosing down, and a minimum of straw. Slurry is thus a liquid manure but by its very nature is anaerobic and no composting occurs. Slurry often contains large numbers of pathogens from infected animals, e.g. salmonellae, and often large numbers of antibiotic-resistant bacteria. Great concern exists over the public health and animal health risks since these liquid manures are often spread untreated onto land. The risks include bacterial and viral pathogens, parasites, and toxic chemicals, such as copper which is included in diets fed to pigs. Minimum requirements, in process of being formulated, include holding the slurry in tanks for 60 days before spreading preferably onto arable land. Where slurry is spread on pasture land, this should not be grazed for 30 days.

3. Control of food supplies

Many diseases may be contracted by the ingestion of contaminated food (Chapter 14). Food may come from infected animals or birds and therefore be infected at source, e.g. meat, milk or eggs, or vegetables may be contaminated during irrigation by polluted water (e.g. watercress beds). Foods may also become contaminated with pathogenic organisms during processing or preparation for eating. Pathogens may be introduced by flies or by the handling of foods by human carriers of food-poisoning organisms. The risk is high with foods prepared for consumption the following day, the organisms multiplying if the food is kept at a temperature conducive to growth.

The increase in urban populations during the present century, and improvements in methods of food preservation, have led to large-scale transport of basic foods from the producer to the consumer areas. This has inevitably increased the risk of infection of many people from a common food source. This risk can be reduced by suitable precautions. Methods employed for improving the keeping quality of a food (i.e. avoidance of spoilage) are often adequate to render a food safe for eating. They include the addition of an acceptable germicide, the application of heat, and storage at low temperature.

(a) *Preservatives added to foods*

Organisms can be inhibited by the addition of small quantities of a chemical preservative. These include acids, esters, phenols, quinones, nitrofurans, nisin, and subtilin. Acid foodstuffs are more easily conserved chemically and sulphurous acid (sulphur dioxide) is used to control acid-tolerant bacteria. All substances used as antimicrobial food preservatives must pass very stringent statutory tests, including feeding to experimental animals, before being used in foods for human consumption.

Smoking meat, fish, cheese, etc. (normally in conjunction with salting), is an ancient method of food preservation brought up to date. The active components of vapour include volatile fatty acids which have both a bactericidal and residual mild bacteriostatic effect.

A wide range of foods is preserved by dehydration. The lower the moisture content of a product, the less liable it is to support microbial growth. A similar effect is obtained by raising the osmotic pressure of a food or beverage; this can be done by adding salt or sugar.

(b) *Heat treatment of foods*

(i) *High temperature.* In practice the terms 'sterilization' and 'pasteurization' are used to differentiate different levels of heat treatment. Pasteurization usually implies that only some of the spoilage organisms or food pathogens are destroyed. With both processes it is necessary to know the temperature required to inactivate a particular organism or, if spore-forming, its spores. This is particularly important in the canning industry since bacterial spores may survive and subsequently produce spoilage or, in the case of pathogens, toxins able to cause food-poisoning (Chapter 14). Canners have very rigid standards, both in the numbers of organisms permissible in the food before canning and in the control of the heat to which the cans are exposed.

Consumer-milk is usually transported by bulk-collection services; it is therefore most essential that adequate measures of control are observed since contaminated milk from one source may be mixed with a large volume of clean milk. The first requirement is good animal husbandry and dairy technique to produce a clean product of high quality. As an additional safeguard most milks are heat-treated to kill pathogenic bacteria which may be present and, at the same time, to reduce the number of contaminants thereby improving its keeping quality. Three methods of heat treatment are currently used.

(α) *Pasteurization.* This method of partial destruction of the microbial population by heat was first introduced by Pasteur to kill contaminating organisms which interfered with the fermentation processes in the manufacture of wine. Its application to milk was first used in Denmark to safeguard pigs against infection from bovine pathogens, but its widest industrial application today is in heat-treating milk for human consumption. By holding the milk for a defined time at a standard temperature (e.g. 15 seconds at 161°F (71.7°C) in the 'High Temperature Short Time' process) most non-sporing organisms, including pathogens, are killed. This renders the milk safe for drinking and extends its keeping quality.

(β) *Boiling.* Greater numbers of micro-organisms are killed when milk is held at a temperature of at least 212°F (100°C) for a standard time; the bottles are sealed immediately afterwards. This process imparts a caramelized flavour to the milk and homogenization results in no visible cream line. Milk subjected to this process is often called 'sterilized' but total sterility is not achieved and it will not keep indefinitely at normal temperature.

(γ) *Ultra heat treated (UHT).* By this method the milk is exposed to a temperature of not less than 270°F (132.2°C) for at least 1 second and filled aseptically into sterile containers. Usually this treatment renders it sterile and therefore gives it excellent keeping qualities. There is no cream line, the cream being dispersed throughout the milk. From 10 to 20 per cent of vitamins are destroyed and a slight flavour is imparted, but this becomes less upon storage. Because of the excellent keeping quality of the milk it is likely to become very popular since less frequent deliveries to the consumer will be possible and export to other countries is facilitated.

(ii) *Low-temperature storage of food.* Food may be chilled to between 0 and 5°C to delay spoilage while awaiting sale or consumption, or to −18°C for extended storage. Frozen foods are not sterile and most spoilage is due to improper hand-ling prior to freezing. Freezing is not generally bactericidal; hence the commercial freezing temperatures cannot be guaranteed to destroy pathogens, particularly with modern freezing techniques using rapid freezing at −35°C followed by storage at −18°C. Prevention of food-poisoning from frozen foods depends therefore on the use of good-quality foodstuffs, on maintaining impeccable hygiene standards during processing and on the avoidance of contamination and multiplication of micro-organisms in the thawed product. After a frozen food is thawed it does not then spoil any more rapidly than it would have done had it not been frozen unless the cell walls have been damaged excessively by poor freezing techniques. No frozen food should contain salmonellae, irrespective of whether it will be cooked before being eaten or not. This is perhaps an ideal situation, practicable only in pasteurized products and frozen vegetables.

4. Control of airborne cross-infection

Respiratory infections arise by inhalation of pathogenic organisms in droplets of moisture from infected hosts or on contaminated airborne dust.

Cross-infection by 'droplet' infection can be restricted by avoiding overcrowding. This may be achieved by the adequate spacing of beds in hospital wards, of desks in schoolrooms, and of stalls in stables. Subdividing large hospital wards into smaller units is more effective in reducing cross-infection than wide spacing of beds. It has long been held that the common-cold virus is transmitted from patient to patient by droplet infection but prolonged personal contact, such as occurs in the home or school, is more important than the casual contact. It is often difficult to separate a purely airborne route from that of personal contact and invariably both pathways of infection have to be controlled at the same time.

The provision of adequate natural or mechanical ventilation, with filtered air, reduces air contamination. The fact that ventilation is better in summer may account, in part at least, for less respiratory disease occurring at that time of the year. In multi-storied buildings with mechanical ventilation, movement of contaminated air can occur from one floor to another by way of lift shafts or laundry chutes and these present a real hazard, especially in hospitals where a build-up of infection in one ward may be passed to another on a different floor. Positive air-pressure within operating theatres excludes contaminated air entering from adjacent rooms and corridors and also removes air from within the theatre which will become contaminated during the operation. Personnel working in contaminated environments, such as operating theatres and wards of infected patients, may be protected by wearing sterile protective clothing and face masks. Face masks must be frequently changed and, although they filter only the larger salivary droplets, are considered adequate to prevent direct contamination of wounds.

Dust derived from animal and human sources is invariably laden with micro-organisms. This comes from hair, desquamated skin, and fragments of textiles resulting from friction between layers of clothing or activities such as bed-making. This is of particular importance when the bedding has been used by an infected patient. Contaminated dust settles on to horizontal surfaces, but is readily redispersed into the air by air currents, dry sweeping, dusting, and similar activities. By the use of water or, more effectively, oils and other mixtures which cause individual particles to adhere together in large aggregates, the raising of dust can be greatly reduced. Vacuum cleaning is a far more efficient way of removing dust without contamination of the air than other means, but precautions must be taken to prevent fine particles of dust from being blown out of the machine. These preventive measures alone, however, seldom provide complete control.

Organisms may be destroyed in air by the use of ultraviolet radiation, and aerosols or sprays of disinfectants. Ultraviolet radiation of wavelength 254 nm, whilst not being the most bactericidal part of the spectrum, is readily available from mercury vapour type lamps and is used to reduce the count of airborne bacteria and viruses within buildings. Since ultraviolet radiations produce an irritant effect on the conjunctiva of the eye the radiation source must be shielded from occupants in the room. This correspondingly reduces its field of effectiveness. Ultraviolet radiations have a limited range of action which varies as the square root of the distance from source; it is far less effective against bacterial and fungal spores than against vegetative cells. Organisms must be directly exposed to the rays and screening them from direct radiation or protection by thin layers of protein materials reduces the 'kill'. Ultraviolet radiations are more effective against organisms suspended in droplets than against those attached to dry, dust particles.

Chemical disinfectants have the advantage of penetrating to all parts of a room. To be efficient they must make contact with the airborne organisms and, for this reason, those in a gaseous or vapour state are more effective than others but less volatile substances are sometimes effective if sufficiently finely dispersed. They are used either as sprays or, in a volatile form, dispersed as aerosols. Disinfectant sprays consist of droplets larger than 150 nm in diameter and result in wet mists; the drops rapidly settle out, lay the dust, and disinfect horizontal surfaces. Aerosols, of droplet size 5–150 nm in diameter, produce dry mists or fogs, which remain suspended in the atmosphere for long periods of time, travel for considerable distances from the source and, by virtue of their large surface area/volume ratio, rapidly build up high local vapour concentrations of volatile substances. All effective air disinfectants act at comparatively low levels of concentration when dispersed in the atmosphere, a feature of great practical importance since these levels are well below those toxic for man and animals. They can therefore be used in occupied buildings. For instance, 1 gram of sodium hypochlorite dispersed in 40,000 litres of air has been found to be bactericidal in a few minutes. The lethal effects are thought to be due to the vapour derived from the aerosol particle producing a lethal concentration localized around the micro-organisms. The particle therefore acts merely as a mobile source of vapour, aiding its persistence and distribution. Humidity usually plays an important role, providing residual moisture on the organismal surface in which the germicide dissolves.

Not all disinfectants are suitable for air disinfection. The most useful include chlorine, sodium hypochlorite, hypochlorous and lactic acid, propylene, and triethylene glycols and phenols of low volatility such as hexyl resorcinol and resorcinol. Many of these have been used successfully for the sterilization of air in public buildings and for the safeguarding of young poultry against pullorum disease in brooder houses. Formalin vapour, due to its toxicity, cannot be used in occupied buildings but is a most effective means of sterilizing the air of unoccupied animal and poultry houses. Reductions greater than 99 per cent in the coliform counts have been achieved by use of this reagent.

5. Control of insect vectors

Certain infective agents enter the blood or tissues of a susceptible host by the bite of an arthropod vector (Chapter 19). They include protozoa (e.g. the malaria parasite), viruses (e.g. the ARthropod-BOrne or arboviruses), rickettsiae (e.g. the agent of typhus fever) and a few bacteria (e.g. the plague bacillus). The vectors include mosquitoes, fleas, ticks, mites, and lice. Some act as mechanical carriers of infection while in others the infective agent passes through part of its life cycle within the vector.

The control of vector-borne diseases, depends on accurate knowledge of the ecology of each of the vectors, including their breeding grounds, geographical location, and other properties unique to individual species. In general control involves eradication of the vector, avoidance of contact by susceptible host with the vector, the treatment of infected patients to reduce the sources of the infective agent should the host be bitten by the vector, and prophylaxis or immunization or both to prevent susceptibles experiencing the clinical illness should they be exposed to infection.

(a) *Eradication of the vector*

Malaria control is directed at eliminating the breeding places of the mosquito. Since the larvae of many species of mosquito breed in ponds the primary need is to drain, fill in, or treat surface waters. The larvae themselves can be suffocated by an oil film spread on the water surface.

Spraying with insecticides, particularly residual insecticides that persist for long periods (e.g. chlorinated hydrocarbons such as DDT, benzene hexachloride or dieldrin), is most effective. This is done on the inside walls of dwellings and on other surfaces, such as ceilings, upon which vectors habitually rest in the day-time—this procedure generally results in effective malaria control. The widespread use of DDT has led, however, to the emergence of DDT-resistant mosquitoes, and alternative insecticides have to be used.

Insecticides are used widely in the control of most vector-transmitted disease. A specific instance of their use in the control of typhus fever in the concentration camps at the end of the Second World War is described later (Chapter 19). Measures to reduce tick populations for the control of tick-borne diseases are not generally practical. In selected areas direct application of chlordane, dieldrin, lindane, diazonin, or benzene hexachloride have provided good tick control. It is more feasible for susceptible hosts to avoid tick-infested areas or, if infected, for ticks to be removed promptly from the patient.

(b) *Avoidance of contact between vectors and susceptible hosts*

Where possible, susceptible hosts should avoid entering vector-breeding areas such as forests, or geographical locations known to be sources of infection. The application of insect repellents to naked skin and clothing have proved useful. These include diethyltoluamide and dimethylphthalate. In endemic areas mosquito screens over windows or doorways, and the use of bed nets at night, also provide protection.

(c) *Avoidance of contact between vectors and infected hosts*

A modern method to control vector-transmitted diseases is to limit access by vectors to infected hosts. This includes the isolation and treatment of patients to reduce the numbers of parasites in their blood below a level at which vectors become infected. Useful drugs against malaria, for instance, include chloroquine base, amodiaquine base, and quinine sulphate or dihydrochloride. Another approach, especially for travellers, is the use of chemoprophylaxis using, for example, proguanil hydrochloride (Paludrine) throughout the period of exposure and for at least 6 weeks after leaving the area.

6. Prevention of cross-infection in hospitals

Cross-infection is a most important problem in hospitals and because of its importance is considered at length in Chapter 12.

FURTHER READING

Burnet, F. M., and White, D. O. (1972) *Natural History of Infectious Disease*, 4th edn. Cambridge University Press (278 pp.).

Hawker, L. E., and Linton, A. H. (eds) (1979) *Micro-organisms—Function, Form and Environment*, 2nd edn. Arnold, London.

Stoenner, H., Kaplin, W., and Torten, M. (eds) (1979) *Bacterial, Rickettsial and Mycotic Diseases*, vols I and II. In Steele, J. H. (chief ed.), *Handbook Series In Zoonoses*. CRC Press, Inc., Florida (643 pp. and 568 pp.).

Chapter 10

The control of infectious disease— II: Protecting susceptible hosts by immunization

In an endemic situation measures to prevent infectious agents reaching susceptible individuals, as considered in the previous chapter, will reduce the incidence of infection. Inevitably these preventive measures fail from time to time. To counter this possibility susceptible members of a community may be protected by making them more resistant to an infection. The best way to do this is to artificially immunize individuals against specific infections to which they may subsequently be exposed. Another approach is to treat infected patients with chemotherapeutic drugs or even to treat individuals who may be exposed to infection—chemotherapy is the subject of the next chapter.

The purpose of artificial immunization is to provide the host with specific protective antibody either by supplying it preformed in serum from another host (passive immunization) or by stimulating the individual to produce his own (active immunity). *Passive immunity* provides immediate protection but, since the antibody is raised often in a different animal species, it is 'foreign' to the host and hence eliminated over a relatively short period of time. There are also certain risks attached to the use of antisera, the most important being anaphylactic shock due to hypersensitivity reactions to foreign sera. Less risky in man is the use of hyperimmune globulins harvested from convalescent and vaccinated humans; pooled human immunoglobulin has also been safely used to provide passive immunity.

Active immunization has been widely used in man and animals for many years. The agents include living bacteria or viruses which have been attenuated to modify the clinical response, killed or inactivated micro-organisms, and modified products or cellular components of microbial cells. Such vaccines have been successfully used against tetanus, diphtheria, whooping cough, anthrax, tuberculosis, brucellosis, poliomyelitis, and yellow fever. Others against cholera, plague, typhoid and related enteric fevers, typhus and related rickettsial diseases have proved less effective or even of doubtful value. For other diseases including dysentery, infections by *Staphylococcus aureus* and *Streptococcus pyogenes*, gonorrhoea, syphilis, many protozoal infections, mycoses, and those due to multiple biotypes of viruses, such as the common cold, effective vaccines have not yet been devised.

A.   WHAT MAKES A VACCINE EFFECTIVE?

The effectiveness of an immunizing agent in providing protection against an infectious disease depends on whether or not the agent is able to stimulate protective antibodies. All pathogenic microbes are antigenic and include many different antigens in their structure each of which, if exposed at the cell surface, will stimulate specific antibody formation. However, unless these antibodies can neutralize the components of pathogenicity, they are valueless in giving protection. Furthermore, assays of antibody levels in the blood serum following vaccination may give no indication of the degree of protection unless they are the 'right' protective antibodies. Before proceeding to discuss ways of producing better vaccines the essential requirements of a vaccine will be considered.

An effective vaccine must be 'safe', 'immunogenic', and 'right'. Whilst the vaccine must simulate the natural infection in producing active immunity it must not produce the clinical illness. Killed or inactivated vaccines are usually much safer than living ones, but even these may produce unwanted local or generalized side-effects, such as endotoxic shock (Chapter 6). Toxoids derived by formalin treatment of exotoxins are relatively safe while producing effective immunity if the disease is produced by a single exotoxin as in diphtheria and tetanus. With living vaccines greater control is necessary. Living vaccines are usually attenuated but they are not avirulent and therefore must be handled with care. The S19 vaccine for calfhood vaccination against brucellosis is able to produce mild infections in man and guinea pigs and can cause pregnant cattle to abort. This strain is very stable and many attempts to increase or decrease its virulence have failed. Also it has never been shown to be transmitted from one animal to another, even in animals in contact with the heavily infected foetal membranes from an animal that had aborted consequent upon receiving the vaccine strain. A number of living vaccines (e.g. the S19 brucella vaccine for cattle (Chapter 15), the BCG vaccine for tuberculosis (see later) and many viral vaccines) stimulate cell-mediated immunity and consequently may produce hypersensitivity (but in vaccination the stimulation of hypersensitivity should be avoided as much as possible). Before living vaccines can be used safely it is essential to prove beyond reasonable doubt that a strain of lowered virulence will not mutate to one of enhanced or full virulence *in vivo*.

The second requirement of an effective vaccine is that it must include good immunogenic antigens. Not all virulence factors associated with a microbial pathogen are good antigens. For instance, the hyaluronic acid capsule of *Streptococcus pyogenes* and the polyglutamic acid capsule of *Bacillus anthracis* are very poor antigens. Where this applies immunogenicity may be increased

ie use of adjuvants (usually suspensions of dead mycobacteria or
Chapter 6) or by modification of the antigens to render them more
lowever, apart from attempts to chemically link 'poor' immunogens to
good antigenic carriers, e.g. proteins, no known method is available to improve
the antigenic quality of these factors by alteration of their chemical structure.

Even if a vaccine is 'safe' and 'good' it will only be effective if the antigens are
'right'. A vaccine must include protective antigens, i.e. those which contribute to
the production of disease in the natural host and which stimulate the production
of protective antibodies or cell-mediated immunity when used in immunization.

All micro-organisms, whether pathogenic or non-pathogenic, and many of their
products, are antigenic but only certain ones produce protective responses. Even
viruses may possess extra-viral antigens, such as protein envelopes which are
accompaniments of the virus rather than determinants of pathogenicity. To be of
value, therefore, a vaccine must stimulate immune responses against virulence
factors.

In Part I a survey of present knowledge on components of virulence is considered. This type of knowledge should lead to a more rational approach to the
preparation of better vaccines. Virulence factors, after being rendered harmless
without altering their immunogenic properties, should be incorporated in vaccines.
This is exemplified by use of toxoids, cell wall substance of *Brucella abortus* and
the capsule of *Bacillus anthracis* (both of which interfere with phagocytosis) and
the surface antigens of poliovirus (which determine the attachment of the virus to
susceptible cells). Some virulence factors are antigenic but not all are protective.
For instance, the α-toxin of *Staphylococcus aureus* produces good titres of anti-α-
toxin but recurrent attacks of infection frequently occur. It is possible that other
factors are involved in the primary lodgement of the organism and anti-α-toxin
does not inhibit this stage of the infection.

The virulence of many pathogens depends on a number of factors. Where a
single factor only is involved it is comparatively easy to prepare a good vaccine.
Where multiple factors are involved it is not easy to determine which is the more
important or whether the use of a single immunogenic antigen will effectively
prevent the infection. For example, in anthrax, virulence involves the capsule and
at least three virulence factors but vaccines prepared against the toxic factors
alone will immunize effectively. In other infections where all the virulence factors
act synergistically full protection can only be achieved when all are present in the
vaccine.

Conventional methods of preparing vaccines involve growth of the vaccine
strains in artificial culture or, in the case of viruses, in tissue culture. Under these
conditions antigenic changes may occur such that factors of importance *in vivo*
are not produced. Hence *in vitro* cultures are often deficient in protective antigens.
This applies more to dead vaccines than living ones but even attenuated strains or
strains of close antigenic similarity to the virulent pathogen may be partially
inadequate.

The application of knowledge of the special conditions needed to produce protective antigens under *in vitro* conditions will obviously improve the value of a vaccine. For instance, the various toxins of *B. anthracis* can be produced *in vitro* by growing the organisms in a serum broth containing bicarbonate and bubbling carbon dioxide through the medium. For the production of more effective killed vaccines of *Br. abortus*, bovine foetal fluids may be added to artificial media to stimulate the production of cell-wall substance which interferes with the bactericidal activity of bovine phagocytes. The use of continuous culture systems rather than static batch cultures is likely to resemble the conditions *in vivo* more closely.

VACCINES IN HUMAN MEDICINE

History

More than a thousand years ago the Chinese practised smallpox prevention by exposing susceptible individuals to mild cases. Following Jenner's observations that injections with cowpox gave immunity against smallpox, vaccination was introduced about 200 years ago. Towards the end of the nineteenth century the first human vaccine against rabies was developed by Pasteur. These early successes led to many vaccines being produced in parallel with advances in medical microbiology, particularly in infectious diseases, and in an understanding of the mechanisms of immunity. They resulted in a marked reduction of previously common diseases, e.g. smallpox, diphtheria, tetanus, poliomyelitis and, more recently, whooping cough and measles.

With the discovery of sulphonamides and antibiotics it was hoped that infectious diseases would be conquered and interest in vaccines waned. Yet despite the effects of education, socioeconomic development, public health measures and the introduction of chemotherapy, infectious diseases still remain among the commonest problems encountered in medicine in the 1980s. In the developing countries, where any or all of these factors are frequently deficient, infections remain the commonest cause of childhood mortality. Major epidemics of 'killer' infections, such as meningococcaemia, are still prevalent and diseases such as measles, a mild infection in twentieth-century western society, has been associated with a mortality of up to 24 per cent in the Third World. However, where available (usually as a result of WHO programmes), measles immunization (and to a lesser extent, meningococcal immunization) has all but eliminated such problems. These and similar successes due to immunization, coupled with the failure of chemotherapy to control spread of infection, an increasing incidence of bacterial resistance, the relative failure of progress in the field of antiviral chemotherapy, and the problems of control of tropical parasitic infestation, have ensured continued clinical and research interest in immunization and the development of new vaccines. Many of the older vaccines, with variable efficacy, are being re-examined in the light of recent knowledge of the pathogenesis of disease. Concern about side-effects, as with whooping cough vaccine, has called for more research into the possible use of purified antigens and recent work is indicating success with oral vaccines to stimulate local gut immunity, as in poliomyelitis and enteropathogenic *E. coli* in pigs. Some modern approaches include the use of fractions of infectious agents (sub-unit vaccines), e.g. capsular polysaccharides of meningococci and pneumococci, and ribosomal vaccines (see later). Attempts are also being made to improve the safety and efficacy of vaccines, and to elucidate the *in vivo* sites where immune responses are elicited.

Modern vaccines

Some of the commoner vaccines in use at the present time in human medicine and possible future developments are considered.

Bacterial vaccines—whole organisms

Tuberculosis. BCG (Bacillus of Calmette and Guérin), the live attenuated vaccine strain against tuberculosis, was derived from *Mycobacterium bovis* by Calmette and Guérin at the Pasteur Institute in France. It was first administered to humans in 1921. In Britain the vaccine is available for general use in susceptible individuals who are at risk, such as tuberculin negative contacts in families where a known case of tuberculosis occurs, as may be found in immigrant families, and to medical and laboratory workers.

Whooping cough. Vaccination with killed suspensions of *Bordetella pertussis* is part of the routine immunization schedule for children in combination with diphtheria and tetanus toxoids. Modern vaccines, which include bacterial antigens occurring in wild strains, reduce the attack rate of whooping cough and the severity of the illness. Immunization should be started at 3 months of age to ensure protection during the earliest months of life when infection is usually more serious.

Bacterial diseases of the intestinal tract. Progress in the development of vaccines to protect against bacterial infections of the gastrointestinal tract has been impeded by a lack of detailed understanding of the pathogenic mechanisms of bacterial invasion and toxin production, together with the need to elucidate the immune mechanisms operating in the intestine or in the tissues in its immediate vicinity. The success with the live poliomyelitis vaccine has encouraged a change from systemic (parenteral) vaccination to the investigation of oral vaccines with the aim of enchancing immunity at the gut mucosal surface. The approach has been strengthened by the success obtained by feeding suspensions of killed specific enteropathogenic *E. coli* to pigs in their diet.

Enteric fever. Sporadic cases of enteric fever occur in Britain mainly in holiday-makers or immigrants from countries where the disease is endemic or, less commonly, following contact with chronic carriers. Vaccination is recommended for travellers outside Northern Europe, Canada, Australia, and New Zealand.

For many years the standard TAB vaccine contained whole heat-killed cells of *Salm. typhi*, *Salm. paratyphi* A and B preserved with phenol, but there is little evidence of protective efficacy against paratyphoid fever. Based on studies in mice, an alcohol-killed vaccine of *Salm. typhi* preserving the Vi antigen, was devised. Clinical trials in different countries have shown that the phenolized one gave better results. More recently, TAB vaccine has been replaced by monovalent acetone, heat or alcohol-killed typhoid vaccines which produce fewer side effects. Killed vaccines will protect against death from enteric fever, presumably by delaying the establishment of the organisms in the liver and spleen rather than by preventing their multiplication. Live vaccines may be more effective by limiting the spread of virulent organisms from the intestinal tissues and restricting their growth in these sites. Oral live vaccines, including a streptomycin-dependent strain of *Salm. typhi*, are currently under investigation. Vaccines based on ribosomal extracts of salmonellae have also been shown to be highly effective in animal models.

Bacillary dysentery. Bacillary dysentery, is generally characterized by bacterial invasion of the intestinal wall although attempts over many years to immunize using killed vaccines, given systemically, have failed to produce demonstrable protection against *Shigella* infection. However, studies, both in animals and man, have shown the development of type-specific protection following the oral administration of live attenuated *Shigella* organisms.

Cholera. Killed cholera vibrio vaccines have been available since the end of the last century, but have proved ineffective in preventing the introduction of cholera into a country, in restricting its spread, and in reducing mortality when adequate treatment facilities are available. Vaccination for travellers is still required, however, by several countries in the Middle East and Africa.

New approaches for immunization, using cholera toxoid, whole-cell vaccine toxoid, purified antigen with or without adjuvant and live attenuated oral vaccines are under consideration, their development having followed advances in the understanding of pathogenic and immunological mechanisms in the disease process. Cholera toxin has been fractionated to provide a subunit vaccine, free from any toxic component, containing the antigenic component responsible for attachment of the vibrio to the gut wall.

Hospital infections by Gram-negative bacilli. The increase in hospital infections caused by Gram-negative bacilli, often resistant to several common antibiotics, has prompted investigation of vaccines directed against prevalent hospital strains. Future developments may result in polyvalent vaccines containing protective antigens for a variety of Gram-negative bacilli. However, it is in the development of a vaccine against *Pseudomonas aeruginosa* that most progress has been made.

Pseudomonas aeruginosa. Infections caused by *Pseudomonas aeruginosa*, increasingly common in hospital practice, present difficulties in management using conventional antibacterial chemotherapy. The protection by vaccination of high risk groups, such as burned or immunodeficient patients, or those with specific metabolic disorders, as in cystic fibrosis, would be an attractive alternative approach. Inoculation with a polyvalent vaccine of *Ps. aeruginosa* conferred immunity, in burned mice, to challenge by virulent organisms. Controlled clinical trials in man using the vaccine PEV-O1, containing surface antigens of 14 different O-serotypes of *Ps. aeruginosa*, have been undertaken successfully in burns units.

Bacterial vaccines—products

Diphtheria. The virtual eradication of diphtheria in the United Kingdom is one of the outstanding successes of immunization. Prior to the national campaign of active immunization begun in 1940 diphtheria was the main cause of death in children of school age but within a few years the disease had almost disappeared. It is still possible that the disease may be imported into Britain as diphtheria is endemic in many parts of the world or may be encountered whilst travelling abroad.

Active immunization against diphtheria is achieved by using a toxoid preparation (formalin- and heat-treated toxin) which is usually administered with tetanus toxoid and pertussis vaccine (or with tetanus alone) as part of the routine immunization schedule for children. The purified toxoid elicits only a poor antibody response in children but its action is greatly enhanced by aluminium adjuvants (aluminium hydroxide or aluminium phosphate) or in combination with other antigens (*Bord. pertussis*) as in the triple vaccine.

Temporary protection (of about 2 weeks) for contacts exposed to infection may be obtained by passive immunization using an intramuscular injection of 1000–2000 units of antitoxin but carried the risk of anaphylaxis or serum sickness. It is recommended that administration of the antitoxin should be combined with active immunization.

Tetanus. As with diphtheria and poliomyelitis, preventive medicine has greatly reduced the incidence of tetanus in Britain. Continued programmes of active immunization against tetanus are justified in that *Clostridium tetani* spores are often present in the soil, dust, and soiled clothing. This constitutes a possible risk of infection in wounds and the disease has a high mortality despite intensive care.

The disease is prevented by active immunization using an adsorbed toxoid as part of the routine immunization programme in childhood. The toxoid is administered as part of the triple vaccine or with diphtheria toxoid alone.

Human antitetanus globulin (Humotet) should be given both for prophylaxis against tetanus, in addition to penicillin and booster toxoid, and for treatment of the established disease.

Bacterial vaccines—cellular components

Polysaccharide capsular antigens of meningococci. *Neisseria meningitidis* can be classified into serogroups on the basis of capsular polysaccharides. Group A, B, and C polysaccharides have been tested for immunogenicity and safety in man, with marked variations in humoral immune response. This may be due to a variety of factors such as increased antibody levels with increasing age, the molecular size of the antigen, and the assay technique used. Field trials, in a number of countries, are underway.

A further development has come from the study of *N. meningitidis* outer membrane proteins. A high percentage of strains causing epidemics of both Group B and Group C meningococcal disease contain the type II outer membrane protein antigen and it is possible that a vacccine containing the single type II protein could provide broader protection than is achieved currently using the group specific A and C polysaccharides.

Polysaccharide capsular antigens of pneumococci. Attempts to prevent pneumococcal pneumonias by vaccination have been made since the beginning of this century. Wright and colleagues (1914) used a killed whole cell vaccine in field trials in South African gold miners with some success but later vaccines containing specific polysaccharides have been shown to be effective.

Despite widespread antibiotic therapy, infections caused by *Streptococcus pneumoniae* continue to make a significant contribution to morbidity and mortality. Penicillin-resistant strains of pneumococci have been reported, notably in South Africa and Papua New Guinea where pneumococci remains a major cause of respiratory illness and deaths, and the use of penicillin is therefore compromised.

Recent studies in South Africa have demonstrated the efficacy of a vaccine containing capsular polysaccharides from 14 selected serotypes (Pneumovax). This has been shown to be safe, immunogenic and to give 78.5 per cent efficacy in prevention of type specific pneumonia and bacteraemia in adults.

M-protein antigens of group A streptococci. Immunity to infection of the upper respiratory tract with Group A streptococci is serologically type specific, mediated by serum bactericidal (opsonic) antibodies specific for the M-protein antigens on the cell wall of Group A streptococci. Some success in clinical trials has been obtained using purified M-protein vaccines administered parenterally or intranasally (topically). Intranasal administration, by means of an aerosol sprayed directly onto the pharyngeal mucosa, was shown to be effective in protecting against clinical illness and colonization after challenge with homologous streptococci. It is possible that use of the vaccine would reduce the number of potential carriers and thereby reduce the community reservoir of infection.

Ribosomal vaccines. Youmans and Youmans (1965) first observed that ribosomes and ribosomal extracts of the avirulent strain *Mycobacterium tuberculosis* H37 Ra in incomplete Freund's adjuvant administered to mice resulted in high level protection against challenge with the virulent H37 Rv strain. Treatment of the ribosomal preparation with RNA-ase resulted in the loss of protective activity indicating that the active fraction is RNA. Vaccines have been prepared from ribosomes or ribosomal extracts of a number of other microorganisms including *Salmonella typhimurium*, *Neisseria meningitidis*, *Streptococcus pneumoniae*, *Streptococcus pyogenes*, *Vibrio cholerae*, *Histoplasma capsulatum*, and *Francisella tularensis*.

Possible advantages of ribosomal vaccines over conventional killed or live cell vaccines include: (1) Ribosomal vaccines, like live vaccines, may induce macrophage activity—this has been reported for ribosomal RNA of mycobacteria which induces macrophage immunity without eliciting a delayed hyper-sensitivity response. This vaccine would therefore have an advantage over the BCG vaccine which converts to a positive tuberculin test. (2) Ribosomal vaccines produce cross-protection against other serotypes within the species in contrast to polysaccharide vaccine, e.g. pneumococcal vaccines, which protect only against the serotypes included in the vaccine. (3) Ribosomal vaccines may provide a means of immunizing against infection for which no conventional vaccine is available.

Bacterial vaccines of the future

Dental caries. Dental caries is one of the most prevalent world-wide diseases and great interest has been aroused in the possibility of conferring protection by vaccination. Carious lesions are initiated by the action on tooth enamel of acid following fermentation of dietary carbohydrates by oral bacteria, primarily *Streptococcus mutans* (Chapter 2). It has been shown recently that inoculation with intact cells or cell wall fractions of *Strep. mutans* confers some protection in monkeys when fed on a caries-promoting diet and challenged with this organism. Before immunization of humans becomes a feasible proposition, further studies are required to assess possible adverse side effects including those associated with cross-reactivity of vaccine antigens with human tissue, and to elucidate the complex immunological interaction (local secretory IgA and serum antibodies) in selecting the most effective route of administration.

Viral vaccines

Smallpox. Prevention of smallpox by vaccination has been possible since the middle of the eighteenth century, following the work of Edward Jenner on the passage of cowpox from person to person, and the practical observation by farm workers that inoculation of lymph containing cowpox prevented not only an attack of cowpox but also smallpox. Vaccinia virus, used in the vaccine, is distinct from both smallpox (Variola major or Variola minor) and cowpox viruses although conferring cross-immunity. The vaccine is prepared by inoculation of the shaved and scarified (abraded) skin of the abdominal wall of sheep and calves with vaccinia virus. The fluid collected from subsequent vesicles may be freeze-dried retaining its potency without refrigeration for six months. The vaccinia virus vaccine is administered intradermally, a vesicle appearing on the 7th–8th day (maximum on 12th day) in primary vaccination. Immunity against smallpox infection lasts for about three years with a longer lasting protective effect against death from smallpox.

The international eradication campaign, which involved active surveillance and a vaccination programme in communities where immunity was inadequate, has proved successful in that no cases of smallpox have been reported since October 1977 (except the two laboratory associated cases of Variola major in Birmingham, 1978). Since all countries have now been recognized as free from smallpox by the International Smallpox Eradication Commission, vaccination is no longer practised.

Poliomyelitis. Twenty-five years ago, poliomyelitis was one of the most severe infectious diseases in Western countries. The earliest developed safe vaccine, the Salk inactivated virus vaccine, contained the three poliovirus types inactivated by formaldehyde. The vaccine is still used in Sweden, Finland, and the Netherlands but has been replaced for routine use in Britain and the USA by the oral Sabin live attenuated vaccine. One important aspect of the effectiveness of the Sabin vaccine is in the production of gut immunity (IgA) following multiplication of the organism in the intestinal wall. The virus is excreted in faeces and extensive use of the vaccine since the early 1960s has led to the establishment of the vaccine-derived strains of poliovirus as the most common in the community and causing, on rare occasions, serious polio-like illness. A small number of poliomyelitis cases recently seen in Britain showed that many of the polio strains causing disease (commonly types 2 and 3) are vaccine-derived strains reverted to neurovirulence. Within a couple of years of the introduction of mass vaccination in infancy, poliomyelitis has become a rare disease in Britain although in much of Africa and Asia it is still common.

Influenza. Influenza vaccines have been available as inactivated preparations and used in Eastern European countries for nearly 40 years but the successful control of the disease has become more likely following the recent development of attenuated vaccines. Inactivated virus vaccines confer protection in the order of 60–70 per cent, when the current epidemic strain is included in the vaccine and there is a good coverage of the exposed population. The composition of the vaccine has therefore to be changed constantly, placing severe limitations on the amount of vaccine available during any epidemic. Immunity is short term usually lasting only a few months.

The development of attenuated vaccine strains gives the potential for longer term immunity as the non-virulent virus multiplies in the upper respiratory tract. The method used for production of the vaccine strain is a recombination of the wild virulent current antigenic strain with an attenuated strain and the selection of progeny which contain genetic material coding for the haemagglutinin and neuraminidase of the wild parent strain with the attenuation of the other parent. For example, the strain WRL 105 has been selected with the antigenic characteristics of A/Fruland/74 and the growth characteristics of A/Okuda/57. The vaccine has no capacity to spread to contacts and has been shown to be effective in preventing the disease. The vaccine may be given by drops, or as a spray, increasing local antibody levels on the respiratory mucosa. A further development for influenza vaccination is the subunit (purified surface antigen) vaccine, conferring good protection at a lower antigenic concentration.

Yellow fever. Yellow fever vaccine consists of an attenuated virus passaged in chick-embryo tissue culture. International travel regulations require that the vaccine must be approved by the WHO and administered at a designated yellow fever vaccination centre. A certificate of vaccination is required for travellers entering or passing through infected areas, currently parts of East and West Africa and South to Central America.

Measles. Measles is generally a mild childhood disease in developed countries whereas in under-developed countries, it is associated with high mortality (5–10 per cent). Routine immunization using an attenuated virus is recommended in Britain during the second year of life. Protective immunization at a younger age is less reliable due to the inbitory presence of maternal antibodies.

Rubella. Rubella is a common childhood viral infection which is usually subclinical (90 per cent) and the clinical disease may be misdiagnosed. The most important consequences of rubella are foetal abnormalities resulting from rubella infection in early pregnancy, especially in the first trimester (Chapter 8). The prevention of infection in the foetus and consequent congenital rubella syndrome is the major objective of the rubella immunization programme in the UK.

The rubella vaccine in Britain contains an attenuated virus grown either in primary rabbit kidney cells or human diploid cell culture. A single subcutaneous dose confers long-lasting immunity and is recommended for schoolgirls aged 11–13 years (pre-pubertal girls) and all non-immune women of childbearing age. Pregnant women should not be vaccinated as the attenuated virus vaccine is capable of crossing the placenta to infect the foetus.

Mumps. An effective vaccine has been introduced in America using an attenuated mumps virus (the Jeryl Lynn strain). There has been little demand in this country for routine vaccination against mumps. It may be offered to susceptible post-pubertal males in whom orchitis is a possible complication of the infection.

Rabies. Vaccination of persons considered to have been exposed to rabies has been practised since the time of Pasteur, but it is only comparatively recently that pre-exposure immunization for those at particular risk has been recommended. For domestic animals the reverse is true; post-exposure treatment is seldom practised as the danger to people involved would be too great should the treatment fail.

Pre-exposure vaccination of animals has been used for many years, notably in rabies eradication campaigns in Japan, Malaya, Rhodesia, and Israel. Countries where the disease is endemic often use attenuated vaccines for veterinary purposes. Rabies-free countries use only inactivated vaccines.

In the past, the most widely used modification of the original Pasteur vaccine was an inactivated rabbit-brain virus vaccine developed by Semple. This was accompanied by the risk of complications including permanent damage to the nervous system (immunological response to the high myelin content of rabbit brain tissue). In the 1950s, two new vaccines relatively free of myelin were introduced, prepared in the brains of suckling animals, usually mice (SMBV) or duck embryo (DEV). Both vaccines have been widely used for pre- and post-exposure vaccination. SMBV is widely used in South America although associated with some neurological complications. DEV produces fewer cases of allergic encephalitis but is less antigenic than the Semple vaccine.

Until 1976, the DEV vaccine was the vaccine of choice in Britain and the USA although with major limitations, including low antigenicity. A safe inactivated virus vaccine has now been produced from virus grown in a human diploid cell line W1-38 a technique widely used for the production of polio and measles vaccine. The product is relatively pure and of higher antigenicity than older types of vaccine with a recommended basic schedule for post-exposure vaccination of five doses spaced at intervals (0, 3, 7, 14, 30, and 90 days). Passive immunization using anti-rabies human immunoglobulin or anti-rabies serum prepared in horses is usually started in conjunction with local cleansing and treatment of the bite wound simultaneously with active immunization.

Pre-exposure vaccination should be offered to all persons at particular risk of contracting the disease, for example staff employed at quarantine stations, veterinary staff, laboratory workers handling rabies virus; the vaccine is also available for travellers to high risk areas abroad.

Herpesvirus vaccine. A number of herpesvirus diseases of domestic animals can be prevented by vaccination using live attenuated virus vaccines, for example infectious laryngotracheitis of chickens, infectious bovine rhinotracheitis, and Marek's disease of chickens. Humans have also been inoculated with experimental herpesvirus vaccines. A live human cytomegalovirus (CMV) vaccine has been developed for possible use in preventing *in utero* infection and subsequent mental retardation of the child and recently a live varicella-zoster virus (VZV) vaccine has been shown to protect schoolchildren in Japan against both primary infection and the development of herpes-zoster.

However, the use of live attenuated vaccines for herpesvirus infections in man poses serious problems. Herpesvirus infections are characterized both by persistence and latency, thus live vaccine strains may manifest themselves later as infections if, for example, the patient is on immunosuppressive therapy. Also, recent evidence has linked some members of this virus group with malignant disease. Interest has therefore been intensified in the use of killed vaccines or, ideally, in the use of sub-unit vaccines free of viral genetic material.

Hepatitis B. Infection with hepatitis B is a major public health problem with an estimated 200 million persistent carriers world-wide. The virus is transmitted parenterally, for example during blood transfusions via hypodermic needles or syringes, by homosexual contact, blood-sucking arthropods, tattooing, and by direct transmission from carrier mothers to the newborn infant. A vaccine against hepatitis B has been urged as an important control measure in conjunction with improved diagnostic measures and further identification or risk factors and the carrier state.

The inability to culture hepatitis B virus in tissue culture and the lack of small laboratory animals susceptible to the virus has impeded progress towards the development of a conventional vaccine. Heat killed serum from patients infected with hepatitis B has been shown to protect susceptible individuals challenged subsequently with infective serum. However, as long as uncertainty remains as to the reliable destruction of infectious components in the vaccine serum it is unlikely that this type of vaccine would be generally acceptable.

Sub-unit vaccines containing 22 nm hepatitis B surface antigen treated with heat or formalin has been used in chimpanzees and in man. These vaccines lack nucleic acid and are non-infectious. However, concern has been expressed that associated host cell components in the vaccine may be involved in the pathogenesis of liver damage and further studies must be undertaken to assess the effects of the vaccine on the immune system.

An alternative approach has concerned the characterization of antigenic polypeptides derived from hepatitis B surface antigens, some of which induce a cell-mediated immune response in animals and whose synthesis may be possible commercially.

A recent significant and novel advance has been made by research workers in Edinburgh and Porton Down who have succeeded in inserting hepatitis B viral DNA into an *Escherichia coli* plasmid. Cells containing the recombinant plasmid are able to synthesize antigenic material that reacts specifically with antisera to hepatitis B viral antigen (HBsAg and HBcAg). The potential therefore exists now for the detailed characterization of antigens involved in the infective process and for the large-scale production of antigens for diagnostic purposes and vaccine development.

Hepatitis A. The development of hepatitis A vaccine remains a prospect for the distant future as growth of the virus in cell culture remains unsuccessful but human immunoglobulin has a protective effect in preventing jaundice in people exposed to hepatitis A. A single dose of immunoglobulin provides protection for up to 6 months.

Vaccines against tropical parasitic infections

Parasitic infections are endemic throughout the underdeveloped world, and also in parts of the developed world, affecting many millions of people. A special programme has been established by the World Health Organization researching the most important parasitic diseases—malaria, schistosomiasis, trypanosomiasis, filariasis, leishmaniasis, and leprosy. One of the main objectives of this programme is to develop effective vaccines. However, their development may not be as successful and so well justified as that of bacterial and viral vaccines. The pathology of infection and persistence of the parasite in the body are unique, and the immune response to infection is poorly understood. Countries which should theoretically benefit most from vaccines are often those with the most severe economic problems, rudimentary health care, poor communications, and adverse climatic conditions. Protection by immunization can only be one of several control measures under active consideration.

Malaria. Development of a vaccine, in addition to the biological control of the vector mosquito, *Anopheles bellator* (Chapter 9), is one of the more promising methods currently being investigated for the control and eradication of malaria. Each development stage in the complex life-cycle of the parasite has specific antigenic determinants and vaccines against the sporozoite, merozoite, or gametocytes are all possible points of attack. Prevention of sporozoite entry into, or growth within, the liver or interruption of asexual erythrocytic development would eliminate disease in humans and transmission to the mosquito vector. Blocking of gametocyte spread to mosquitoes would stop transmission of the infection although the individual would not be protected from the consequences of the disease. A fully effective vaccine against any one of these stages would ultimately eradicate malaria.

A sporozoite vaccine for *Plasmodium falciparum* has been used in volunteer studies in humans, with immunity persisting in some for 3 months. The immunity is species specific and lacks cross-immunity, for example, to *P. vivax*. The severe limitation of this vaccine is the availability of antigen, which can only be obtained from sporozoites in infected mosquitoes. Further work is being undertaken on vaccines for the merozoite and the gametocyte but as yet a practical vaccine for this complex disease remains elusive.

FURTHER READING

General

Dick, G. (1978) *Immunisation.* Update Publications Ltd, London.
Robbins, J. B., and Hill, J. C. (eds) (1977) Symposium on current status and prospects for improved and new bacterial vaccines. *J. Infect. Dis.*, **137**, (Suppl.), 1.

Tuberculosis

D'Arcy Hart, P., and Sutherland, I. (1977) BCG and vole bacillus vaccines in the prevention of tuberculosis in adolescence and early adult life. Final report to the Medical Research Council. *Br. Med. J.*, **1**, 293–5.

Enteric infections

WHO Technical Report Series No. 500. (1972) Oral enteric bacterial vaccines. Geneva.
Tapa, S., and Cvjetanović, B. (1975) Controlled field trial of the effectiveness of one and

two doses of acetone-inactivated and dried typhoid vaccine. *Bull. Wld. Hlth. Org.*, **52**, 75–80.

Mel, D. M., Gangarosa, E. J., Arsić, B. L., Radovanović, M. L., and Litvinjenko, S. (1977) Effectiveness of oral live attenuated Shigella vaccines. In: *Proceedings of the Symposium on Bacterial Vaccines*, Zagreb.

Report Philippines Cholera Committee. (1973) A controlled field trial on the effectiveness of the withdrawal and subcutaneous administration of cholera vaccine in the Philippines. *Bull. Wld. Hlth. Org.*, **49**, 389–94.

Polysaccharide vaccines

Austrian, R. (1977) Prevention of pneumococcal infection by immunization with capsular polysaccharides of *Streptococcus pneumoniae*: current status of polyvalent vaccines. *J. Infect. Dis.*, **136** (Suppl.), 38–42.

Emmerson, A. M. (1980) The need for a pneumococcal vaccine. *J. Antimicrob. Chemother.*, **6**, 301–9.

Gold, R., Lepow, M. L., Goldschneider, I., Draper, T. L., and Gotschlich, E. C. (1975) Clinical evaluation of group A and group C meningococcal polysacchoride vaccines in infants. *J. Clin. Invest.*, **56**, 1536–47.

Zollinger, W. D., Mandrell, R. E., Alteri, P., Berman, S., Lowenthal, J., and Artenstein, M. S. (1978) Safety and immunogenicity of *Neisseria meningitidis* type 2 protein vaccine in animals and humans. *J. Infect. Dis.*, **137**, 728–39.

Whooping cough

Report (1977) *Whooping cough vaccination*. Review of the evidence on whooping cough vaccination by the Joint Committee on Vaccination and Immunisation. DHSS, HMSO, London.

Dental caries

Bowen, W. H., Cohen, B., Cole, M. F., and Colman, G. (1975) Immunisation against dental caries. *Br. Dental. J.*, **139**, 45–58.

Hepatitis B

Burrell, C. J., Mackay, P., Greenaway, P. J., Hofschneider, P. H., and Murray, K. (1979) Expression in *Escherichia coli* of hepatitis B virus DNA sequences cloned in plasmid pBR322. *Nature*, **279**, 43–7.

Gerety, R. J., Tabs, E., Purcell, R. H., and Tyeryar, F. J. (1979) Summary of an international workshop on Hepatitis B vaccines. *J. Infect. Dis.*, **40**, 642–8.

Parasitic infections

Miller, L. H. (1977) Current prospects and problems for a malaria vaccine. *J. Infect. Dis.*, **135**, 855–64.

Hospital Gram-negative infections

Jones, R. J., Roe, E. A., and Gupja, J. L. (1979) Controlled trials of a polyvalent pseudomonas vaccine in burns. *Lancet*, **ii**, 977–83.

Chapter 11

The control of infectious disease— III: Chemotherapy as an aid to host defence

The treatment of an individual with an antimicrobial drug provides the body defences with a further aid against microbial attack. Chemotherapy of established infections has proved of immense value in relieving suffering (even of avoiding death) and, in addition, has countered enormous potential economic losses due to infectious diseases in animals. Unlike most antimicrobial chemicals (disinfectants, antiseptics, etc.) chemotherapeutic agents possess unique properties which make them suitable agents to be administered to living animals and man. First, they exhibit selective toxicity, i.e. they are highly active against particular micro-organisms while being relatively non-toxic to the host and the host tissues. This property may be due to one of several factors. For instance, the drug may inhibit a metabolic pathway present in the microbe but not in the host, or it may be able to penetrate to the sensitive target in the micro-organism but not to a similarly sensitive site in the host's cells. Different modes of action exhibited by the various chemotherapeutic agents are considered below. A second property unique to these agents lies in the fact that they are not seriously antagonized by body tissues, fluids, or secretions and therefore remain active *in vivo*.

Some chemotherapeutic agents are synthesized artificially (e.g. sulphonamides, trimethoprim, and chloramphenicol); others are produced as natural products of the metabolism of micro-organisms, the so-called antibiotics. Strictly an antibiotic is a substance produced by one micro-organism against another (antibiosis) but the term is often used loosely to include all antimicrobial agents. Many antibiotics are chemically modified to give them an advantage over the parent compound (see below). These compounds are called semisynthetic antibiotics (e.g. methicillin, cloxacillin, and carbenicillin; Figure 11.1).

Of the thousands of antibiotics so far discovered, isolated, and characterized, only relatively few have proved clinically useful. Examples of these, and their producing micro-organisms, are shown in Table 11.1.

Clinically useful antibiotics may be classified in different ways. They can be divided on the basis of whether they are bacteriostatic or bactericidal (Table 11.2). Bacteriostatic agents inhibit multiplication of micro-organisms; their destruction

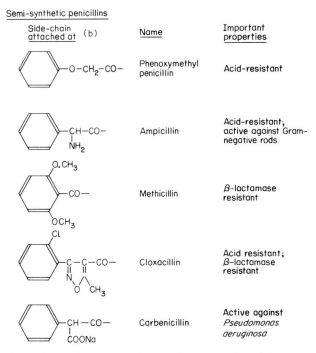

Figure 11.1.   Structure and properties of benzyl penicillin and
some semi-synthetic penicillins

and elimination from the body depends exclusively on the body defences. Bactericidal agents actually kill micro-organisms and thereby play a more active role in host defence. Some agents are bacteriostatic at high dilution and bactericidal at higher concentration. Whatever their activity antibiotics act in collaboration with body defences and bacteriostatic drugs usually are as useful as bactericidal ones, unless the defences are impaired.

Another possible grouping of clinically useful antibiotics is based on their range of antimicrobial activity (Table 11.3). Generally particular species of micro-organisms demonstrate natural susceptibility; others which do not possess a sensitive target site exhibit natural insusceptibility. Some antibiotics are active against Gram-positive bacteria but relatively inactive against Gram-negative ones; others are active, to a greater or lesser degree, against most species of bacteria including the rickettsiae—these are referred to as the broad-spectrum antibiotics. Other smaller groups of antibiotics are active against the tubercle bacillus, fungi, or viruses respectively. Useful antibiotics for the treatment of infections caused by specific organisms are indicated in Table 11.2. Many naturally susceptible microorganisms may acquire resistance to one or more antibiotics by either of a number of mechanisms. For this reason it is not possible to predict the range of antibiotic sensitivity of these organisms based solely on their identification; in these species isolates from patients must be tested for their sensitivity/resistance pattern to decide the most suitable antibiotic for treatment.

Table 11.1.   Date of discovery and source of the more important antibiotics (after Garrod and O'Grady, 1981, and reproduced by permission of Churchill Livingstone)

Name		Date of discovery	Microbe
Penicillin		1929–1940	*Penicillium notatum*
Tyrothricin	{ Gramicidin { Tyrocidine	1939	*Bacillus brevis*
Griseofulvin		1939	*Penicillium griseofulvum Dierckx*
		1945	*Penicillium janczewski*
Streptomycin		1944	*Streptomyces griseus*
Bacitracin		1945	*Bacillus licheniformis*
Chloroamphenicol		1947	*Streptomyces venezuelae*
Polymyxin		1947	*Bacillus polymyxa*
Framycetin		1947–1953	*Streptomyces lavendulae*
Chlortetracycline		1948	*Streptomyces aureofaciens*
Cephalosporin C, N and P		1948	*Cephalosporium* sp.
Neomycin		1949	*Streptomyces fradiae*
Oxytetracycline		1950	*Streptomyces rimosus*
Nystatin		1950	*Streptomyces noursei*
Erythromycin		1952	*Streptomyces erythreus*
Novobiocin		1955	*Streptomyces spheroides* *Streptomyces niveus*
Cycloserine		1955	*Streptomyces orchidaceus* *Streptomyces gaeryphalus*
Vancomycin		1956	*Streptomyces orientalis*
Kanamycin		1957	*Streptomyces kanamyceticus*
Paromomycin		1959	*Streptomyces rimosus*
Fusidic acid		1960	*Fusidium coccineum*
Lincomycin		1962	*Streptomyces lincolnensis*
Gentamicin		1963	*Micromonospora purpurea*

Table 11.2. Antibiotics which may be useful in treatment of specific infections based on common patterns of sensitivity and resistance

	Erythromycin	Novobiocin	Vancomycin	Bacitracin	Benzyl penicillin	Methicillin (Celbenin)	Cloxacillin (Orbenin)	Cephaloridine (etc.)	Carbenicillin (Pyopen)	Ampicillin (Penbritin)	Tetracyclines	Chloramphenicol	Neomycin/kanamycin	Gentamycin	Streptomycin	Polymyxin (Colistin)	Griseofulvin	Nystatin	Amphotericin
Gram-positive bacteria																			
Staphylococcus aureus	+	+	+	+	+	+	+	+	+	+	+	+	+	+	+	–	–	–	–
Staphylococcus aureus (penicillinase producer)	+	+	+	+	–	+	+	+	–	–	+	+	+	+	+	–	–	–	–
Streptococcus pyogenes	+	+	+	+	+	+	+	+	+	+	+	+	–	–	–	–	–	–	–
Enterococci	+	–	+	+	+	+	+	+	–	+	+	+	–	–	–	–	–	–	–
Gram-negative bacteria																			
Haemophilus influenzae	+	+	–	–	–	–	–	+	+	+	+	+	+	+	+	+	–	–	–
Escherichia coli	–	–	–	–	–	–	–	+	+	+	+	+	+	+	+	+	–	–	–
Salmonella spp.	–	–	–	–	–	–	–	+	+	+	+	+	+	+	+	+	–	–	–
Shigella spp.	–	–	–	–	–	–	–	+	+	+	+	+	+	+	+	+	–	–	–
Pseudomonas aeruginosa	–	–	–	–	–	–	–	–	+	–	–	–	+	+	+	+	–	–	–
Rickettsiae	–	–	–	–	–	–	–	–	–	–	+	+	–	–	–	–	–	–	–
Lymphogranuloma group	–	–	–	–	–	–	–	–	–	–	+	+	–	–	–	–	–	–	–
Fungi																			
Candida albicans	–	–	–	–	–	–	–	–	–	–	–	–	–	–	–	–	–	+	+
Dermatophytes	–	–	–	–	–	–	–	–	–	–	–	–	–	–	–	–	+	–	–
Viruses*	–	–	–	–	–	–	–	–	–	–	–	–	–	–	–	–	–	–	–
c = bactericidal; s = bacteriostatic	s	s	c	c	c	c	c	c	c	c	s	s	c	c	c	c	s	s	s

*Viruses are unaffected by the various antibiotics in clinical use. The considerable variety of antibiotics which do affect DNA and RNA synthesis, and are capable of inhibiting virus replication in vitro (e.g., actinomycin, ribavirin), are far too toxic for normal clinical use and chemotherapeutics are employed instead.

+ = sensitive; – = resistant.

Table 11.3. A classification of some antimicrobial agents based on their range of activity (after Hawker and Linton, 1979, *Micro-organisms—function, form and environment*, and reproduced by permission of Edward Arnold)

	Narrow spectrum*	Broad spectrum†
Antibacterial agents	Benzyl penicillin Phenoxymethyl penicillin Methicillin Flucloxacillin	Ampicillin Amoxycillin Carbenicillin
		Sulphonamides
	Cephalosporins Erythromycin Lincomycin Clindamycin	Trimethorprim Tetracyclines‡ Chloramphenicol
		Streptomycin
	Vancomycin	Kanamycin Neomycin
	Fucidin	
		Gentamicin
	Bacitracin	Amikacin
Antituberculous agents	Streptomycin Para-aminosalicylic acid Isoniazid Rifampicin Ethambutol	
Antifungal agents	Nystatin Griseofulvin Amphotericin B 5-fluorocytosine Miconazole Clotrimazole	
Antiviral agents§	Methisazone Idoxuridine Cytosine arabinoside Adenine arabinoside Amantadine	

*Usually Gram-positive bacteria and Gram-negative cocci.
†Gram-positive and Gram-negative bacteria.
‡Tetracycline is also the drug of choice for many mycoplasma, rickettsial and chlamydial infections.
§Drugs have very limited uses—often best given prophylactically.

Mode of action of antibiotics

Antibiotics may also be grouped on the basis of their mode of action—this classification will be used to summarize the range and properties of antibiotics in common use. In choosing the most appropriate antibiotic for clinical use, properties other than the sensitivity of the pathogen also must be considered, such as the pharmacology and toxicity of the agent. For instance, the size and frequency of dose to maintain an effective concentration, the site of infection and the route of administering the agent to reach that site, whether or not the agent is destroyed by gastric acid, the rate of absorption or penetration into tissue, and the degree of excretion into the urogenital system. In considering the mode of action, therefore, some of these properties will be noted.

Drugs acting on the folic acid pathway

Sulphonamides. The synthesis of these compounds in the 1930s was one of the earliest and most important advances in the treatment of infectious disease. They are still useful and cheap drugs for the treatment of urinary tract infections but have been superseded in most other infections. A large number of sulphonamides have been synthesized; the structural modifications confer differences in pharmacological properties but do not change the spectrum of organisms sensitive to them; a bacterium sensitive to one is sensitive to all.

The folic acid pathway is vital in the production of co-enzymes involved in nucleic acid synthesis. Since sulphonamides have a similar molecular structure to para-aminobenzoic acid they act as competitive analogues and inhibit the first stage of the synthesis.

Trimethoprim. This is another synthetic compound with wide antibacterial action; it has higher affinity for bacterial dihydrofolate reductase (rather than mammalian) and acts later in the synthesis of bacterial folinic acid.

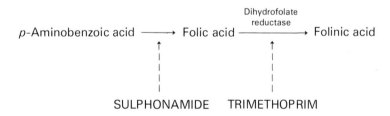

Figure 11.2.   Sites of inhibition by analogues in the metabolic pathway leading to nucleic acid synthesis

Cotrimoxazole. This is the name given to a fixed combination (sulphamethoxazole) and trimethoprim. The two drugs act synei combined action is greater than the sum of their independent activi

Antibiotics acting on cell wall synthesis

Penicillins. Penicillins are a large group of effective, bactericidal, general non-toxic antibiotics, produced by species of *Penicillium.* The molecule of benzyl penicillin can be split by amidase to yield penicillanic acid; the addition of different side-chains to this molecule provides very useful clinical agents, the semisynthetic penicillins (Figure 11.1). Penicillins act by preventing cell wall synthesis of many Gram-positive and some Gram-negative bacteria, and are most active against growing organisms. The presence of bacteriostatic drugs therefore antagonizes their activity. Interference with cell wall synthesis exposes the protoplast to osmotic shock and death.

Benzyl penicillin (often called penicillin), the first antibiotic to be widely used clinically, is active in high dilution against cocci, both Gram-positive and Gram-negative, and most Gram-positive bacilli. It is still the best and the cheapest antibiotic for the treatment of infections caused by pyogenic streptococci (e.g. *Streptococcus pyogenes* and mastitis streptococci), with few recent exceptions most pneumococcal infections and, at present, all *Neisseria meningitidis.* Resistance is common in *Staphylococcus aureus*, due to the ability of strains to produce the enzyme β-lactamase which destroys the antibiotic. Since 1976 clones of *Neisseria gonorrhoeae*, also producing β-lactamase, have been experienced. One disadvantage in the use of benzyl penicillin lies in the fact that it cannot be given by mouth since it is readily destroyed by gastric acid, and must be administered parenterally. Certain of the semisynthetic penicillins have been devised to be acid-resistant, to have broader spectra of activity or to be β-lactamase resistant (Figure 11.1). Carbenicillin is unique among the penicillins in being useful in the treatment of *Pseudomonas aeruginosa* infections commonly experienced in burn lesions. Despite many advantages the semisynthetic penicillins often have a lower activity than penicillin which is still the most effective antibiotic against sensitive organisms.

Cephalosporins. Cephalosporins are produced by species of *Cephalosporium.* They act similarly to penicillin, have a wider spectrum of activity (including Gram-negative bacilli and β-lactamase-producing staphylococci) but bacteria readily develop resistance to them and they are more toxic than the penicillins.

Antibiotics affecting the cell membrane

Polyenes. Polyenes, e.g. nystatin and amphotericin, are water-insoluble drugs which complex with lipids. They are antifungal drugs. Amphotericin remains the drug of choice for a number of systemic mycotic infections; it is administered intravenously in a colloidal sol and binds to sterols in cell membranes—it exerts its antifungal action at the cell membrane by allowing release of cellular components. This often results in severe toxicity for the patient. Amphotericin has a very high affinity for ergosterol, the major sterol in fungal membranes, as against cholesterol in mammalian cells.

These membrane-active drugs share the disadvantages of poor penetration to some sites of infection and yield comparatively disappointing results in practice. Fungal resistance to them rarely arises.

Antibiotics affecting protein synthesis

Many antibiotics interfere with protein synthesis in susceptible bacteria: this results in an imbalance of cellular processes which may inhibit cell propagation and may cause death. Antibiotics differ as regards the stage in protein synthesis at which they interfere with protein synthesis, as indicated under each group discussed below.

Aminoglycosides. Aminoglycosides prevent ribosomes joining with the mRNA and may also cause the mRNA code to be misread so that the wrong amino acid is incorporated into the protein. Aminoglycosides are a group of important antibiotics which act against staphylococci and aerobic Gram-negative bacilli (but not streptococci and anaerobic bacteria). They must be administered parenterally and are known to demonstrate toxic effects on the vestibular or cochlear organs, and kidneys. For these reasons some can only be used topically.

Streptomycin was the first aminoglycoside to be introduced, and second in time to penicillin. Its discovery was an important advance since the antibiotic was effective against Gram-negative bacilli and *Mycobacterium tuberculosis*. It has now been superseded except as a combined treatment with penicillin in special circumstances, and is no longer the principal agent in the treatment of tuberculosis.

Gentamicin exhibits a wide range of activity among the Gram-negative bacilli and, in addition, is active against *Pseudomonas aeruginosa* and very resistant strains of *Proteus*. Resistance has unfortunately been experienced in previously sensitive species.

Tetracyclines. These antibiotics bind to ribosomes in such a way that the tRNA is unable to interact with the ribosome. This occurs in sensitive bacteria but, although similar sensitive pathways are present in eukaryotic cells, the drugs seem unable to penetrate into these cells. Tetracyclines are bacteriostatic antibiotics having a wide spectrum (including the Gram-positive cocci, aerobic Gram-negative bacilli, both enterobacteria and parvobacteria), bacteroides and even rickettsiae and chlamydiae. They are the treatment of first choice for brucellosis,

Q-fever, etc., but their usefulness today is limited due to the
resistance and certain side-effects.

Chloramphenicol. Chloramphenicol prevents transfer of the
chain in protein synthesis. It is another wide-spectrum, bacteri
which, being a small molecule, penetrates deeply into tissues and
the rare but serious side-effect of bone-marrow aplasia, it is reserved for the treatment of very serious infections for which it is the most effective drug, e.g. typhoid
fever.

Clindamycin. This antibiotic prevents translocation of the functional ribosome on
the mRNA strand. It is active against staphylococci and against bacteroides
(many of which are resistant to other commonly used agents such as the
penicillins). Again side-effects (colitis) limit its use to serious infections with
relevant bacteria.

Antibiotics affecting nucleic acid replication

Rifampicin. Rifampicin is the most important member of a group of antibiotics
which inhibit the bacterial enzyme concerned with RNA replication. It has a wide
spectrum of activity but, as it is one of the best antibiotics for the treatment of
tuberculosis and as resistance to it readily develops, its use is being restricted.
Metronidazole inhibits DNA synthesis. It is effective against protozoal parasites
(such as *Entamoeba*, *Giardia*, and *Trichomonas*) and anaerobic bacteria (including *Bacteroides* spp.).

Antimicrobial prophylaxis

Chemotherapeutic agents are sometimes used to prevent infection becoming
established in the host (i.e. prophylaxis). In prophylaxis the agent is administered
to individuals who are exposed to increased risk of infection. It is more likely to
succeed where the individual is exposed to a single microbial pathogen over a
relatively brief time period. To avoid the potential disadvantages consequent on
the use of antibiotics, prophylaxis should only be practised where favourably
indicated, and these instances are relatively few in human medicine; prophylaxis,
rightly or wrongly, is more widely practised in veterinary medicine. Examples of
the prophylactic use of antibiotics against bacterial infections are considered in
Chapter 12; prophylaxis in malaria is considered below.

Malaria

Antimalaria prophylaxis is strongly indicated for travellers entering an endemic
area. The drug of choice is chloroquine sulphate which suppresses clinical

symptoms and eliminates erythrocytic parasites. It does not prevent infection since it has no effects on sporozoites infected by mosquitoes or on malaria parasites within the liver. Treatment is started 1 or 2 weeks before exposure and is continued throughout the duration of the visit and for 6 weeks after departure from the endemic area. Not all *Plasmodium* spp. are equally sensitive and other drugs, e.g. primaquine, are sometimes taken after ending a course of chloroquine. For prophylaxis against chloroquine-resistant *P. falciparum*, a combination of pyrimethamine and sulphadoxine is used.

The emergence of antibiotic resistance

Antibiotics have revolutionized the treatment of infectious diseases but not without a number of serious potential hazards. These include the occurrence in a few patients of hypersensitive reactions and of superinfections by insusceptible species (Chapter 12). The most important hazard, however, is the development of antibiotic resistance. In the last analysis, the success or failure of antibiotic therapy depends on the sensitivity of the infectious agent to the antibiotic of choice. To limit these hazards care must be taken in the prudent use of antibiotics.

Contrary to popular opinion antibiotics do not cause bacteria to become resistant. Antibiotic resistant organisms were already present in limited numbers in collections stored before antibiotics were used in clinical medicine. These organisms already possessed genes which coded for resistance to specific antibiotics. It is not known how they originated but it is assumed that they were the products of random mutations in genes which normally code for enzymes involved in basic cell functions. Where a mutation altered the gene, such that the enzyme which it coded was able to destroy an antibiotic or interfere with its mechanisms of action, then the progeny of the cell in which the mutation occurred exhibited resistance to the corresponding antibiotic. This, of course, was not evident until the antibiotic was discovered and used but, once this had occurred, the resistant strains were selected and grew preferentially to become the dominant components of the population.

The situation has been aggravated by the fact that resistant genes in one bacterium may be transferred to sensitive ones by either of several well defined mechanisms described below.

Transformation

This is a method of transfer by which free DNA is transferred from one cell to another. If the DNA codes for antibiotic resistance the recipient cell becomes instantly resistant. Transformation occurs rarely in nature.

Transduction

This involves a bacterial virus (bacteriophage or phage) which carries the DNA from one bacterium to another. The phage replicates in an antibiotic resistant bacterium. During the stage of phage maturation, when the new phage DNA is assembled within new phage coats, by chance the new phage particle may pick up a small piece of bacterial DNA instead of phage DNA. This is carried by the phage to a sensitive bacterium and the DNA coding for resistance is released into the sensitive cell during the process of phage infection. Transduction is especially important for the transfer of resistance genes between staphylococci and has been shown to occur *in vivo*.

Conjugation (sexual mating)

This occurs in many Gram-negative bacteria, including Enterobacteriaceae, *Pseudomonas* spp., *Vibrio* spp. etc. The resistant donor bacterium synthesizes a surface structure (sex pilus) which can attach to a receptor site on a sensitive (recipient) bacterium. The DNA coding for resistance (often called an R plasmid) changes from a circular structure to a linear form and is transferred across the sex pilus bridge into the recipient cell. Before transfer the R plasmid replicates; one piece remains in the donor cell and the other is transferred thereby imparting resistance to the recipient. The R plasmid replicates in the recipient cell which, in turn, transfers DNA to other sensitive bacteria and to its progeny. Transfer can occur within 10 minutes *in vitro* and resistance spreads rapidly through a bacterial population.

R plasmids may code for resistance to one or many antibiotics. The R plasmid may be transferred in part or the whole; the range of resistance acquired by the recipient cell depends on the number of different R determinants transferred.

Transfer by conjugation occurs not only between bacteria of the same species, e.g. between *Escherichia coli*, but also between species of different genera of Gram-negative bacteria, e.g. between *E. coli* and salmonellae or *Shigella*. Since this can occur between non-pathogens and pathogens it is of great importance in clinical medicine.

It must be stressed that not only do antibiotics not create resistance, neither do they cause transfer of R plasmids. However, they influence transfer by selecting strains which have acquired R plasmids at the expense of antibiotic sensitive bacteria. Transfer rarely occurs in the gut of the normal individual but can be demonstrated readily in patients on oral antibiotics.

Plasmids frequently code for other genetic information including resistance to metals (mercury, tellurite, cadmium, arsenic, etc.), resistance to agents that damage DNA (γ-irradiation, UV irradiation, etc.), metabolic functions (hydrogen sulphide production, carbohydrate fermentations), and pathogenicity factors (colonization factors—K88, K99, etc., haemolysin production, colicin production, enterotoxin production, etc.). Some of these are considered in Chapter 4.

FURTHER READING

Broda, P. (1979) *Plasmids*, Freeman & Co., Oxford.
Gale, E. F., Cundliffe, E., Reynolds, P. E., Richmond, M. H., and Waring, M. (1981) *The Molecular Basis of Antibiotic Action*, 2nd edn. John Wiley, Chichester.
Garrod, L. P., and O'Grady, F. (1981) *Antibiotic and Chemotherapy*, 4th edn. E. & S. Livingstone, London.

The control of infectious disease— IV: Hospital infections

1. Introduction

More than a century has passed since a new word 'hospitalism' was coined to express the dread of gangrene and death which was characteristic of hospitals in the mid-nineteenth century. Mortality rates of over 40 per cent were reported in those days in some of the large city hospitals, whereas patients treated in rural areas were four times less likely to die after the same treatment. During the succeeding years surgical and medical techniques have changed beyond all recognition, conditions have been greatly improved, most of the causal organisms of infections in hospital have been identified, and most infections are now amenable to treatment. Yet infections acquired in hospital still remain one of the main causes of morbidity and mortality, and represent an increasingly costly drain on the provision of health care at the present time. A number of factors which contribute to this problem follow.

(a) *Hospital patients are often highly susceptible to infection*

Those most vulnerable to infection are the elderly and the very young, and the majority of patients in general hospitals fall into these two age extremities. The pre-existing disease or condition, for which the patient was admitted to hospital, may itself have lowered the patient's natural resistance to disease, as may the treatment given—immunosuppressive drugs, irradiation, etc. (Chapter 1). Moreover, the natural defences of the body are often impaired by injury to skin or mucous membranes before admission or during surgery in hospital. As modern biomedical technology improves, so people of greater ages are increasingly subjected to more extensive operations of longer duration, often with multiple incisions, catheters, drainage tubes, etc., all of which make the patient more susceptible to invasion by micro-organisms. Many patients in hospital receive antibiotics and this often suppresses the growth of sensitive strains in the normal body flora, thus removing another barrier against infection by pathogens (Chapter 2).

(b) *Patients are often in close contact with each other*

Staff and patients usually share a common supply of food and water, and so outbreaks of infectious diseases may occur from time to time as in any comparable institution—e.g. schools and hotels—but, in addition, there is the hazard of close proximity to patients admitted to hospital already infected or to those infected during their stay in hospital.

(c) *The hospital environment may be a reservoir of highly virulent organisms*

Not only may patients and staff be the cause of the spread of virulent organisms, but the many items of equipment essential in a hospital may become and remain contaminated by pathogenic organisms for long periods. It is well recognized that the widespread use of antibiotics in hospitals has led to an increase in antibiotic-resistant bacteria endemic in the environment; moreover, the particular conditions within a ward or unit, together with the treatment regimens used there, may select out bacterial strains of particular virulence in that situation, e.g. the hazard of *Staph. aureus* phage type 80 in maternity wards and the multiple resistant *Klebsiella* strains in urological wards.

The frequency and severity of infection depends on the susceptibility of the patient, the treatment given, and the length of stay in hospital, but an average figure may be derived from the many surveys reported in recent years and would indicate an infection rate of 5–10 per cent. This level of infection may be considered as '*endemic infection*', in contrast to sporadic outbreaks of particular types of infection viz. *epidemics*.

Infections acquired in hospital (sometimes called *nosocomial* infections) may be caused by micro-organisms derived from another patient, from the hospital equipment, or from the materials used; this is commonly called *cross-infection*. However, a patient may succumb to infection following a procedure performed in hospital but caused by an organism that he was carrying when he entered hospital, perhaps as part of his normal body flora; this is referred to as *self-infection* (or *auto-* or *endogenous infection*).

2. Causal organisms

Almost any organism may cause infection in a hospital patient, but certain organisms are particularly associated with hospital infections and some of these rarely cause disease elsewhere. This depends partly on the virulence of the strain, and partly on the likelihood of suitable sources of reservoirs of the organism being available in the hospital at the time but, more importantly, on the fact that patients have diminished resistance to infection.

In a recent WHO publication, Parker (1978) suggested three categories as a rough indication of the circumstances under which particular organisms may be expected to cause infection in hospital (Table 12.1): *'conventional'* pathogens (P) that cause clinical disease in healthy persons, often responsible for 'institutional' outbreaks in hospital; *'conditional'* pathogens (C) that cause significant disease only in persons with reduced resistance to infection, or when implanted directly into tissue or a normally sterile body area (this category includes the bulk of infections that are attributable to specific procedures performed on hospital patients); and 'opportunist' pathogens (O) that cause generalized disease only in patients with profoundly diminished resistance to infection, usually those in whom there is severe underlying disease.

3. Sources of infection

Sources of infection are often the breeding places of the micro-organisms and, outside hospital, these may be other human beings, animals, or inanimate objects in the environment. In hospital, animal sources are of little importance, environmental sources are of some importance, but the sources of greatest importance are the patients and staff.

Patients with overt infection—a septic wound or a respiratory infection—are obviously a hazard to their fellows, but so are carriers of organisms that do no harm to the host but are potential pathogens for others. This carriage may be purely mechanical, e.g. on healthy skin; due to colonization, e.g. nasal carriage of staphylococci; or a patient who is incubating a disease, e.g. many viruses are shed before clinical virus infection becomes apparent; or the convalescent patient, e.g. typhoid carriers or a staphylococcal carrier with 'healed' bed-sores; or a patient who may be temporarily colonized or carrying a pathogen against which he is himself immune.

Inanimate sources are likely to be important where multiplication of micro-organisms is possible and this usually entails a warm and moist environment. Most Gram-positive bacteria survive well even under dry conditions, but Gram-negative bacteria do not, and hence it is mainly in fluids, weak disinfectants, distilled water and moist equipment such as humidifiers that these have caused many problems.

Clearly, inanimate sources that can convey pathogens direct to the patient's susceptible site (e.g. weak disinfectants) are more likely to cause infection than more remote sources, such as sinks and sluice equipment; the latter can also act as sources of contamination for intermediate objects which, in turn, become the means of transmission of infection.

4. Routes of infection

Many infections contracted in hospital result from the transmission of micro-organisms by a combination of several routes of infection, but two routes are of special importance—*aerial* and *contact* spread.

Table 12.1. Major microbial causes of hospital-acquired infections

Class of micro-organism	Organism	Pathogenecity in hospital patients*		
Gram-positive cocci	*Staphylococcus aureus*	P	(C)	
	Other staphylococci and micrococci		C	
	Streptococci group A	P		
	Streptococci group B		C	
	Streptococci groups C and G	P	(C)	
	Enterococci		C	
	Other non-haemolytic streptococci		C	
	Anaerobic cocci		C	
Anaerobic bacilli	Histotoxic clostridia		C	
	Clostridium tetani		C	
	Non-sporing Gram-negative bacilli		C	
Gram-negative aerobic bacilli	Enterobacteria: *Salmonella, Shigella,* enteropathogenic *Escherichia coli*	P		
	Other *Escherichia coli, Proteus, Klebsiella-Serratia-Enterobacter*		C	
	Pseudomonas aeruginosa, other pseudomonads		C	
	Flavobacterium meningosepticum		C	
	Acinetobacter		C	
Other bacteria	*Cornyebacterium diphtheriae*	P		
	Listeria		C	(O)
	Mycobacterium tuberculosis	P		
	Anonymous mycobacteria			O
	Bordetella pertussis	P		
Viruses	Hepatitis	P		
	Smallpox, vaccinia	P		
	Chickenpox	P		(O)
	Influenza and other respiratory viruses	P		
	Herpes simplex	P	(C)	(O)
	Cytomegalovirus	P	(C)	(O)
	Measles	P		
	Rubella	P		
	Rotaviruses	P		
Fungi	*Candida*		C	(O)
	Nocardia			O
	Moulds		C	
	Histoplasma, Coccidioides, Cryptococcus	P		(O)
Other	*Pneumocystis*			O
	Toxoplasma	P		(O)

Many common infections are spread by aerial transmission following the release of organisms from the respiratory tract of an infected person, from a person's skin by natural shedding of skin scales, during the dressing of wounds, or from an aerosol formed by moving or opening an inanimate source of infection. The degree of danger depends not only on the virulence of the organisms and the length of time that they can remain alive, but also on the size of the particle on which the organisms are suspended (Chapter 9). Very small particles formed during normal breathing, sighing, or talking and by resuspension of tiny dust particles by movement or dusting (often only 2–15 μm in diameter) can remain suspended for hours or even days, and may transmit infection over quite large distances, for example, when recirculated through an air-conditioning system.

Viral respiratory infections (e.g. influenza, common cold, etc.) may be transmitted as readily in hospital as outside, and in paediatric wards, outbreaks of childhood fevers (e.g. measles, chickenpox, etc.) can be a problem. Aerial spread is an important route of transmission for streptococcal respiratory infections (e.g. tonsillitis, sore throats, etc.) and for meningococcal and *Haemophilus* infections (e.g. meningitis). The spread of tubercle bacilli by this route from patients with respiratory tuberculosis, and of *Staph. aureus* from patients with staphylococcal pneumonia, is sufficiently hazardous to warrant isolation procedures. The main cause of postoperative wound infection is *Staph. aureus*, and aerial spread is often involved whether the infections are due to cross-infection or self-infection. Direct sources of infection include the patient's own nose or skin, other nasal or skin carriers in the ward or operating theatre and patients or staff who may have infected lesions. It may occur indirectly by the resuspension of settled contaminated dust particles during cleaning, dusting, or the movement of people or objects such as clothing, bedding, or curtains. Infection by Gram-negative bacteria is less likely to occur by these means, but aerial spread from respiratory apparatus and air-conditioning equipment is more common. Spores of clostridia may be spread around the hospital in dust from streets and buildings, and hence airborne infection is a possibility although infections such as gas gangrene are more likely to be endogenous.

Direct contact transmission may occur in a few instances in hospital; for example, a *herpetic whitlow* on a nurse's finger, if she does not wear gloves when

---

*P = 'Conventional' pathogen; causes clinical disease in healthy persons.
  C = 'Conditional' pathogen; causes significant disease only in presence of specific predisposing factor.
  O = 'Opportunistic' pathogen; causes generalized disease, but only in patients with profoundly diminished resistance to infection.
(C) = Chance or severity of infection greatly increased in predisposed persons.
(O) = Gives generalized infection rarely except in patients with profoundly diminished resistance.

From Parker, M. T. (ed.) (1978). *Hospital-acquired Infections*, World Health Publications, European Series No. 4.

attending to a patient's oral hygiene, may be derived by direct contact with the patient's infected lips; *ophthalmia neonatorum*, an inflammation of the conjunctiva in the newborn contracted by direct contact with the mother's infected birth canal during delivery; staphylococcal or streptococcal *wound infection* by direct contact with a septic or recently healed lesion on the skin of a member of staff.

More often, however, hospital infections are due to *indirect* contact, often combined with aerial spread. The hospital abounds with objects, known as *fomites*, which may act as an intermediate link between a source of infection and a patient's susceptible site; faeces, urine or pus, as well as contaminated dust particles or fluids, may be carried on cutlery, toys, thermometers, bedding, aprons, bed pans, instruments, etc., but the intermediaries most difficult to control are the clothes and hands of doctors and nurses. This is the route by which many cases of sepsis by staphylococci and streptococci occur, and some cases of infection by Gram-negative bacteria. In old or unhygienic buildings cockroaches, flies, and other insects may act as fomites, and of recent years, minute insects known as 'Pharaoh's ants' have become a considerable problem in that they are well adapted to the warmth of modern centrally heated buildings. They have been shown to be capable of contaminating sterile dressing packs with Gram-negative bacteria by carrying them from sources such as sinks and drains.

In hospital, many procedures involve contact with the patient's body that goes deeper than the skin; infection may thus be caused by *injection* or *insertion*. Catheters and many optical instruments, such as cystoscopes, may become contaminated before they are inserted into the patient, or a sterile instrument may introduce organisms from the patient's own flora in the process of insertion; for example, the introduction of urethral organisms when a sterile catheter is inserted into the bladder. If the site of insertion is normally sterile, then even a small degree of contamination may lead to infection; there is much less danger of this where the instrument is passed into a site in the body where there is a large normal flora, e.g. throat, anus. Where needles or instruments incise the skin, precautions must be taken to avoid, as far as possible, contamination by the patient's skin organisms. Where a patient is particularly susceptible to a potential pathogen, which is likely to be present on the skin in the operation area, it may be necessary to give prophylactic antibiotic cover during the operation, as in the prevention of gas gangrene in patients with a poor blood supply who are undergoing amputation.

Of special note are the occasional but serious infections caused by infected infusion fluids and by hepatitis B virus. Since the infective agent of hepatitis B is present in the saliva, as well as the blood of carriers (approximately 1 in 10^3 of normal people in the UK), problems of cross-infection by this organism are of special interest in dental surgeries and hospitals, and in haematology laboratories.

Infections may occur in hospital by *ingestion* but should not be more frequent than in any institution where there is communal feeding although hospital kitchens are not always subject to the same stringent requirements as commercial catering establishments. Moreover, in hospitals for the mentally subnormal there may be a very low standard of personal hygiene, and in children's wards there are often additional problems of hygiene, such as the sucking of toys. As a result, outbreaks of intestinal infections tend to occur most often in these two types of hospital.

5. Incidence and prevalence of infection

Care must be taken in the interpretation of published data of infection rates in a community, whether in or outside hospital. It is important to distinguish between bacterial colonization of a site, trivial infection, and more serious infection with marked clinical symptoms. For example, indwelling urinary catheters are likely to become colonized; this may cause a mild local reaction which disappears when the catheter is removed, or it may lead to severe and long-lasting infection. In addition, while some reports give *incidence* figures, i.e. the proportion of all patients in a given period for which there is evidence of infection, other studies are *prevalence* studies where samples are obtained from all patients on one or several days chosen arbitrarily and the proportion of those apparently infected is calculated. Figures for prevalence are often higher than those for incidence since confirmation by repeat positive cultures is not required and also, since incidence figures are very dependent on the degree of surveillance and investigation shown by the medical and nursing staff, these may reveal far fewer of those cases where clinical symptoms are not evident. Recent data from surveys in the USA give a range of incidence rates from 3.5 to 7.1 per cent of patients compared with prevalence rates which vary from 11.9 to 15.5 per cent.

Moreover, a single figure for the incidence or prevalence of infection in a hospital is likely to be of little value, or even misleading. Postoperative sepsis, in particular, is greatly dependent on the type of surgical procedure employed. Surgical wounds range from 'clean' where hollow organs are not opened and no inflammation is encountered, to 'contaminated' where the operation areas are known to be heavily contaminated (e.g. the colon) or infected (e.g. an appendix abscess). Recent data from both the USA and the UK give average overall incidence rates of about 5 per cent for clean and 20 per cent for contaminated wounds. In 1980 a prevalence study was organized by a multidisciplinary PHLS committee which embraced 43 hospitals in the UK and included over 18,000 patients. The prevalence of infection was found to be about 20 per cent, of which 9.2 per cent was judged to be hospital-associated infections, and 9.8 per cent acquired in the community before admission to hospital. Of the hospital-acquired infections, the largest group of infections was the urinary tract infections (39 per cent). Respiratory infections accounted for 25 per cent, wound infections 24 per cent, bacteraemia constituted a very important 4 per cent, and various other infections accounted for the remaining 8 per cent.

Most postoperative wound sepsis, is likely to originate in the operating theatre and Kelly and Warren (1978) have shown that culturing a wound swab taken at the end of the operation gave a useful indication of likely sepsis subsequently. In a study of 214 transperitoneal operations, they found that less than 1 per cent of the wounds yielding no growth at the end of the operation subsequently became infected, whereas 37 per cent of the positive swabs were associated with later infection. Prevention or a marked reduction in the ultimate contamination of an operation site has been the goal of surgeons ever since the days of Lister and Von Bergmann and enormous efforts have been made in this direction in postwar years.

In addition, great attention needs to be paid to the period of postoperative care in the wards where many other infections, in particular most of the urinary tract infections, are likely to occur. Methods of prevention, therefore, need to take into account not only the sources of infecting organisms and their routes of spread, but also the wide variety of surgical and medical procedures in theatre and ward to which patients are subjected.

6. Prevention of infection

(a) *Asepsis and aseptic techniques*

The terms 'asepsis' is used to describe attempts to reduce or prevent infection by removing or enclosing potentially harmful micro-organisms. The term is sometimes used when disinfectants or antibiotics are employed to achieve a state as near as possible to sterility but strictly refers to the prevention of infection by physical methods.

Aseptic techniques must take account of all likely sources of infection—instruments and equipment, staff and patients, and the environment—as well as those routes of spread which are likely to be important.

In recent years, much has been achieved by improved methods of sterilizing instruments, dressings, and equipment. In most hospitals, the boiling tanks in each ward and theatre (incorrectly called 'sterilizers') have been removed and, instead, there is a CSSD (Central Sterile Supply Department) or a TSSD (Theatre Sterile Supply Department). The staff in these departments clean and pack instruments, prepare dressings, and then sterilize these packs. This is usually done in high-vacuum autoclaves which are well maintained, operated by experienced personnel, and monitored automatically. Another major contribution to improved asepsis of recent years is the availability of a very wide range of prepacked, disposable instruments, drainage bags, syringes, dressing packs, etc.; most of these are sterilized commercially by γ-irradiation or ethylene oxide. Most intravenous fluids and baby feeds are also prepared commercially where the standards of control and efficiency for bulk processes are much more reliable than in small hospital dispensaries or food kitchens.

It has been less easy to reduce contamination of the environment. Theatres can be designed to have smooth, durable, non-porous surfaces and a minimum of upward-facing horizontal areas (ledges, shelves, window-sills, etc.), where dust will accumulate. Theatre routine also has improved with the realization that any unnecessary movement in a theatre is a hazard and that no unnecessary equipment should be stored in the operating room. A major improvement has been the installation of air-conditioning which not only greatly improves the comfort of the operators and increases safety by removing any escaping anaesthetic gases, but also, by maintaining a positive pressure near the patient, constantly removes airborne bacteria which might otherwise settle on the operation site and cause infection (Chapter 9).

In the wards, however, it is much less easy to make the environment safe for aseptic procedures. The large number of people, the frequent movement, the dispersal of numerous bacteria during bed-making, drawing of cubicle curtains, etc., all contribute to a high bacterial count in the air, especially at certain times of the day. For these reasons, some wards now have 'dressing stations'. These are small side-rooms with forced ventilation and surfaces that can readily be disinfected; into these rooms patients are taken, often in their beds, for changes of dressings, catheterizations, and other minor procedures where aseptic precautions are desirable.

The most difficult sources to deal with prior to carrying out any aseptic procedure are people—doctors, nurses, and patients. All skin surfaces and hair under normal conditions are heavily colonized by a person's own bacteria and also usually contaminated by bacteria picked up casually from other people and from the environment. Many of these bacteria are potential pathogens. The only totally efficient way to prevent doctors' or nurses' bacteria from contaminating an operation site is for the operating team to work in clothing rather like space suits, each with his own air supply and cooling system; this is done in certain highly specialized centres where the prevention of infection is considered of paramount importance (e.g. some surgical units dealing with hip replacements). Normal clothing is porous to bacteria and much research has gone into designing fabrics which are relatively impervious to bacteria but which, unlike waterproof aprons, will give the operator reasonable comfort by allowing evaporation of moisture. Many operating suits are now made in a close-weave ventile fabric which greatly reduces the shedding of bacteria especially if simple additional precautions are taken, such as securing the trousers around the ankles. Hoods for hair and beard, and masks over nose and mouth, are normal wear in operating theatres. However, badly designed or ill-fitting masks, or those that have been worn too long or inadvertently touched, may be more hazardous than no mask at all. For these reasons, masks are often not employed for ward procedures.

General cleanliness of the person and clothing of members of staff is clearly desirable, but two rituals of the past, now recognized as being of possible harm, are the prolonged scrubbing of hands which may seriously damage the skin and lead to increased microbial growth there, and preoperative showers which have been shown to increase the shedding of bacteria subsequently.

One unpopular measure which is necessary in the cause of the prevention of infection is the banning of members of staff with boils or similar lesions. An occlusive dressing cannot be relied on to give protection since skin remote from the lesion may be contaminated with the causal organism, often staphylococci. Healthy carriers of *Staph. aureus* usually have to be considered an acceptable risk, but if the type of organism carried is known to be of particular virulence in that situation, then essential staff who are carriers can be treated to eliminate the staphylococcus.

There is little that can be done about the patient's bacterial flora except in the case of bowel surgery where colonic bacteria are usually reduced in numbers by wash-outs or antibiotics. In most patients it is the site of injection or operation which receives attention. Shaving shortly before operation is advantageous and the skin or mucous membranes are then disinfected (see below).

For all infections, both in theatre and ward, probably the most important routes are the hands of doctors and nurses. Hence, even for simple procedures such as emptying a filled urine drainage bag, nurses are expected to wear sterile disposable plastic gloves; these are cheap and ill-fitting but serve a very useful purpose in providing a barrier, impervious to bacteria, between the nurse's hands and the contaminated material. For operative techniques, sterile disposable gloves of high quality and correct fit are required but a high proportion of these inevitably become perforated during a single operation. For this reason, the operator's hands are carefully disinfected (see below) before putting on his gloves so that even under the otherwise ideal conditions of warmth and moisture, bacterial multiplication will be minimized and hence contamination through a perforation in the glove less likely to initiate infection.

Many surgical and nursing procedures employ 'no-touch' techniques; sterile materials or contaminated swabs may be held and manoeuvred by forceps, thus avoiding contact with hands whether gloved or not. Forceps are often included in dressing packs so that they may be used in this way but also impervious paper wrappers are sometimes included for this purpose, and some sterile plastic urethral catheters are provided in a sterile sleeve of plastic so that the catheter can be held by the fingers but without contaminating it, thus allowing much more sensitive control of its insertion.

One important infection which may be acquired in hospital, and where aseptic precautions are of great importance, is viral hepatitis due to hepatitis B virus. One of the chief problems is that the virus is present in large amounts in the blood (0.0000001 ml of blood is infectious) and smaller amounts in the saliva of patients ill with hepatitis B infection and also in about 1 in 1000 of the normal population in the UK. There are various categories of patient who must be considered as of 'high risk' when handled by doctors, dentists, or nurses; these include people known to HBsAg-positive (hepatitis B surface antigen positive), multi-transfused patients, drug addicts, inmates of mental institutions, and promiscuous homosexuals. The spread of infection may occur by contact with contaminated blood or saliva (as in transfusions—now virtually eliminated, tatooing, the use of non-sterile needles by drug addicts) or by the faecal–oral route (as in mental institutions) and, more recently, new evidence has indicated that other routes such as inhalation may be implicated. Doctors, dentists, and nurses all need to take aseptic precautions—wearing gloves and face masks for all procedures, the wide use of disposable dressings and instruments; autoclaving rather than chemical disinfection should be used wherever possible. Samples of blood for laboratory investigations have to be double-wrapped in specially labelled packs, and laboratory staff must also employ aseptic techniques in handling them.

(b) *Antisepsis*

Since the work of Ignaz Semmelweiss and Robert Lister, the search has been on for suitable chemical agents for use in the medical field. Where physical methods of sterilization and disinfection can be used, they are preferred since chemicals have many disadvantages. Most chemicals with a powerful antimicrobial action are too toxic to use in contact with human tissues and most agents that are safe to use have a restricted spectrum of activity; very few are active against spores, viruses, or fungi and even fewer are capable of destroying all types of micro-organisms. Numerous outbreaks of infection in hospitals have been traced to con-taminated bottles of disinfectants since most chemicals are inactivated by many forms of organic materials and bottles of disinfectant solutions, once opened, are liable to become and remain contaminated.

Yet there are many instances where physical methods of disinfection cannot be used and hence chemicals play a most important part in the prevention of infec-tion. There are so many agents available that many hospitals or hospital groups have found it helpful, on grounds of safety and economy, to formulate a local policy and to recommend specific agents and appropriate concentrations for common situations. One example of such a policy is a publication by the South Western Regional Hospital Authority, entitled *Notes on Disinfection and Sterilisation* (3rd edition, 1975).

For many situations where environmental contamination used to be dealt with by chemical disinfectants, e.g. the cleaning of furniture, floors, walls, etc., it has been found to be equally, if not more, effective to wash thoroughly with hot water and detergent. However, there are some occasions where a chemical disinfectant is appropriate for 'domestic' hospital hygiene, such as where floors or surfaces have become heavily soiled by spillage; one of several clear soluble phenolic pre-parations is often recommended for this. Many wards have facilities for disinfect-ing bedpans and urinals, and for pasteurizing crockery and cutlery, but baths, plastic covers of mattresses, mops, and trolley surfaces are examples where chemical disinfection is usually needed to reduce contamination.

Most instruments and medical equipment can be sterilized or disinfected by physical means (usually autoclaving or pasteurizing) but some types of plastic cannot be heat-treated. Where catheters or other items which cannot be treated as disposable are made of these materials, chemical disinfection with ethylene oxide or glutaraldehyde is often used. (*Note:* it is important to ascertain the type of plastic and the manufacturer's instructions before treating plastic goods with chemicals.) Many optical instruments (e.g. cystoscopes, bronchoscopes, etc.) may be damaged by high temperatures; some can withstand boiling, and most can be disinfected by pasteurization temperatures, but chemical methods (e.g. with ethylene oxide or glutaraldehyde) are often employed for these delicate instruments. Clinical thermometers cannot be disinfected with heat and are best wiped with spirit after use and stored dry at the bedside.

It is for the disinfection of skin that chemical disinfectants are of greatest importance. Provided fingernails are kept short, doctors' and nurses' hands are best washed in the wards with plain soap and warm running water followed by thorough drying with a paper or other individual towel. Hospital staff may need to wash their hands very many times each day and a hand cream (with or without an added antibacterial agent) should be used to prevent soreness, but only during outbreaks of cross-infection in ordinary wards is it usually considered necessary to apply disinfectant skin preparations repeatedly as in theatre practice. There, at least 2 minutes should be spent in scrubbing the nails, hands, and forearms with a sterilized nail brush, and with 'Hibiscrub' (containing chlorhexidine and a detergent) or a povidone iodine surgical scrub (e.g. Betadine or Disadine); then, before putting on sterile gloves, the hands should be dried with a sterile cloth. These skin disinfectant preparations have a cumulative effect so that by repeated usage during an operating session, the bacterial content of the skin can be kept at very low levels.

Patients' skin before injection may be rubbed with a swab impregnated with alcohol (e.g. commercially available impregnated swabs such as 'Medi-swab') or where there is a higher risk of infection or contamination (e.g. taking blood for culture, or before lumbar puncture) an alcoholic solution of iodine or chlorhexidine may be used. These latter solutions are also used, depending on hospital practice and possible sensitivity of the patient, for the treatment of an operation site after shaving shortly before being prepared for theatre. In Accident and Emergency departments, the skin of patients on arrival may be heavily ingrained with dirt as well as bacteria. Where the skin is unbroken, a detergent jelly (e.g. Swarfega) may greatly help before disinfection with chlorhexidine or povidone iodine; broken skin may be cleaned with chlorhexidine and cetrimide (e.g. Savlon) or with a povidone iodine surgical scrub.

Patients who are admitted with an infected skin lesion or who develop a wound or urinary infection postoperatively, may pose a serious hazard to others, and chemical disinfection is of value in reducing this hazard. Infected wounds are swabbed with a disinfectant solution such as chlorhexidine; infected urine may be rendered harmless by emptying into a vessel containing a disinfectant such as a clear soluble phenolic, and where baths are used to speed up wound healing, disinfection of the bath between patients is essential.

(c) *Prophylactic antibiotics*

Since antibiotics first became widely available, they have been used not only to treat infections but also to *prevent* infection developing. There are several situations where the use of prophylactic antibiotics has proved its worth, usually where the infection to be prevented is caused by a single species of bacterium which has a known sensitivity to the antibiotics available or where the patient is so seriously ill that any additional infection would be highly dangerous. Examples include prophylaxis just before dental treatment or tonsillectomy of patients who have a history of rheumatic fever or other heart abnormality which would predispose to bacterial endocarditis, and penicillin prophylaxis to prevent gas gangrene in elderly people undergoing leg amputations because of poor blood supply in their limbs. With recent advances in surgery, such as the field of transplantation and the insertion of arterial prostheses or joint replacements, and also with many uses in modern medicine for immunosuppressive drug therapy, the number of people who need protection from infection has increased dramatically. Where they are aimed at preventing special known risks, prophylactic antibiotics are generally thought justifiable although formal evidence of their efficacy is not easy to obtain.

In practice, prophylactic antibiotics are given much more frequently than is necessary or advisable, especially to patients undergoing general surgery. In a recent survey in an American hospital, 27 per cent of patients admitted for medical treatment and 29 per cent of those for surgical treatment, were given antibiotics; in the surgical patients, 58 per cent of the antibiotics were given in the hope of preventing infection, compared with only 6 per cent of those given to non-surgical patients. In the opinion of the medical staff overseeing infection control in that hospital, more than half of all the antibiotic therapy was inappropriate. Similar surveys and reviews in the UK have led to the general conclusion that apart from cases of special known risks (such as those suggested above) prophylactic antibiotics are likely to do more harm than good; where conditions predispose to postoperative infection, antibiotics are unlikely to prevent it but merely to change the type of the infecting organism.

During the past 30 years of antibiotic chemotherapy, the patterns of hospital-acquired infections have changed; first from the mainly sensitive staphylococci and β-haemolytic streptococci in the 1940s to a much smaller number of streptococci in the 1950s with predominantly staphylococci resistant to most of the common antibiotics then available. Then in the 1960s, for reasons that are still not clear, staphylococci subsided in importance and were replaced by enteric Gram-negative bacilli and yeasts. Subsequently it has become clear that antibiotic resistance can be transferred among different strains and genera of Gram-negative bacilli by conjugation, and a hospital where antibiotics are widely used has proved to be an ideal environment for the emergence, selection, and spread of multiple-resistant enterobacteria. As a result, organisms which used to be considered as normal commensals of the human body or, at worst, very low-grade pathogens, are now the *opportunistic pathogens* (Chapter 2) responsible for many serious outbreaks of hospital infections. Table 12.2 is an extract from the findings of the 1980 National Prevalence Survey in UK hospitals, and it will be seen that Gram-negative bacilli accounted for more than half of all hospital-acquired infections. Moreover, those strains of *Klebsiella*, *Proteus*, *Pseudomonas*, and *Serratia* which

Table 12.2. The distribution by percentage of the named organisms recorded as causing hospital-acquired infections (from the 1980 National Prevalence Survey). Meers *et al.*, *Journal of Hospital Infection*, 1981, Supplement. Reproduced by permission of Academic Press Inc. (London) Ltd.

	Percentage named organisms
Staphylococcus aureus	17.6
Other staphylococci	3.0
Streptococcus pneumoniae	1.8
Streptococci, group A	0.5
Streptococci, other groups	3.4
Faecal streptococci	4.0
Viridans streptococci	0.1
Escherichia coli	26.1
Proteus spp.	11.2
Klebsiella spp.	7.2
Pseudomonas aeruginosa	7.0
Enterobacter spp.	1.1
Serratia spp.	0.5
Salmonella spp.	0.1
Haemophilus influenzae	1.7
Other Gram-negative bacilli*	5.2
Bacteroides spp.	1.7
Clostridium spp.	0.6
Other anaerobes	0.6
Mycobacterium tuberculosis	0.1
Other bacteria	1.6
Viruses etc.	0.8
Candida spp.	4.1

*Including those not fully identified.

have been the cause of recent outbreaks of infection have been very difficult to eradicate, and one of the contributing factors to this has undoubtedly been their multiple-resistance to antibiotics.

Prophylactic antibiotics, therefore, have an important and sometimes essential part to play in the prevention of hospital infection, but a measure of control must be exercised by clinicians to avoid increasing the problems associated with infections which the indiscriminate use of antibiotics might aggravate.

(d) *Isolation*

Isolation procedures and facilities within a hospital may be needed to prevent the spread of infection. These may include techniques such as barrier nursing and reverse barrier nursing, administrative arrangements such as cohort isolation in maternity wards, or special facilities ranging from plastic enclosures in a ward to specially designed isolation buildings equipped with air-locks and very complex isolators.

The aims of isolation may be either 'containment (or source) isolation', where a barrier of some form is maintained around an infected patient who is a potential source of serious infection to others, or 'protective isolation' where a barrier is kept around a highly susceptible patient (such as one receiving immunosuppressive drugs or cytotoxic agents) to prevent pathogenic microbes from other people or from the environment reaching him in sufficient numbers to initiate infection. Many hospital wards have no special isolation facilities and the best that can be done is to employ elaborate nursing procedures—barrier nursing for containment isolation and reverse barrier nursing for protective isolation. Only one or a few nurses deal with the patient, and they follow strict procedures for skin decontamination and the wearing of gowns, gloves, and masks; special arrangements have also to be made for toilet facilities, food, bedding, books, etc. Such measures are unlikely to be very effective within an open ward and are better carried out with the patient in a single room, preferably with an ante-room so that the patient's room is not in direct communication with the ward or corridor and where the nurse can prepare herself and the materials she is to take in to the patient and to bring out. However, it is well recognized that isolation in single rooms does little to prevent infection by the aerial route unless mechanical ventilation is provided—negative pressure for containment isolation and positive pressure for protective isolation. Where isolation rooms are purpose-built, they can be equipped with an air-lock and with facilities for both positive or negative pressure, and hence can be used for either type of isolation. Another form of ventilation which may be used in small rooms for patients where their protective or containment isolation is less critical, is *laminar air flow* in which recycled, filtered air is continually passed across the patient. For temporary, emergency arrangements and where isolation rooms are not available, a mobile, transparent plastic canopy can be fitted over the patient in his bed, or over one part of the patient (e.g. a burned limb) and the air inside can be mechanically ventilated.

Areas where there is a high risk of infection within a hospital are often designed to incorporate many isolation features. Thus *intensive care units* and *special care baby units* are usually divided into cubicles which can be used as small isolation units whenever necessary. In these, highly susceptible patients are nursed and many items of equipment, such as ventilators and incubators, pose an additional hazard because of the difficulties of disinfection and the risk associated with reservoirs of fluids.

In maternity wards, the danger of cross-infection from baby to baby can be greatly reduced by abandoning the large communal nursery. *Rooming-in* is one alternative in which each baby is kept at all times in the same room as its mother. If this is not possible but several small nurseries are available, *cohort isolation* can be adopted; babies born within a period of two or three days are segregated into one small nursery and no more added until all have been discharged.

Isolation facilities would undoubtedly be an advantage to more patients than are normally treated in this way, but they are very expensive in equipment, space, and staff. Inevitably, a selection has to be made of those patients for whom it is essential; it is usually the responsibility of the Hospital Infection Control Committee to advise doctors and nurses as to the degree of isolation recommended for each infectious disease and to advise when it is mandatory to transfer a patient to an infectious diseases unit for strict isolation.

(e) *Hospital design*

Antiquated buildings and their poor maintenance are often blamed for outbreaks of infection, and while this may sometimes be unjustified, the reduction in hospital infection has been one of the aims of postwar design of new hospitals and the modernization of old ones. Some of the changes introduced have been advocated as improvements towards this goal, such as increased spacing between beds and the subdivision of large wards by high partitions, but investigations have shown no change in infection rates following these alterations; their value has been mainly of a social character. Real improvements in wards have come with the provision of more washbasins spaced throughout the ward, better siting of sluices and toilet facilities, and with new surfaces on floors, furniture and window ledges, etc., that are more easily kept free of dust. It is doubtful whether the installation cost and additional heating costs needed for mechanical ventilation in general wards would be justified in terms of reduced infections, but this facility has become an essential part of the design of special-care wards, recovery rooms, dressing stations, etc. Air-tight doors in corridors, and the dangers of the circulation of contaminated air via lift shafts and laundry chutes, are among the problems to be solved by hospital architects.

Some dangers can be averted almost entirely by careful design, for example, where the rooms of a CSSD are so arranged that it is physically impossible for any objects to reach the *sterile* store from a *dirty* or *clean* area, without first passing through an autoclave chamber. Zoning of this sort has also been employed in the design of modern operating theatre suites. The Medical Research Council Sub-Committee (1962) proposed the definition of four zones:

A. *Protective zone:* entrance lobby, recovery room and changing rooms. Unchanged staff have access to this zone.

B. *Clean zone:* anaesthetic rooms, scrub rooms, rest rooms, inner lobbies and clean corridor, sterilizing department, and sterile store. Only changed staff have access to this zone.

C. *Aseptic or sterile zone:* operating room, and sterilizing or lay-up rooms.

D. *Disposal zone:* sink-rooms and disposal corridor. This zone has independent communication with the outside.

Such zoning, shown diagrammatically in Figure 12.1, requires a separate outer corridor, clean corridor and disposal corridor. Where a series of operating theatres are incorporated in a suite, the zoning arrangements may be very complex and full implementation of these ideas may be too costly to justify solely for the purpose of improving infection rates alone. Mechanical ventilation and air-conditioning are essential; positive pressure should be greatest in the aseptic zone (C), gradually reducing through the other zones, clean (B) to protective (A) to disposal (D). Even the positioning of the theatre(s) in relation to the wards may have some effect on hospital infection; for many years the top floor has been the

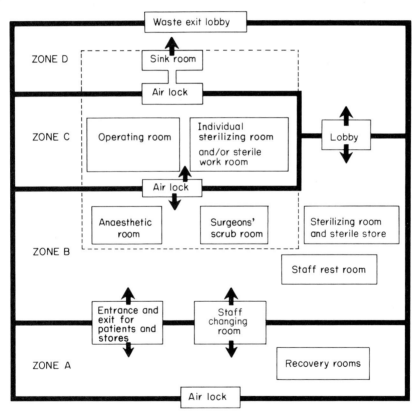

Figure 12.1. Ventilation zones in an operating suite. Zone A: protective zone; Zone B: clean zone; Zone C: sterile zone; Zone D: disposal zone. The rooms within the dotted line constitute the operating-room suite. The whole of this group should be replicated in multiple operating-room suites, except that individual sterilizing rooms are not required if a central sterilizing room is included

usual site for operating theatres but the 'stack effect' of contaminated warm air rising from the wards below via the stairways and lift shafts is a possible hazard. Rooms on the top floor also have other disadvantages, e.g. greater exposure to solar heat, and nowadays theatre suites in new hospitals are positioned at as low a level as possible in the building.

(f) *Surveillance and infection control*

Many large hospitals have an Infection Control Team or Committee which usually includes a microbiologist, at least two senior consultants (a surgeon and a physician), a pharmacist, and an infection control nurse (ICN). This latter post is often filled by a senior nurse who is based in the laboratory but whose advice and cooperation is readily accepted by the medical and nursing staff because of her own nursing and microbiological experience; the ICN often forms an important link between clinical, nursing, and laboratory staff.

Surveillance requires the daily scrutiny of laboratory records by a senior member of the laboratory staff who involves the ICN in any items of significance. It also entails frequent visits to the wards by the ICN and depends very greatly for success on good personal relationships with ward staff so that information is readily volunteered and can be assembled from many sources. Other functions of the Infection Control Team are to monitor hygienic practices, to advise various hospital committees on matters of policy relating to the prevention of infection (e.g. policies for the use of antibiotics, and of disinfectants). There is also a need for considerable participation in the education of all categories of staff where hygienic practice is important—students, nurses, kitchen workers, autoclave operators, cleaners, etc.

When an outbreak of infection does occur, it is the responsibility of the Infection Control Team to consult with the clinical and nursing staff and to institute the appropriate control measures—recommended treatments, special nursing procedures, sometimes a restriction on antibiotic usage, a transfer to an isolation hospital and sometimes, in extreme cases, the closure of a ward.

It is only in recent years that workers have begun to take the trouble to keep adequate records of their rates of infection and this is now a major contribution made by the Infection Control Team. Only by the study of carefully collected data and a detailed knowledge of the running of the hospital—what procedures and what materials are actually employed in every ward for every treatment—can faults be exposed and put right, and limited resources concentrated on those areas which will contribute most to the prevention of hospital infection.

FURTHER READING

Lowbury, E. J. L., Ayliffe, G. A. J., Geddes, A. M., and Williams, J. D. (1975) *Control of Hospital Infection—a Practical Handbook.* Chapman & Hall, London. (306 pp.).
Maurer, I. M. (1973) *Hospital Hygiene*, Arnold, London. (116 pp.).

Medical Research Council Report. (1962) Design and ventilation of operating-room suites for control of infection and for comfort. *Lancet*, **ii**, 945–951.

Medical Research Council Report. (1968) Aseptic methods in the operating suite. *Lancet*, **i**, 705–709, 763–768, 831–839.

Parker, M. T. (1978) *Hospital-acquired infections: Guidelines to Laboratory Methods.* WHO Regional Publications, European Series No. 4. WHO, Copenhagen (61 pp.).

South Western Regional Hospital Authority. (1975) *Notes on Disinfection and Sterilisation.* 3rd edn. (37 pp.).

Williams, R. E. O., Blowers, R., Garrod, L. P., and Shooter, R. A. (1966) *Hospital Infection*, 2nd edn. Lloyd-Luke, London (386 pp.).

PART III

SELECTED STUDIES IN EPIDEMIOLOGY, BASED ON VARIOUS ROUTES OF TRANSMISSION OF INFECTION AND TOPICS OF SPECIAL INTEREST

Chapter 13

The zoonoses

The deviation of man from the state in which he was placed by nature seems to have proven to him a prolific source of disease. From the love of splendour, from the indulgences of luxury, and from his fondness for amusement, he has familiarised himself with a great number of animals, which may not originally have been intended for his associates. The wolf, disarmed of ferocity, is now pillowed in the lady's lap. The cat, the little tiger of our island, whose natural home is the forest, is equally domesticated and caressed. The cow, the hog, the sheep and the horse, are all for a variety of purposes brought under his care and dominion (Edward Jenner, London, 1796).

The literal translation of the word *zoonosis* (Gr., *zoon*, animal, and *nosos*, disease) is *animal disease*. However, by wider usage it has come to mean those diseases and infections which are naturally transmitted between vertebrate animals and man and, although this use of the word is etymologically inexact, the WHO Expert Committee on Zoonoses adopted this definition in 1958.

By this definition man is singled out as the unique vertebrate animal involved in the epidemiology of the infections which, in some instances, exaggerates his functional role. On the other hand his central placement has provided considerable opportunity for medical–veterinary cooperation in all aspects of zoonotic disease control. It was stated that zoonoses comprise 'the most significant group of communicable diseases', and this emphasizes their prime importance in public health. Zoonoses are therefore a biologically heterogeneous collection of infections and infestations which have little in common. At present more than 100 zoonoses are recognized (Table 13.1). The importance ascribed to a particular zoonotic disease is a reflection of its incidence in human populations as well as its virulence and severity. On this basis, the ones which justify strict surveillance and rigorous control measures include rabies, plague, anthrax, and bovine tuberculosis.

A very wide variety of infectious organisms are classified as *zoonoses*; they include micro-organisms (viruses, rickettsiae, bacteria, fungi, and protozoa) and helminths, and arthropods. Each possess the common ability to infect both man and other vertebrate animals during natural transmission between them. With the continuing exploitation of the more exotic vertebrates by man it may be expected that others will be added to this list in the future.

Table 13.1   A selected list of zoonoses, taken from the list of the Joint FAO/WHO Expert Committee on Zoonoses, 3rd Report, Rome (1967). Reproduced by permission of WHO

Disease	Causative organism	Animal reservoirs
1. Viral diseases		
Arbovirus infections	Various arboviruses	Rodents, birds, equines, goats, sheep, monkeys, swine, marsupials
Herpes B virus disease	Herpes B virus	Monkeys
Influenza	Influenza virus type A	Swine (?), birds (?)
Lymphocytic chorio-meningitis	Lymphocytic chorio-meningitis virus	Mice, dogs, monkeys
Newcastle disease	Newcastle disease virus	Chickens
Poxvirus infections		
Cowpox	Cowpox or vaccinia virus	Cattle
Orf (contagious ecthyma)	Contagious ecthyma virus	Sheep and goats
Paravaccinia (milkers' nodules)	Paravaccinia virus	Cattle
Rabies	Rabies virus	Carnivores, bats, other wild animals
2. Rickettsial diseases		
Flea-borne:		
Murine (endemic typhus)	*R. mooseri*	Rats, mice
Mite-borne:		
Rickettsial pox	*R. akari*	Mice
Scrub typhus (tsutsuga-mushi)	*R. tsutsugamushi*	Rodents
Tick-borne:		
(North) Queensland tick typhus	*R. australis*	Bandicoots, rodents
Spotted fever (including Rocky Mountain, Brazilian, and Colum-bian spotted fevers)	*R. rickettsiae*	Dogs, rodents, other animals
Fièvre boutonneuse	*R. conorii*	Dogs, rodents
North Asian tick-borne rickettsiosis	*R. sibirica*	Rodents
Q fever	*R. burnetii*	Cattle, sheep, goats, wild and domestic mammals, birds
3. Chlamydia infection		
Psittacosis (ornithosis)	Psittacosis	Psittacines and other birds (including poultry)

Table 13.1. (*continued*)

Disease	Causative organism	Animal reservoirs
4. Bacterial diseases		
Anthrax	*Bacillus anthracis*	Ruminants, equines, swine
Brucellosis	*Brucella abortus, Br. suis,*	Cattle, swine, goats, sheep,
	Br. melitensis	horses
Salmonellosis	*Salmonella* spp.	Mammals, birds, rodents
Erysipeloid	*Erysipelothrix rhusiopathiae*	Swine, poultry, fish
Leptospirosis	*Leptospira* spp.	Rodents, dogs, swine,
		cattle, bandicoots
Listeriosis	*Listeria monocytogenes*	Rodents, sheep, cattle,
		swine
Pasteurellosis	*Pasteurella multocida*	Mammals, birds
Plague	*Yersinia pestis*	Rodents
Pseudotuberculosis	*Yersinia pseudotuberculosis*	Rodents, cats, fowls
Relapsing fever (tick-borne)	*Borellia* spp.	Cattle
Tuberculosis	*Mycobacterium tuberculosis*	
	var. bovis	Cattle, goats, swine, cats
	var. hominis	Dogs, swine, monkeys
	var. avium	Poultry, swine, cattle
Tularaemia	*Francisella tularensis*	Rabbits, hares, sheep, wild
		rodents
Vibriosis	*Vibrio parahaemolyticus*	Fish
5. Fungal diseases		
Dermatophytosis ringworm,	*Microsporum* spp.	Cats, dogs, horses
favus	*Trichophyton* spp.	Horses, cattle, poultry,
		small mammals
6. Protozoal diseases		
Amoebiasis	*Entamoeba histolytica*	Dogs, lower primates
Balantidiasis	*Balantidium coli*	Swine
Coccidiosis	*Isospora* spp.	Dogs
Leishmaniasis		
Kala azar	*Leishmania donovani*	Dogs
Oriental sore	*Leishmania tropica*	Dogs, rodents
American	*Leishmania* spp.	Dogs, wild mammals
Malaria	*Plasmodium knowlesi*	Monkeys
	Plasmodium simium	Monkeys
	Plasmodium cynomolgi	Monkeys
Toxoplasmosis	*Toxoplasma gondii*	Mammals, birds
Trypanosomiasis	*Trypanosoma cruzi*	Dogs, small mammals
	Trypanosoma rangeli	
	Trypanosoma rhodesiense	Antelope, cattle

A seventh category—Helminthic diseases—is not included.

1. General considerations

Early man was initially a hunter and subsequently a domesticator of the wild animals with which he co-inhabited the land. It may be deduced, therefore, that zoonoses have been a continuing part of man's evolution from those early days of domestication when he had considerable contact with animals. Man has long been resourceful at adapting to and modifying his environment to give himself better chances for survival, and there is evidence to suggest that he exercised this resourcefulness against zoonoses in earlier times. Classical records indicate that comparisons were certainly made at the time of Aristotle and Hippocrates, and they have been continually made ever since, though it was necessary for the new science of bacteriology, developed by Pasteur and Koch, to emerge before any real scientific progress was made towards checking the spread of zoonotic diseases. Prior to this man was slow to recognize the crucial role of animals in the spread of zoonoses.

2. Historical associations between animals and health

In the past, ancient laws, religious writings, folklore and superstitions served to emphasize the need for caution when dealing with particular animals or animal meat. This led to modern interpretations that, in ancient times, direct associations with human disease were established. For example, the Mosaic Law taught that pork and the flesh of the camel, rabbit, and hare were 'unclean': 'Of their flesh shall ye not eat, and their carcase shall ye not touch; they are unclean to you' (Leviticus 11.8). Early biblical interpreters appraised the religious content of such scriptures more than any public health content, a view entirely rational and consistent with the climate of scientific knowledge at the time.

In his book, *Eat Not This Flesh*, Simoons (1961) discussed the reasons why the flesh of the pig, ox, horse, camel, dog, and fowl are tabooed by various ethnic groups in the world. It seems clear that public health or hygienic reasons for declaring a designated flesh as 'unclean' are slight or non-existent. In the case of the dog kept as a working animal there would appear to be stronger reasons for linking its 'uncleanliness' as an animal rather than as flesh, and these are undoubtedly based upon previous ethnic experiences with this most common of domesticated animals. The zoonosis most commonly transmitted by dogs in the Middle East is echinococcosis, and it is claimed that the Prophet Mohamed recognized this. Indeed, the high incidence of the disease in Middle Eastern shoemakers may be attributed directly to the ancient practice of bating (tanning) hides with dog faeces. Since the dog has been domesticated for at least 60,000 years it is certain that echinococcosis is an ancient zoonosis.

It is of interest to note that in the United Kingdom at the present time, with its 20 million household pets of which some 25 per cent are dogs, echinococcosis is a rare disease. The emerging dog zoonosis in the UK is toxocariasis, caused by roundworms, and the disease will continue to remain a problem until sufficient public awareness has been generated.

3. The origins of zoonoses

There is reason to suppose that most, if not all, of man's infections and infestations originated in animals. Prolonged interaction of infectious agents and animal hosts have induced mutual survival adaptations and mutations to take place. Those of the parasite have been directed towards increased infectivity to combat host immunity mechanisms rather than towards increased virulence which inevitably leads to a depletion of the hosts upon which survival depends; those of the host animal towards increased resistance against overburdening by the parasite. Optimal survival for each obtains when a state of equable symbiosis is achieved, as is observed with the intestinal flora and the skin flora of the animal body. Hence, from an ecological standpoint it can be argued that the most lethal diseases are those produced by parasites which are still in a state of disequilibrium with their animal hosts. These parasites are essentially in a state of continuing adaptation, and their ability to adapt themselves to an increasing variety of hosts to perpetuate survival will increase their threat to man.

Theobald Smith, in 1934, writing about malaria disease, stated that 'the ancestry of the malaria parasite belongs to the insect because the most important, viz., sporogonic phase takes place in it'. Bruce-Chwatt in 1965, in his deliberations on the palaeogenesis and palaeoepidemiology of primate malaria, further developed this concept, indicating that insects date back some 200 million years, primates only 20 million years, and man a mere 2 million years. The most likely possibility concerning the evolution of malaria is that it developed in the gut of the insect (mosquito) and later became adapted to various primates (including man) through the feeding habits of the insect. Most other zoonoses may have developed similarly and, as adaptation to new animal hosts took place, the hosts themselves would become excretors and reservoirs of the disease, with or without overt clinical signs of infection. The greater the contact of man with original animal hosts, the greater the risk of establishment of a zoonotic epidemiology.

Zoonoses are the most frequent, and often the most dreadful, infections to which man is exposed. This is especially true for the tropical and sub-tropical areas of the world where arthropod-borne diseases are rife. The most frequent zoonotic exposures occur in those for whom close animal contact is a particular occupational hazard e.g. in veterinarians, slaughterhouse workers, farmers, forestry workers, factory food-processors, etc. The next in frequency of exposure are those having close contact with soil and water during normal occupational or leisure activities, e.g. agricultural workers, sewage workers, fishermen, hunters, bathers, etc. The least degree of exposure is by those who consume improperly processed animal foods, and who may suffer salmonellosis or helminthiasis as a result. The clinical and pathological characters of such zoonoses will be the same for man and animal alike, though the virulence is likely to be different in each according to the degree of immunity which the host expresses.

Man's continuing expansion of his 'lebensraum' and his disruption of stable natural conditions by land reclamation, with or without accompanying irrigation, can culminate in either eradication or exacerbation of natural zoonoses. When a potential human pathogen exists within the animals of an affected area, man's disruptive actions may make him a terminal link in the chain of dissemination and thereby create a zoonosis. Under most natural conditions soil, water, and arthropods constitute the usual indirect vehicles of spread; also, the wild vertebrates themselves may occasion transmission.

Such a situation pertains at present with the opening up of the Amazonian rain forests by the trans-Amazonian road-builders. These workers can expect to fall victims to prevailing forest zoonoses simply as a result of the more intimate contact with exotic animal populations. For various reasons, particularly an active immunity due to regular exposure to infections, the forest zoonoses have had little effect upon the native inhabiting the forest.

Wherever zoonotic control and eradication schemes are implemented the veterinary profession has played a key role. It has successfully implemented eradication programmes against the bovine diseases of tuberculosis and brucellosis (Chapter 15) in the UK and elsewhere on the European and North American continents.

4. Shifts of hazard

In the countries of the Western world modern advances in the treatment and control of animal diseases, coupled with changes in the patterns of agricultural and industrial practices and man-made alterations of his environment, have contributed towards a shift in the epidemiological patterns of certain commoner zoonoses. In general, the shift has been away from the social or adventitious hazard towards the occupational hazard. Brucellosis is now contracted less from the ingestion of contaminated milk and milk-based foodstuffs and more from direct contact with infected animals by veterinarians (especially when implementing eradication schemes), abattoir workers and meat-packers (Chapter 15). Similarly, psittacosis has become less of a zoonosis associated with pet psittacine birds (such as the budgerigar, macaw, or parrot) and much more associated with the factory processing of poultry (especially turkeys).

A further example of a severe zoonosis which has undergone an unexpected shift of hazard, sufficient to prompt speedy Governmental action in Great Britain, is rabies. In the mid-1970s an epizootic of rabies in feral foxes commenced to sweep across France from the north-east towards the south-west, and this gave rise to considerable unease in Great Britain. The last enzootic of rabies in England in 1918 was successfully controlled by a vigorous slaughter policy. The rapidly fluctuating Orders of 1970 reflected the equally fluctuating fears of a concerned

Government in respect of the inadvertent importation of canidae and felidae which might be incubating rabies, and the finding that the incubation period could extend beyond 12 months in the dog.

From January 1973, when Great Britain entered the European Economic Community (EEC), the continuing French epizootic posed a more immediate threat because of the increase in traffic between the two countries due to EEC entry and the difficulty of adequate Customs surveillance. The potentially explosive situation induced the rapid drafting of a new Rabies Act and Control Order in 1974 as a countermeasure.

In view of the continuing changes in agriculture, industry, and the environment it is likely that the list is incomplete. Indeed, there is an increasing implication of the domestic pig as a reservoir of human viruses, and it may well turn out that other domesticated animals may also be implicated in the continuing shifts of hazard of the present-day zoonoses.

5. A classification of zoonoses

As stated earlier, the term 'zoonoses' encompasses a biologically heterogeneous collection of infections and infestations which, although shared in nature by man and the lower animals, have little in common with each other. Classification schemes have been built around the nature of the aetiological agent involved (bacterial, viral, protozoal, etc.), or the groupings of the animal hosts (wild, domesticated, domiciled) which share the disease with man, and hence around the separate control and eradication problems which these groups generate. Possibly the most useful classification is based on the various epidemiological patterns (see Schwabe, 1969). While this scheme is not perfect it is at least workable.

In Schwabe's scheme there are four primary subdivisions: direct, cyclo-, meta- and saprozoonoses. Direct zoonoses are those in which the epidemiological pathway involves a single vertebrate host, cyclozoonoses in which the pathway involves more than one vertebrate host, metazoonoses in which the pathway involves an alteration of vertebrate and invertebrate hosts, and saprozoonoses in which the pathway involves an alternation of vertebrate host and non-animal vehicle.

Three secondary subdivisions are recognized: anthropozoonosis, zooanthroponosis, and amphixenosis. Anthropozoonoses comprise those diseases with animals as the reservoirs and man as a secondary host, zooanthroponoses those with man as the reservoir and animals as secondary hosts, and amphixenoses those with both man and the lower vertebrates as reservoirs independent of each other.

It is of interest to note that the largest body of infections and infestations are classified as metazoonoses, the second largest as direct zoonoses, the third as saprozoonoses, and the smallest as cyclozoonoses. Of the bacterial and viral zoonoses, 96 per cent (84 out of 87) fall within the classifications of metazoonoses

(51 per cent) or direct zoonoses (45 per cent). Within the metazoonoses the causal agents are not totally adapted to their vertebrate hosts, so that in the invertebrate vector there may be either further development or multiplication, thereby requiring an extrinsic incubation period and what is in effect an extrinsic host in which to undertake it. In terms of palaeogenesis, therefore, the metazoonoses should be considered amongst the oldest of infections, as has already been discussed for malaria. The arthropod-transmitted viruses are, on this argument, the oldest viruses still extant. Within the direct zoonoses the causal agents are sufficiently host-adapted and have sufficiently defined propagation requirements that non-vertebrates, plants, or soil serve only as passive transmission agents. In terms of palaeogenesis, these must be considered the younger infections.

(a) *Major zoonoses*

Five zoonoses are properly regarded as major in view of the scale of human suffering and mortality which they produce in the world. All are metazoonoses and all have arthropod vectors save one—schistosomiasis—which has a snail vector. It is noteworthy that no bacterial or viral zoonoses occur on the same enormous scale as these. Consideration of the principal reservoirs indicates that the dog constitutes an important mammalian reservoir for each. Early domestication of this animal has undoubtedly created more human suffering than any other.

(i) *Filariasis* (Wucheriasis). This affects 300 million humans worldwide annually, and has primates, dogs, cats, rats, and other mammals as reservoirs. The disease is caused by a variety of filarial nematode worms which usually come to reside in the lymphatics and generate elephantiasis of the affected tissues. The worms also cause 'river blindness' when they invade the ocular tissues, and in parts of Africa this disease affects the sight of 1 in 10 persons. The disease was no doubt carried from the African continent to the New World by the Negro slave traffic, but eventually declined except for a few localized areas in South America.

(ii) *Malaria.* This disease affects 200 million humans annually and has primate, dog, and rodent reservoirs. It is the most geographically widespread of the zoonoses. Around 25 per cent of African adults suffer from it at one time or another. Worldwide, $\frac{1}{4}$ million children die of it each year. In India and Sri Lanka, where it regressed for a time, it is now once more on the increase. The problem of eradication has been made more difficult by the need to maintain irrigation canals in dry areas which also maintain the aquatic larval forms of the anopheline mosquito vectors, and also by the emergence of DDT- and dieldrin-resistant mosquitos. However, substantial reduction of the vector population has effected cessation of transmission, as was shown in Sardinia in the years immediately following the Second World War.

(iii) *Schistosomiasis* (Bilharzia). This disease affects 200 million humans annually and has primates, dogs, cattle, sheep, and rodents as reservoirs. The disease is caused by trematode worms which reside in the blood vessels and affect bowel, bladder, and intestines. It is often a disease of rural development where artificial lakes and canals are constructed and become the living areas of the water snail vectors. Spread from the African continent to the New World followed the slave traffic, and it is now established in broad areas of the Caribbean and South American continent. The disease would be less of a problem in Egypt if the population was less dependent upon the same irrigation canals for washing, ablution, and recreation.

(iv) *Trypanosomiasis*. Trypanosomiasis affects 10 million humans annually and has dogs, cattle, pigs, and many other mammals as reservoirs. It is found in South America as Chagas' disease—with the heart as target organ, and in Africa as 'sleeping sickness'—with the central nervous system as target organ. Attempts to eradicate the disease by attacking its vector, the tsetse fly, have been similarly hampered by the emergence of insecticide-resistant vectors.

(v) *Leishmaniasis*. Leishmaniasis affects around 9 million humans annually and has dogs and rodents as reservoirs. The disease may either attack the oral and nasal tissues in man through invasion of the skin by infected reticuloendothelial cells (bush or forest yaws) or attack the viscera by the same means (kala-azar. Its epidemiology in Kenya is poorly understood since the reservoir there has not been identified.

(b) *Other zoonoses*

As pointed out earlier, the importance attached to a particular zoonotic disease may reflect either its incidence in human populations or its virulence and severity in man. The major zoonoses, considered above, obviously fall into the former category. In addition a number of viral, bacterial, and rickettsial diseases are also of outstanding importance to man due to their virulence rather than their prevalence. A number of these are selected for special consideration in later chapters, partly because of their present-day or historic importance, and partly to illustrate important principles of epidemiology and control.

FURTHER READING

Andrewes, C. H., and Walton, J. R. (1976) *Viral and Bacterial Zoonoses*. Baillière Tindall, London (161 pp.).

FAO, Rome (1967) *Third Report FAO/WHO Expert Committee on Zoonoses*. WHO Technical Report Series 378, or FAO Agriculture Series 74 (127 pp.).

Schwabe, C. W. (1969) *Veterinary Medicine and Human Health*, 2nd edn. Williams & Wilkins, Baltimore (713 pp.).
Simoons, F. J. (1961) *Eat Not This Flesh; Food Avoidances in the Old World.* University of Wisconsin Press, Madison (241 pp.).
Steele, J. H. (1979) *The Epidemiology and Control of the Zoonoses.* In Hobson, W. (ed.), *The Theory and Practice of Public Health*, 5th ed., Oxford University Press (785 pp.).
Steele, J. H. (ed.) (1979, 1980) *CRC Handbook Series in Zoonoses.* Section A: Bacterial, Rickettsial and Mycotic Diseases, vols I and II. CRC Press, Inc., Florida (643 pp.).

Chapter 14

Foodborne infections

A. HUMAN FOOD-POISONING

Many natural infections of animals are common to man (Chapter 13) and some of these may be contracted by the ingestion of contaminated food. This may occur when the food comes from an infected animal or bird and includes meat, milk, or eggs. Foods may also be contaminated, such as vegetables grown in contaminated soil or irrigated by sewage-polluted water (e.g. watercress beds), or during processing for preservation or preparation for eating. Pathogens may be introduced onto food by flies but more commonly by human carriers, during food handling. Food-poisoning is not limited to microbial pathogens but may be caused by poisons derived from plant or animal sources, or by chemicals added inadvertently or as preservatives at too high concentrations.

1. Food-poisoning defined

A special group of diseases contracted by the ingestion of food are designated 'food-poisoning'. In Great Britain the term 'bacterial food-poisoning' is legally restricted to poisoning resulting from the ingestion of food containing certain bacteria (or their toxins) which are not included among other specific infectious diseases notifiable to the Minister of Health. Thus, although the dysentery bacillus may be ingested in food and cause gastroenteritis, it is not officially classed as a food-poisoning organism since it causes a disease notifiable in its own right.

The types of foods involved in food-poisoning are those favouring growth of the causative organisms, so that large numbers of bacteria or their products are present in the food at the time it is ingested. Outbreaks of food-poisoning frequently involve a large proportion of the people who have eaten a particular food or a specific meal. For these reasons they are usually explosive in character. They occur most frequently after eating foods prepared the previous day; the organisms multiply in food kept at a temperature conducive to growth.

The source of an outbreak is not always easy to discover. The distribution of cases among people eating the same food, particularly a communal meal, may not be even. Not everyone eats the same amount or article of food, and often the food-poisoning organism is unevenly distributed throughout the food, causing some people to ingest a larger dose than others. The clinical response is dose-related.

Investigations to detect the offending food involve lengthy questioning of patients to determine the types and sources of foods eaten over the previous 24 hours or so, and sometimes questionnaires are used to determine precisely which dishes of a meal were not eaten by each individual partaking of the same communal meal.

Two groups of microbiological agents are recognized: pathogenic bacteria which can infect the alimentary canal and produce symptoms of gastroenteritis and bacteria (or their toxins) which do not infect the gut but produce various toxic symptoms when ingested with food. Where living pathogenic micro-organisms can be demonstrated in large numbers in the patient's gut diagnosis is relatively easy. However where numbers are small it is not so easy to distinguish between an infective and a toxic form of food-poisoning.

The symptoms produced by different bacterial toxins may be similar and the final diagnosis thus depends upon laboratory tests. The time from the consumption of the food to the onset of symptoms also varies according to the causal agent. With some forms of chemical poisoning symptoms may arise over a period of 10 minutes to 2 hours, with bacterial toxins around 6 hours, with living bacteria around 12–72 hours, and 2–3 days where poisonous toadstools are the cause.

2. Infective food-poisoning

In this form of food-poisoning the infective agent is able to establish itself in the gut, multiply rapidly, and produce disease over a number of days—often about a week or longer. The commonest causes of this form of poisoning are species of *Salmonella* and, because of their importance and extensive epidemiology, these will be considered at length later in this chapter.

A more recently described cause of infective food-poisoning is *Vibrio parahaemolyticus* (synonyms: *Pasteurella parahaemolytica* and *Beneckea parahaemolytica*). This is frequently associated with the consumption of raw fish and shellfish, particularly in Japan. It has been isolated from fish, shellfish, and coastal waters in Asia, Europe, the United States, Australia, and South America but outbreaks originating outside of Japan have been infrequent. Cases among air travellers flying on routes from Japan have been reported, the incriminated foods having been prepared in Japan.

The organism is present in greatest numbers in coastal waters during the summer months. This may be the chief reason why the disease in man occurs mainly during the warmer months of the year. The incubation period is variable but is most frequently 14–20 hours. Symptoms closely resemble those of salmonellosis and dysentery and for this reason, in the absence of laboratory investigations, an infection may be erroneously diagnosed. Symptoms usually commence with violent epigastric pain, accompanied by nausea, vomiting, and diarrhoea. In severe cases mucus and blood may be present in the stools. Often mild headaches and fever are experienced. In fatal cases there may be mild oedema and hyperaemia of the small intestinal wall but, as in cholera (Chapter

16), the mucosa is intact and no histological changes are evident. These observations suggest that the principal symptoms are caused by an enterotoxin produced in the intestine by the vibrios.

The organism is halophilic but can be isolated from stools of patients without difficulty on media containing 3 per cent NaCl. It is much less easy to isolate from foods and environmental materials because many vibrios resembling *V. parahaemolyticus* are widely distributed.

Campylobacter enteritis in man has been diagnosed with increasing frequency over recent years (in 1979 over 10,000 strains were isolated from diarrhoeic patients in the UK). The causal organism is vibrio-like but is classified in the genus *Campylobacter* on the basis of its microaerophilic growth requirement. A spectrum of species have been isolated, ranging from *C. jejuni* through intermediate types to *C. coli*. The onset of diarrhoea is preceded by fever, malaise, headache, and sometimes aching of limbs and colicky abdominal pains, with nausea but rarely vomiting. The organisms are regularly found in the small intestine. There is good evidence that humans contract infection from food, particularly chickens, and a few milk-borne outbreaks have been recorded. Similar organisms have also been isolated from dog faeces.

3. Bacterial cells or their products as causes of food-poisoning

In this form of food-poisoning no infection of the gut occurs but bacterial cells or exotoxins of bacterial origin produce symptoms of food-poisoning when ingested in foods. Since these toxins are absorbed by the gut wall they are called enterotoxins. They are preformed in the food by growth of the causative organism and although the vegetative cells may be killed by heating the food prior to eating, some toxins, being moderately resistant to temperatures below 100°C, remain unchanged and cause symptoms of food-poisoning when ingested. The most important bacterial toxic food-poisonings are caused by the toxins of specific strains of *Staphylococcus aureus* and *Clostridium botulinum*, the latter frequently causing a form of food-poisoning (with a high mortality rate) known as botulism (see later). Strains of *Clostridium perfringens* also cause toxic food-poisoning and certain non-pathogenic bacteria, such as *Bacillus cereus* and *Proteus* spp., not generally considered to be toxigenic, produce acute gastroenteritis when consumed in sufficiently large numbers.

(a) *Staphylococcal food-poisoning*

This form of toxic food-poisoning is the consequence of specific strains of staphylococci which produce enterotoxins as they multiply in food. Five antigenically distinct enterotoxins have been identified (A–E) and more than one may be produced by a single strain. The enterotoxins are polypeptides which act as emetic toxins upon ingestion. The larger proportion of food-poisoning strains produce the

A-toxin (77.5 per cent has been reported); the other are rarely incriminated. Enterotoxin-producing strains are usually coagulase-positive, of human origin, and the majority belong to a limited number of phage types (group III and a few types, e.g. 42D, in group IV). Although milk is incriminated in a proportion of outbreaks, staphylococci which cause mastitis in the cow do not produce enterotoxin type A and therefore cattle strains are not usually the cause of food-poisoning in man. The enterotoxins are heat-resistant (e.g. they can stand boiling for 30 minutes) and are stable for months in filtrates held at 4°C. They often survive in foods which have been heated.

The foods commonly incriminated are shown in Table 14.1. The staphylococci multiply in foods held in a temperate climate or a warm kitchen. They grow exponentially between 7 and 45.5°C and the toxin is produced during or just after multiplication. It is generally thought that 10^5–10^6 organism per gram of food must have grown up in the food to produce a clinical dose of enterotoxin. The amount of enterotoxin produced is proportional to the rate of multiplication and 1 µg of pure type A can cause a clinical response in man. Symptoms may arise 2–4 hours after ingesting the food but more usually from 6 hours onwards. Symptoms of staphylococcal food-poisoning include violent vomiting, diarrhoea, and prostration but there is no fever and recovery is rapid, often within 24 hours. This type of poisoning is only rarely fatal. Most incriminated foods have been contaminated from human sources at some stage after cooking. Cooked meats are

Table 14.1. Number of outbreaks of food poisoning where vehicle of infection was recorded, 1980 (1979 figures)

Food	Salmonella spp.		Cl. perfringens		Staph. aureus		B. cereus	
Chicken	20	(29)	3	(3)	–	(2)	–	(–)
Turkey	30	(26)	3	(8)	–	(–)	–	(–)
Beef	4	(6)	9	(11)	2	(8)	–	(–)
Pork/ham	7	(6)	10	(7)	2	(2)	–	(–)
Lamb	–	(–)	2	(–)	–	(2)	–	(–)
Other meat	9	(4)	14	(18)	2	(–)	–	(–)
Gravy	–	(–)	1	(3)	–	(–)	–	(–)
Rice	–	(–)	–	(–)	1*	(–)	13	(5)
Milk	9	(2)	–	(–)	1	(–)	–	(–)
Other foods	5	(3)	1	(1)	2	(2)	–	(–)
All outbreaks where vehicle of infections was recorded	84	(76)	43	(51)	10	(16)	13	(5)
Total outbreaks	439	(455)	55	(56)	11	(17)	13	(5)

*In this incident the prawns were also positive. Sardines (1) and tinned crabs (1).

Data from *PHLS Communicable Disease Report*, CDR 80/28, 1981. Reproduced by permission of the Communicable Disease Surveillance Centre, London.

often handled during the processes of trimming, pressing, the addition of gelatine, etc., and are then allowed to cool at room temperature. Where a large bulk of food is involved it often remains warm for a long time, allowing multiplication of the micro-organisms to occur.

The toxins may be assayed and antigenically identified by feeding to susceptible animals or volunteers. Kittens have been widely used but are unreliable because other substances can cause emesis. Monkeys are more specific, and very sensitive, the enterotoxins of staphylococci alone causing emesis. Human volunteers have proved most useful but are not always available! It is now possible to assay amounts of enterotoxin in excess of 1 μg by a double-diffusion agar gel precipitation test using standard antisera against the five identified toxins in parallel with suitable positive controls.

(b) *Botulism*

The disease botulism is an intoxication rather than an infection. *Clostridium botulinum*, an anaerobe found in soil, directly or indirectly contaminates foods. Botulism is a form of toxic food-poisoning with a high mortality rate but fortunately, whilst cases are dramatic, they are rare in man. This is particularly so in Great Britain but four cases occurred in 1978 following the eating of tinned salmon.

In the past, the type prevalent in Europe was type B often associated with contaminated pork products. Type A more common in the USA, was associated with canned vegetables and fruit. The heat-resistant spores survive in home-canned foods which may be inadequately heated in the sterilization process. Under the anaerobic conditions prevailing in the cans the spores germinate and toxin is produced, the presence of the growth being inapparent since often there is an absence of putrefactive changes. Since botulism is a toxic food-poisoning, outbreaks are sporadic with no secondary cases.

In man botulism has never reached epidemic proportions but with modern trends to produce foods in very large quantities the potential risk is always present. A particular problem has arisen with frozen birds for the table. *Cl. botulinum*, if present, could multiply in the anaerobic conditions prevailing within the sealed plastic container while the bird is being thawed. To offset this potential risk perforated plastic bags, allowing access of air, avoid anaerobic conditions arising at least at the surface.

Diagnosis of botulism is based on the demonstration of the toxin in food or gut contents of affected hosts or in serum before death. Isolation of the organism in the absence of the toxin is of no significance. A saline extract of the toxin is injected intraperitoneally into suitable animals, e.g. the guinea pig or chick. Animals protected with type-specific antitoxin survive; animals not protected, or given antitoxin of a different type, develop paralytic symptoms and die.

Many animal species are naturally susceptible to botulinum toxins, particularly types C and D (Table 14.2). Within an animal species there is usually a prevalence of a particular type which, in part, corresponds to the distribution of the type in local soils, and to dietary habits. Differences between types are based on the antigenic specificities of the toxins they produce but all toxins have the same pharmacological activity. Thus irrespective of the antigenic type the symptoms produced in different animals are essentially the same. That certain animal species are susceptible to particular types and not to others may be because of different haemagglutinins produced by the various types.

Outbreaks have occurred among chickens, pheasants, etc., feeding on fly larvae which, in turn, were infected by feeding on putrefying carcasses contaminated with *Cl. botulinum* type C. Evidence has been presented that the titre of toxin, which is very low in the larvae, increases in the birds—indicating that growth of the organisms must occur, probably in the caecum. The birds exhibit flaccid paralysis of the neck muscles, a condition accounting for the descriptive name limberneck. A devastating disease in migratory birds has often been reported in the USA. Ducks and other birds, during the hot months, drink from shallow pools in which decaying larvae provide a favourable medium for *Cl. botulinum* type C to grow and generate toxin. Thousands of birds die each year as a consequence of drinking and feeding at these pools—the disease is called western duck sickness.

Table 14.2. Some epidemiological properties of the various types of *Clostridium botulinum*

Type	Host species affected or susceptible	Common source	Geographical location of high incidence
A	Man, chicken	Home-canned vegetables and fruit	Western USA
B	Man	Meat products, especially pork	Europe, eastern U.S.A.
C	Aquatic wild birds (western duck sickness, limberneck, alkaline disease)	Rotting fly larvae in ponds	Western U.S.A., Canada, South America, Australia
	Cattle (midland cattle disease), horses (forage poisoning), mink	Forage contaminated by decomposing rodent	Australia, South Africa, North America
D	Cattle (lamziekte)	Carrion (osteophagia)	South Africa, Australia
E	Man	Uncooked fish and marine foods	Japan, Northern U.S.A., Labrador, Alaska, Great Lakes, Sweden, Denmark, USSR, Iran
F	Man	Cooked meats, dried venison	Denmark, U.S.A.

Animal foods, such as meat for mink or fodder for horses and cattle, may give rise to botulism should the foods be contaminated with botulinum toxin. Forage disease of horses arises following the eating of fodder contaminated by a decomposing carcass, e.g. of a rodent. Lamziekte in cattle in South Africa is a form of botulism with a fascinating epidemiology. This disease occurs in areas known to be deficient in phosphorus. To satisfy their craving (osteophagia) animals chew bones from animals that have died on the open veld. Should these have been contaminated with *Cl. botulinum* type D, the toxin is generated in the decomposing carcass, the bone marrow becomes contaminated with the toxin, and animals chewing these bones develop lamziekte (lame sickness). Prevention is by vaccination or providing salt licks for the cattle, which satisfy their craving for minerals and hence the need to depend on bones.

(c) *Clostridium perfringens*

Particular strains of *Cl. perfringens* type A cause a relatively mild form of food-poisoning following the ingestion of cooked foods, usually meat containing large numbers of the vegetative cells of the organism (Table 14.1). Symptoms of abdominal pain and diarrhoea arise from 6 to 24 hours after ingesting the food and recovery is usually complete 24 hours following the onset of symptoms. Complications and fatalities occur, but rarely. The illness is caused by an enterotoxin produced by the organism in the gut. The spores of this strain of *Cl. perfringens* type A survive modest temperatures of cooking, being able to withstand boiling for 1–4 hours, and are able to multiply rapidly in the resultant meat preparation in relatively pure culture since other vegetative contaminants are likely to have been killed. The removal of oxygen in the process of cooking meat, the presence of reducing substances (such as glutathione) and the absence of competition by other organisms, are all conducive to growth of the clostridia—especially if the food is kept warm.

Unlike most *Cl. perfringens* type A the food-poisoning strains are usually non-haemolytic, produce only small amounts of α-toxin and no θ-toxin (see Table 5.6). The enterotoxin is a heat-sensitive, antigenically distinct protein with a molecular weight of 36,000. It is produced both *in vitro* and *in vivo*. The toxin induces excess fluid movement in the intestinal lumen, resulting in diarrhoea. Large numbers of the organisms are present in human gut faeces of sick patients (usually in excess of 10^6 per gram: in a normal person *Cl. perfringens* type A would be present in numbers around 10^3–10^4 per gram). In establishing a diagnosis, the numbers of these organisms in the patient's faeces must therefore be determined.

(d) *Bacillus cereus*

Two distinct clinical forms of food-poisoning are caused by this organism. The classical form is similar to that caused by *Cl. perfringens*. Symptoms, which occur

10–13 hours after eating the food, include acute colitis or enterocolitis, abdominal pain and profuse diarrhoea; nausea is only moderate and vomiting rare. The foods involved differ between countries according to local diets. The commonest include meat and meat products, puddings, vanilla sauce, cream pastry, vegetables, etc., often prepared and stored in a moist state under conditions conducive to both contamination and growth of *B. cereus*. The other form of food-poisoning occurs between 1 and 5 hours after eating the contaminated food and closely resembles staphylococcal food-poisoning. Symptoms include acute gastritis or gastro-enteritis in which nausea and acute vomiting occur. This form frequently follows the eating of boiled or fried rice, in which the spores survive and multiply as the prepared food is kept for several days at room temperature before being served, e.g. in Chinese restaurants.

The relatively rapid onset of symptoms, with an absence of fever, suggest that *B. cereus* food-poisoning is an intoxication. An enterotoxin has been shown to be produced in food consequent upon growth of the bacilli. Symptoms of food-poisoning only arise where the numbers of *B. cereus* in the food exceed 10^7 per gram or millilitre.

(e) *Non-specific bacterial causes of food-poisoning*

Occasionally outbreaks of food-poisoning have been reported in which the foods were heavily contaminated with species of *Proteus, Escherichia*, or *Streptococcus*. It is generally thought that large numbers of these organisms have an irritating effect on the gastrointestinal mucosa and set up symptoms of toxic food-poisoning. Gross contamination of food with these organisms, generally considered to be non-pathogenic, followed by storage at temperatures conducive to bacterial propagation, presents a potential hazard to the consumer.

(f) *Poisonous fungi*

Some fungi cause poisoning if eaten as a food or as contaminants of human or animal foods. Only few mushrooms are poisonous but unfortunately there is no simple test by which they can be distinguished, since they contain different toxins. Some mushrooms cause gastrointestinal upsets, others cause disorientation, fever and changed pulse rate, or attack the central nervous system or damage the liver and kidneys. Most patients recover but occasionally death occurs.

The species most commonly causing human deaths in Europe is the 'death cap', *Amanita phalloides*, and in America, *A. virosa*. In *A. phalloides* two classes of poisonous cyclopeptides are found, the amatoxins and the phallotoxins both cause liver damage i.e. are hepatotoxins. The gastrointestinal symptoms have been attributed to the amatoxins but there is doubt whether the phallotoxins cause natural poisoning since they are not fatal when given orally. The high proportion of fatalities resulting from ingestion of *A. phalloides* is due, at least in part, to the first symptoms, vomiting and diarrhoea, not appearing till 10–24 hours after eating the toadstool, by which time damage to the liver has already commenced. Transient improvement follows in those who survive the gastrointestinal disturbances, but eventually progressive liver damage leads to death.

The consumption of certain poisonous fungi as part of rituals performed by certain tribes in Central America leads to hallucinogenic effects. These 'sacred mushrooms', including species of *Conocybe, Panaeolus, Psilocybe,* and *Stropharia*, induce chiefly visual hallucinations; two species of puffball, *Lycoperdon mixtercorum* and *L. marginatum*, produce auditory hallucinations. The active ingredients are psilocybin and psilocin.

Fungal contaminants of food, leading to hallucinations, occasionally occur when bread made from rye, infected with the ascomycete *Claviceps purpurea*, is eaten. More often ingestion by man or animals of cereals or grass heavily contaminated by sclerotia (or ergots) of the fungus leads to constriction of the vascular supply to the extremities of the body. This results in lameness and the development of gangrenous lesions of the feet, especially of the hind limbs in animals, and less frequently the tips of the ears and tail. Ergot contains ergine, ergonovine, and ergotamine—potent alkaloids of pharmacological importance. Ergotism is one example of a group of diseases caused by toxic metabolites (mycotoxins) of certain fungi. This group of diseases are generally called mycotoxicoses.

Aflatoxicosis, caused by *Aspergillus flavus* in many animal species, is another example. Where foods—including maize, cottonseed, palm kernels, and groundnuts—are stored after harvesting at temperatures around 30°C and at a relative humidity of 80–85 per cent, the fungus grows and aflatoxins are produced. An outbreak in turkeys in the UK occurred in 1960 when thousands died from a mysterious disease initially called 'Turkey X disease' before the fungal cause was discovered. The acutely ill birds showed inappetance, lethargy, a rapidly developing weakness, convulsions, and death within 5–7 days of the onset of symptoms. Necrosis and haemorrhage of livers and kidneys were the main post-mortem findings. Birds and fish (trout) are more often affected than domestic animals.

The high incidence of primary liver cancer among certain African tribes, whose diet includes groundnuts, led to the deduction that aflatoxins may be carcinogenic for man. More recently an additional aetiological agent, hepatitis-B virus, has been incriminated. It is postulated that the aflatoxins, rather than acting as primary carcinogens, suppress the host's cell-mediated immune responses. This results in the host being less able to rid itself of infection by the hepatitis-B virus; the virus, in consequence, persists endemic in a large proportion of the population. The resulting chronic liver infection—with associated cirrhosis—leads in the long term to a high incidence of liver carcinoma.

Facial eczema in sheep and cattle, found mainly in New Zealand and Australia, is caused by sporidesmin, a mycotoxin present in the spores of *Pithomyces chartarum* which may grow on pasture grasses. The toxin causes liver damage resulting in jaundice, oedematous inflammation, and photosensitization of areas of skin not protected from sunlight by wool or hair. The photosensitization is due to a build-up of phylloerythrin (a breakdown product of chlorophyll) in the peripheral blood due to failure of the liver to excrete it.

Species of *Fusarium* growing on various grains also cause mycotoxicoses in animals and man. The mycotoxin can cause vomiting in pigs and an oestrogenic syndrome involving shedding of the vaginal and cervical mucosa. Dairy cattle show inappetance, loss of milk yield, scouring, and staggering gait. Horses develop nervous symptoms, incoordination of limbs, and occasionally lesions on the lips and in the mouth. Man may be affected, producing nervous symptoms. Similar illnesses are produced in horses, cattle, sheep, and pigs by *Stachybotrys atra* which may grow on damp hay and straw used for fodder when the animals are housed in autumn and winter. Another disease of sheep involving liver damage is consequent upon the ingestion of lupin fodder contaminated with *Phomopsis leptostromiformis*. Moulding of food and feedstuffs thus raises many questions concerning hazards to animal and human health.

B. SALMONELLOSIS IN MAN AND ANIMALS

Almost 2000 internationally recognized serotypes of *Salmonella* have been isolated from man and animals, and of these, a large number are known to cause disease in warm-blooded animals. With so many serotypes it is not surprising that the range of animal species, and the forms the disease takes in different animal species, are considerable.

In man, two main types of disease are recognized. The *enteric fevers*, caused by *Salm. typhi, Salm. paratyphi* A, *Salm. paratyphi* B, and *Salm. paratyphi* C, initially produce gut infection which is followed by invasion of the gut wall and subsequently a septicaemia occurs; in fatal cases, the numbers of organisms in blood usually increase prior to death. During the septicaemic phase the patient presents a high fever. The mortality rate in typhoid fever is generally higher than in paratyphoid fever, as is also the severity of the disease.

Infections by other *Salmonella* serotypes produce an *infective type of food-poisoning*, chiefly a gastroenteritis. Symptoms of headache, nausea, vomiting, diarrhoea, and abdominal pain, sometimes accompanied by fever, arise within 8–24 hours after ingesting contaminated food but may occur as late as 72 hours. In most individuals, symptoms slowly abate and the patient is better within a week but, in a few cases, invasion of the blood stream occurs and this is accompanied by more severe symptoms of cramp, extreme thirst, coma, and sometimes death. Some patients, after recovery from enteric fever or salmonella food-poisoning, continue to excrete the pathogen for months in the absence of symptoms. These carriers are a continuing source of infection to others in the community, and should not be employed in the preparation or handling of food.

In animals, salmonellae cause a wider range of disease than that experienced in man. Many animals exhibit symptoms similar to those of the human enteric fevers which are described by the clinical term paratyphoid. The causal organisms are not, however, those responsible for enteric fevers in man. Many salmonellae which are confined to gut infections in man are able to invade the gut wall of animals and produce septicaemia in a wide range of animals, particularly in calves and cattle. These include serotypes such as *Salm. typhimurium, Salm. enteritidis, Salm. newport,* and *Salm. bredeney,* but also host-adapted strains. *Salm. dublin,* for example, causes acute infection in calves which spreads rapidly between animals housed in groups. Infection during the first 2 weeks of life causes a sudden loss of appetite, weakness, diarrhoea, and frequently death. Septicaemia is often followed by symptoms of dysentery (i.e. inflammation of the large intestine) and sometimes pneumonia and liver abscesses (often termed yellow liver disease of calves). In adult cattle acute infection, which often arises in late autumn when animals are brought in from pasture or under the strain of pregnancy and the associated heavy milking, is accompanied by fever (40–41°C), inappetance, sudden fall in milk yield and diarrhoea. Some serotypes cause abortion in pregnant animals; *Salm. abortus-equi* in mares, *Salm. abortus-ovis* in ewes but other strains, such as *Salm. dublin,* which usually cause paratyphoid, can also cause pregnant cows to abort.

Pigs and poultry constitute important reservoirs of salmonellae and a wide variety of serotypes are regularly isolated from them. Until recently the most common cause of salmonellosis in the pig was *Salm. cholerae-suis,* causing pig paratyphoid which often proved fatal or left unthrifty survivors. This important porcine strain is now experienced less frequently but other serotypes, particularly *Salm. typhimurium,* have produced high rates of morbidity and mortality in herds of specific pathogen-free (SPF) pigs. Many pigs are symptomless carriers and excrete a wide range of serotypes which may be pathogenic for man. The same situation prevails in poultry but specific infections are also encountered. Diseases in poultry include infections with *Salm. pullorum* (pullorum disease; bacillary white diarrhoea—BWD) and *Salm. gallinarum* (fowl typhoid). These two organisms are not found so frequently as in the past, due to the selection of disease-free chicks to stock large poultry-rearing stations but infections continue to persist on conventional farms.

Apart from infections in farm animals and poultry, most other animal species may also be infected by salmonellae. These include domestic pets, such as dogs and cats, but also more exotic animals such as tortoises and other reptiles. Salmonellae have been isolated frequently from wild rodents and birds, especially pigeons, starlings, and seagulls. The reservoirs of infection are therefore varied and far-ranging, which indicates the immense problem of control. This is made more difficult by the frequent occurrence of infected animals which are clinically normal, and not suspected as being infected. The environment is being continuously contaminated by the infected exreta of these animals.

1. Host and non-host-adapted *Salmonella* serotypes

Measures to control the spread of salmonellae must take into account the degree of host-specificity of certain serotypes. Some are very strongly host-adapted, i.e. they usually cause disease in a limited range of animal species. In man, they include the causal agents of enteric fevers which rarely cause natural disease in animals. A small number of limited outbreaks due to *Salm. paratyphi* B have been reported in cattle drinking water contaminated with effluent from inadequately processed human sewage. In animals, *Salm. dublin* is most commonly found in cattle but may occur in sheep and man. *Salm. abortus-equi* and *Salm. abortus-ovis* are host-specific for the horse and sheep respectively, and *Salm. pullorum* and *Salm. gallinarum* are mainly found in poultry, although a small number of human cases have been reported. In host-adapted infections control is mainly aimed at interfering with the animal-to-animal cycle of cross-infection and often it is not necessary to consider outside sources of infection.

Many non-host-adapted strains infect a wide range of animal species. For instance, *Salm. typhimurium* can infect man, cattle, pigs, poultry, rodents, etc. Consequently non-host-adapted salmonellae are more difficult to control by virtue of the multiplicity of potential sources of infection.

2. Routes of infection

The mode of transmission of infection is most commonly by the faecal/oral route. Clinically sick animals and those subclinically infected excrete salmonellae in their faeces and, to a lesser extent and more variable, in other body fluids (e.g. urine) and secretions (e.g. milk). Apart from *Salm. typhi*, a waterborne infection arising from sewage-contaminated water supplies (Chapter 9), the majority of salmonella infections are foodborne. This is true of human and most animal infections. Cross-infection also occurs between animals. Calves and cattle, for instance, habitually lick each other's coats and if these are contaminated with faeces from an infected animal oral infection occurs. Where foods are incriminated these may be contaminated at source, e.g. milk, eggs, or the foods may be contaminated with infected human or animal excreta.

Compounded animal feedingstuffs are commonly contaminated with salmonellae. The protein supplements of animal origin are the components usually responsible. Peruvian fish meal, for example , is often grossly contaminated with many salmonella serotypes. Fish caught at sea become progressively contaminated with human sewage in the bilge water of the fishing vessels and by bird droppings which contaminate the fish on the quayside during landing. Another source of potentially contaminated animal protein is feather meal and poultry litter, byproducts of the poultry industry, which are often recycled to animals by incorporating them in compounded foods.

Animal foods may be contaminated, particularly on the farm, by contaminated droppings from infected rodents and birds. In the Netherlands, flyborne cross-infection has been shown to occur, doubtless from exposed faecal pats to susceptible animals.

In man the faecal/oral route is the usual route of infection. Foods form the most common intermediate vehicle of cross-infection. They may be contaminated at source, that is, derived from a salmonella-infected animal. Outbreaks have occurred due to the drinking of raw, bovine milk from infected herds, usually with *Salm. dublin*. The eating of contaminated and inadequately cooked eggs have also caused salmonella food-poisoning. Meat, whilst the most common vehicle of infection for man, is rarely contaminated in the muscle of the live animal. Where meat is implicated, therefore, it is usually the consequence of carcass contamination with faeces from infected but clinically healthy animals or from human carriers who handle the food in preparation or cooking. The preparation of thawed poultry for cooking is particularly hazardous since the carcass is often contaminated with salmonellae.

3. The incidence of salmonellosis in animals

Information on the incidence of salmonellosis in animals in England and Wales is based on returns by local Veterinary Investigation Centres to the Central Veterinary Laboratory. Data have been collected over many years and the incidence of infection from 1958 to 1979 in the principal farm animals and poultry is shown in Figure 14.1. Fluctuations in incidence within the same animal species is closely linked with changes in husbandry methods. For example, the increased incidence of calf infections in the UK from 1965 onwards was directly related to a change in husbandry practice which involved the rearing of large numbers of calves under intensive systems. This resulted in concentrations of calves from many sources with inevitable cross-infection. In these calves infection occurs rapidly and extensively since they are already considerably stressed by lengthy transport and thereby rendered unusually susceptible. A rapid rise in clinical disease follows. The incidence of clinical salmonellosis in poultry, in contrast, has fallen to insignificant levels.

The incidences of different serotypes in animals varies from time to time and reflects the numbers and variety present in animal feeds and the environment. Analysing the return for the period 1968–73, Sojka *et al.* (1975) noted that four serotypes (*Salm. dublin, Salm. typhimurium, Salm. cholerae-suis*, and *Salm. abortus-ovis*) caused 91.6 per cent of all outbreaks reported in animals. The commonest was *Salm. dublin* (70.1 per cent), the most frequent cause of salmonellosis in cattle (78.9 per cent). Despite the large proportion of outbreaks caused by the four serotypes, 134 other serotypes were isolated over the same period. This indicates the large number of 'exotic' serotypes brought into the country in animal proteins food supplements and annually this number has increased over the last 15 years.

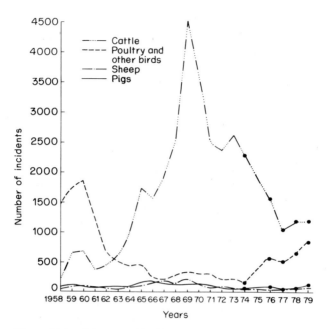

Figure 14.1. The annual incidence of salmonella infections in England and Wales, 1958–79 (all serotypes). Data from Sojka *et al.* (1975) and Salmonellosis Reports (Zoonoses Order, 1975), Annual summaries for 1976–79. MAFF, Weybridge, England. Reproduced by permission of the Veterinary Record

The pattern of salmonellosis in the pig is of particular interest. By far the commonest serotype associated with porcine pathology in the past was *Salm. cholerae-suis*. This, in fact, was the first species to be described in association with hog cholera (Salmon, 1885), and the genus is named after him. In recent years a steady transition has occurred; the incidence of *Salm. cholerae-suis* has decreased and given place particularly to *Salm. typhimurium* and other serotypes. Despite the historical interest, the incidence of clinical illness is very low in the pig, only about 100 incidents a year have been reported over the last 10 years in the UK.

However, numerous workers have reported that up to 12 per cent of pigs on farms in the UK may excrete salmonella in their faeces, indicating that whilst clincial illness is rare, many are infected by eating contaminated foods and become persistent carriers. They constitute a hidden reservoir of infection for man (see below) and are a public health risk. An important conclusion must be drawn from these observations. The data processed at the Central Veterinary Laboratory are a summary of clinical illness occurring in the national herds but does not give a true reflection of the incidence of infection which must include vast numbers of subclinical cases, over and above the clinical ones.

These comments also apply to the situation in poultry—many birds are sub-clinically infected by eating contaminated foods and their droppings are affected with salmonellae. Contamination of the carcass with gut contents at slaughter creates an important public health risk.

4. Incidence of salmonellosis in man

The incidence of salmonellosis is constantly changing as sources of infection in food and the environment change from year to year. The following four statements set out the position at the present time:

(1) over 80 per cent of all identified causes of food-poisoning in the United Kingdom are due to salmonellae;
(2) of all foods, meat is most frequently implicated as a source of infection;
(3) poultry meat and pork are most commonly implicated; and
(4) there has been a steady increase in the number of cases of salmonellosis caused by 'exotic' serotypes of *Salmonella*.

The number of incidents in which the food responsible was established is relatively few, but analyses of identified sources of human infections over many years indicate significant trends. By far the larger proportion of outbreaks have originated from animal meats or meat products; other foods (e.g. vegetables and fruit) are of little importance. Whereas eggs and egg-containing foods (e.g. sweetmeats) constituted an important source of salmonellae in earlier years, these have gradually assumed less importance. This is chiefly due to legislation which, in 1964, required that all liquid egg, both home-produced and imported, should be pasteurized. This controlled one of the most important routes by which eggborne infection was reaching man. On the other hand, meat as a source of salmonellae has steadily grown in importance. During recent years there has been a marked increase in outbreaks associated with poultry meat (Table 14.3), with a reduction in outbreaks linked with beef, pork, and meat products. Beef, as a source of infection, has declined but poultry, and pork, continue to be the principal sources; there are two important reasons for this. With the derationing of animal feeds and the resumption of imports the broiler industry was reconstituted on modern intensive lines and by the early 1960s assumed its present form. By 1964 chicken meat was the cheapest animal protein available and consumption has risen steadily since. For instance, the estimated consumption of poultry meat increased from 15.8 lb per head of population in 1964 to 23.6 lb in 1971—an increase of 49 per cent; upwards of 300 million chickens were consumed in the UK in one year (1975). The second reason concerns the feeding practice. Most compounded foods fed to cattle are pelleted, a practice which virtually pasteurizes the food and renders it free of salmonellae; in contrast only about 50 per cent of feedingstuffs for pigs and poultry are pelleted. Also, since cattle graze on pasture, they do not require so much added protein to their diet and most compounded foods

Table 14.3. Type of meat implicated in salmonella food poisoning outbreak in England and Wales 1956–60, 61–66, 66–70, 71–75

Vehicle of infection	Outbreaks in 5 yearly periods (%)				
	1956–60	1961–65	1966–70	1971–75	1976–80
Beef	2.8	1.6	5.0	2.9	4.3
Pork and ham	16.0	16.4	10.0	17.5	9.2
Poultry	3.8	13.1	42.1	47.5	68.1
Other	77.4	68.9	42.9	32.1	18.4
Total	100.0	100.0	100.0	100.0	100.0

Data reproduced by permission of the Communicable Disease Surveillance Centre, London.

incorporate mainly vegetable proteins. Pigs and poultry are usually reared under intensive conditions of husbandry and are dependent on foods incorporating animal proteins, such as meat, bone meal, and fish meal, which are frequently contaminated with salmonellae.

The incidence of host-adapted serotypes of salmonella (e.g. *Salm. dublin, Salm. cholerae-suis, Salm. pullorum*) in man is relatively low; the majority of outbreaks are caused by strains with a wide host range and wide geographical distribution. The incidence of a particular serotype in man often closely parallels a contemporary or previous high incidence of the same serotype in animals. Hence, the high frequency in animals of 'exotic' serotypes, often imported in animal protein supplements, is followed by a similar trend in man.

There is now strong evidence that the same serotype follows the food chain from animal feedingstuffs, to pigs and poultry, to contamination of their carcasses at slaughter and thence to man. This is supported by recent work which confirms direct carcass contamination with gut organisms (*E. coli*) at slaughter, and the subsequent colonization by these same organisms of the human gut by individuals handling the meat (Figure 14.2). The evidence is very strong that it is the handling of the raw meat and contamination of the kitchen, rather than the occasional eating of inadequately cooked meat, which constitutes the greater risk.

5. Control of salmonellosis

Salmonellosis in animals causes serious economic loss to farmers and food producers and at the same time constitutes the chief source of human infections. Attempts to reduce the level in animals will therefore benefit the producer and at the same time improve the public health situation. The ultimate objective of the farmer must be to produce salmonella-free animals and, of the food-producer, to produce salmonella-free products for human consumption. Control measures to these ends will be considered under three headings: the live animal, the abattoir environment, and the consumer end of the food chain.

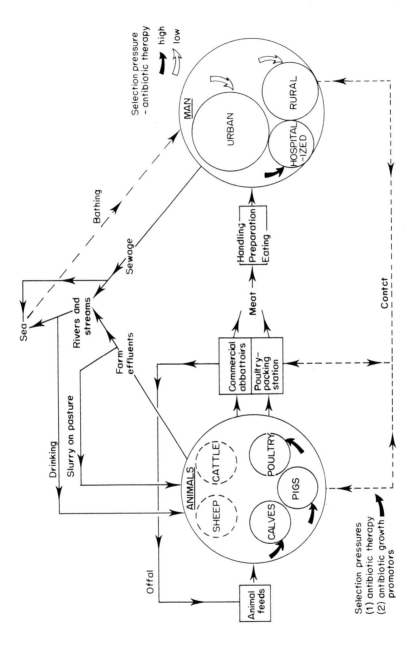

Figure 14.2. Flow of enteric organisms between animals and man. Reproduced by permission of the Veterinary Record

(a) *Control of salmonellosis in the live animal*

Where the salmonella serotype is adapted to a particular animal species cross-infection occurs between animals of the same species and only rarely are other sources incriminated. Attempts must therefore be made to break the cycle of infection. This has been done very successfully in the poultry industry by testing sera for agglutinins against pullorum disease. Reactors are removed from the flocks and the flock retested every 3–4 weeks until no further reactors are found. By this means the number of infected birds is reduced to a minimum. Eggs used for hatching are obtained from disease-free flocks but, as a further safeguard against the possibility of introducing infection into a hatchery, all eggs are fumigated with formaldehyde vapour after being placed in the incubator. This should kill salmonellae which may be present in faecal contamination of the egg-shell. If an infected egg escapes these preventive measures, infection can spread rapidly to the newly hatched chicks and the outbreak must be dealt with speedily and drastically. Survivors must be slaughtered and their carcasses incinerated together with all food and litter. The incubator must be fumigated with form-aldehyde vapour. Only by these stringent measures has pullorum disease in commercial flocks and on poultry-raising farms, been reduced to the present low incidence of recent years.

Most *Salm. dublin* infections in a closed herd of cattle follow recognized patterns of cross-infection. The usual source of infection is the clinically normal adult carrier cow. Newborn calves become infected by sucking an infected dam and by licking her faeces-contaminated coat. So long as the calf runs with the dam repeated infection will take place. Calves frequently show clinical illness and, if they survive, are often unthrifty and of poor quality. Many become infected without showing clinical illness; this may be due in part to receiving subclinical doses of infection, to passive immunity received in the colostrum of the dam or to immunity resulting from prior vaccination. Infected animals, if removed from their dams and raised in calf-rearing units, often throw off infection and, under good husbandry conditions, become free of infection in under 6 weeks. By raising calves in isolation it is possible to build herds with salmonella-free animals.

In recent years modifications in the calf industry have produced conditions highly conducive to the spread of salmonellae. Intensive methods now frequently employed necessitate the purchase of large groups of calves at frequent intervals. This has resulted in many calves from different breeding farms being transported over long distances, often hundreds of miles, through markets and dealers' pre-mises, and eventually to a calf-raising farm (Figure 14.3). The chance that one calf in a group collected from multiple sources may be infected is far greater than if all originated from one or two sources. Also the process involves at least one collecting centre, and several hundred calves per week may pass through this centre which may become progressively contaminated with the possibility of cross-infection occurring. To this must be added the stress of long journeys by

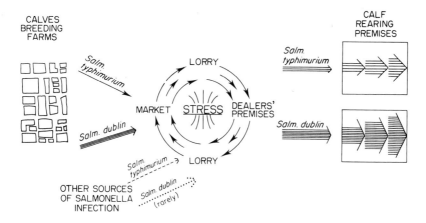

Figure 14.3. The calf 'merry-go-round' (after Stevens *et al.* (1967) *Vet. Rec.*, **80**, 154–61. Reproduced by permission of the Veterinary Record)

road or rail and the changes in feeding routine. It is little wonder that unprecedented outbreaks of salmonellosis· have occurred and attempts at control with antibiotics have only precipitated the additional hazard of drug resistance arising. Until recently most salmonellae in the UK were sensitive to the majority of broad-spectrum antibiotics. Since 1977 a clone of *Salm. typhimurium* arose which was resistant to streptomycin, sulphonamide, tetracycline, kanamycin, ampicillin, and, most serious of all, chloramphenicol. Aided by the calf industry, this strain spread to many parts of the country (Figure 14.4). The same antibiotic resistant *Salm. typhimurium* has also been isolated subsequently from humans, indicating that the build-up of infection in one situation can influence the incidence of infection in another.

Heat treatment of separate food components (e.g. animal protein supplements) before foods are compounded, or of the finished product, renders them free of viable salmonellae. Steam pelleting of compounded foods has been shown to eliminate salmonellae. It is anticipated that the adoption of the Protein Processing Order in the UK, requiring all imported protein supplements to be heat-treated, will limit the flow of 'exotic' serotypes to animals; this has been successfully carried out in Denmark.

The value of vaccination against specific salmonellae (e.g. *Salm. dublin*, *Salm. typhimurium*, and *Salm. cholerae-suis*) and chemotherapy is mainly in safeguarding the individual animal against clinical disease. Neither significantly affects the incidence of alimentary infection and antibiotic therapy often extends the duration of excretion.

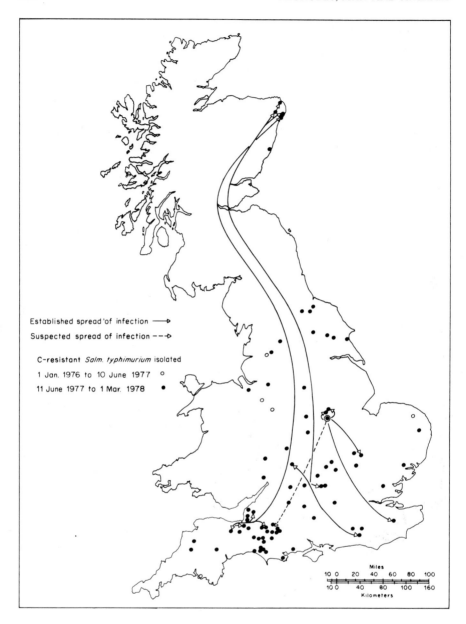

Figure 14.4. Premises on which chloramphenicol-resistant *Salmonella typhimurium* have been isolated from calves, and routes of spread in Great Britain, 1 January 1976 to 31 March 1978 (from *Animal Disease Report* (1978), vol. 2, no. 2. MAFF, Weybridge, England)

(b) *Control of salmonellosis at the abattoir*

Clinically sick animals are rarely slaughtered for human food, and meat inspectors reject fevered or septicaemic carcasses. The abattoir, however, has no means of detecting the healthy salmonella carrier animal sent for slaughter. Contamination of carcasses by faeces from healthy carriers constitutes the chief route of infection for man. Abattoir practice must, therefore, be directed at reducing the level of carcass contamination to a minimum. After slaughter good hygiene is essential to avoid carcass contamination, especially during evisceration.

(c) *Control of salmonellosis in man*

Whilst an integrated effort on the part of animal producers and meat processors is essential to reduce the flow of salmonella to man, this must be followed by similar efforts by food-handlers and kitchen staff. Since the main source of salmonella for man is meat contaminated at slaughter attempts must be made to break the chain of infection for man. Greater efforts are needed to educate food-handlers, kitchen staff, and housewives in the risks involved, and in improved food hygiene. A number of simple expedients meticulously carried out would go a long way to reduce the high incidence of food-poisoning in man. Raw meats and cooked foods or foods eaten raw should never be handled or sold by the same personnel, nor should the same knives, etc., be used for both purposes. Similarly, in the kitchen, the risks of handling raw and cooked foods without due precautions must be appreciated. The greatest risk appears to be in the contamination of hands, table surfaces, knives, etc., that results from the handling of raw meat. Poultry are frequently contaminated with salmonellae and the abdominal cavity, giblets, and defrost fluids are potentially dangerous. Spillage of fluids onto table surfaces, mopping up with dish cloths that have multiple uses and contamination of hands, lead to the frequent contamination of other foods and kitchen utensils. Refrigeration of prepared foods to avoid multiplication of sub-infective doses of salmonella which may be present, or to avoid contamination from other sources, is most important. Thorough thawing—particularly of large frozen joints such as chicken, but more particularly the larger turkey—before cooking ensures that at least a pasteurizing temperature penetrates to contaminated areas. The more meat is divided, e.g. minced meats, the more deeply will surface contamination be carried throughout the whole mass, and these products should be used only if fresh and adequately cooked. Foods prepared one day but used subsequently must be refrigerated to maintain bacteriological safety or be adequately reheated to destroy subsequent contamination or increase in numbers of micro-organisms.

The kitchen staff may themselves be a source of infection and contaminate food by handling. Strict personal hygiene, especially washing of hands before and after handling food, is essential. Medical checks on food-handlers, especially any who may be suffering with intestinal disturbance, is strongly advised. This is

particularly important in persons returning from holidays or visits abroad where
they may have picked up infection.

FURTHER READING

Beech, F. W. (1979) Microbiology of food and beverages. In Hawker, L. E., and Linton,
A. H. (eds), *Micro-organisms—Function, Form and Environment*, Arnold, London,
pp. 340–53.

CDC Report (1979) *Botulism in the USA 1899–1977*. US Department of Health, Education and Welfare, Public Health Service (41 pp.).

CDC Report (1977–78) *Annual summaries. Foodborne and Waterborne Diseases Surveillance*. US Department of Health, Education and Welfare, Public Health Service.

Hobbs, B. C., and Gilbert, R. J. (1978) *Food Poisoning and Food Hygiene*, 4th edn.
Arnold, London (366 pp.).

International Commission on Microbiological Specifications for Foods (ICMSF) (1978)
Microorganisms in Food, No. 1: Their significance and methods of enumeration, 2nd
edn. University of Toronto Press (434 pp.).

Lutwick, L. I. (1979) Relation between aflatoxin, hepatitis-B virus, and hepatocellular
carcinoma. *Lancet*, **1**, 755–7.

Sojka, W. J., Wray, C., Hudson, E. B., and Benson, J. A. (1975) Incidence of salmonella
infection in animals in England and Wales, 1968–73. *Vet. Rec.*, **96**, 280–4.

Sojka, W. J., and Field, H. I. (1970) Salmonellosis in England and Wales, 1958–1967.
Vet. Bull., **40**, 515–31.

Reports (1976–79) Salmonellosis Reports (Zoonoses Order, 1975) Annual summaries.
ADAS. MAFF, Weybridge, England.

Reports (1977–79) Animal Disease Reports, vols. 1–3. ADAS. MAFF, Weybridge,
England.

Report (1980) Animal Disease Report, new series, vol. 1. ADAS. MAFF, Weybridge,
England.

Chapter 15

The epidemiology of brucellosis

Brucellosis is a zoonosis which can be transmitted to man by a number of routes including foodborne spread. In man, the disease can assume many clinical forms; these include intermittent, recurrent, or undulant fever associated with weakness (asthenia), nocturnal sweats, muscular pains, enlargement of the spleen, nervous irritability, and chills or rigors. In the acute form the irregular pattern of the fever is characteristic. The fever lasts, as a rule, for 1 to 3 weeks, followed by a temporary abatement lasting several days. A chronic, milder form may be difficult to diagnose and the patient is often tried beyond endurance since protracted ill health may be attributed to 'neurasthenia', i.e. imaginary ills.

A wide of range of disease is experienced also in many species of domesticated animals. The most common disease in the UK is in cattle in which an acute placentitis in pregnant cows leads to abortion. In bulls infection causes an orchitis. In cattle it can also cause infections of joints, e.g. hygromas of the knee, and in horses inflammation of the bursa beneath ligaments, as in 'fistula withers' or 'poll evil'. The infection in goats and sheep may give rise to an acute illness accompanied by abortion but, more frequently, produces chronic infections detectable only by bacteriological examinations of milk, blood, or urine. Pigs are also affected; occasionally abortion occurs.

Prior to 1906 the association between undulant fever in man and contagious abortion in animals was not recognized. The historical developments which led to an understanding of the epidemiology of human and animal infections evolved independently and, for this reason, will be considered separately. Eventually the disease was recognized as a zoonosis and the convergent epidemiological strands were given added synthesis when Alice Evans demonstrated the close systematic relationships between the various causal agents from both human and animal sources.

1. The history of brucellosis in man

The history of brucellosis in man from the beginning of the nineteenth century up to the present time is both fascinating and instructive. It provides a classical example of an epidemiological investigation involving the application of basic principles which led to the control of this important disease first in man and subsequently in animals. It is also an example of an infection involving more than one route of transmission from animals to man.

In the absence of serological or bacteriological confirmation, the symptoms of acute brucellosis are not sufficient of themselves to identify the disease as undulant fever. It is not surprising, therefore, that in the past this disease was confused with other fevers, such as typhoid fever and malaria, often prevalent in the same geographical areas. It was not until Marston (1859) published a description of the disease, which he called Mediterranean or gastric remittant fever, that the disease was distinguished. His clear description rescued 'undulant fever from the rubbish heap of the continued fevers' (Dalrymple-Champneys, 1960). He was aided in this by observing the various fevers at a military base in Malta during the Crimea War and also by the use of the recently improved clinical thermometer.

The aetiology of the disease was not known before Bruce commenced his studies in 1884. Stationed with the Mediterranean Garrison on the Island of Malta, he recognized the importance of this disease in reducing the strength of the garrison. In 1886 he observed the presence of minute 'cocci' in smears of spleen pulp, and later in histological spleen sections from patients who had died of the disease. In 1887 Bruce succeeded in growing the organism from the spleen removed from a fatal case within 10 minutes of death and, subsequently, from the blood of living patients. Pure cultures of the organism produced a similar disease in monkeys. Considering the organism to be a coccus Bruce and his wife (a skilled microscopist) named the organisms *Micrococcus melitensis*.

These early observations were rapidly extended. Almroth Wright (1897) published a note on the occurrence of Malta fever in India, in which he reported the presence of agglutinins to Bruce's organism in the blood of experimentally infected animals and also in newly infected patients.

The importance of the disease among the army and naval forces prompted the Royal Society in 1904 to appoint the Mediterranean Fever Commission under the chairmanship of Bruce. The terms of reference were to investigate the prevalence of Mediterranean fever among the forces of the crown on Malta. Circumstances favoured their investigations; many hundreds of cases under strict military control were available to them. At the time the Commission was appointed the source and route of infection was unknown. The first breakthrough was in 1905 when Zammit was serum-testing six goats which he intended using for experimental infections. To his surprise he demonstrated high titres of agglutinins to the organism already present in the blood of these animals. Within a week the animal source of the infection was confirmed when Horrocks found the organism in large numbers in the milk of goats. Further investigations revealed that 40 per cent of the goats on the island had agglutinins in their blood and many apparently healthy animals were excreting the organisms in their milk. This finding resulted in the now famous order of 1906 forbidding the drinking of unboiled goats' milk by British naval and military personnel stationed in Malta. During the 5 years preceding the work of the Commission an average of 355 cases of Mediterranean fever were reported annually among the forces with many deaths, whereas in 1907 only nine cases were reported and, in each instance, the patients had drunk raw goats' milk (Table 15.1). Such an immediate decline in incidence of infection has been described as 'one of the most dramatic demonstrations of an epidemiological truth on record' (Topley and Wilson). The civil population on the island did not follow the example set by the forces and the incidence of undulant fever continued unabated. It was not until 1939 that Valetta, the capital city, commenced pasteurization of goats' milk and prohibited the drinking of raw milk. A similar dramatic fall in incidence followed in Valetta but no decline occurred among the populace living outside the city and who were not controlled by legislation (Table 15.2).

The early history of this disease demonstrates a number of important principles of epidemiology. The discovery of the causal agent and identification of the organism did not of themselves affect the incidence of the disease. The knowledge of the reservoir of infection and its mode of transmission were required before measures of control could be implemented. Once this was discovered the control of human brucellosis was achieved by rendering the milk safe by heat treatment. Another feature of interest was the difficulty in recognizing the importance of an article of food (e.g. goats' milk) which was consumed generally by the whole population with little or no preference by any social class. The first indication was the discovery of agglutinins in the blood of large numbers of goats, and this led to the discovery that goats' milk was the source of human infections. Later it was shown that ewes' milk played a similar role in France, Italy, Algeria, and many Mediterranean countries. Cows' milk was not immediately suspected but strong presumptive evidence from 1912 onwards led eventually to its incrimination in the 1920s.

Table 15.1. Undulant fever in Malta (after Eyres, 1908, 1912 and reproduced by permission of *The Lancet*)

	Civil		Navy		Army	
	Cases	Deaths	Cases	Deaths	Cases	Deaths
1901	642	54	252	3	253	9
1902	624	45	354	2	155	6
1903	589	48	339	6	404	9
1904	573	59	333	8	320	12
1905	663	88	270	7	643	16
1906	822	177	145	4	163	2
1907	714	78	12	0	9	1
1908	502	?	6	0	5	0
1909	456	?	10	0	1	0
1910	318	?	3	0	1	0

Table 15.2.   Undulant fever in the civilian popu-
lation of Malta (after Agius, cited by Topley and
Wilson, *Principles of Bacteriology, Virology and
Immunity*, 1975, published by Edward Arnold)

	Valetta		Rest of Malta	
	Cases	Deaths	Cases	Deaths
1934	146	6	1763	82
1935	85	2	1228	78
1936	43	2	830	50
1937	58	4	976	56
1938	72	3	913	47
1939	10	2	865	50
1940	1	0	955	38

In 1939 pasteurized milk was supplied to the capital
city (Valetta) and the sale of raw milk was prohibited
in this town but not in the rest of the island.

2. The history of brucellosis in animals

Records of enzootic disease in cattle in Britain began to appear at the beginning of
the nineteenth century, and abortions, causing extensive losses to stock-owners,
were reported. In 1805 Lawrence wrote: 'cows are well known to be much given
to abortion, slinking or slipping their calves; in an early period of gestation it is
sometimes epidemic and thence some people have supposed it even contagious'.
Two years later the following was written in *The Complete Farmer*: '. . . it is considered certainly contagious and when it happens the abortion should immediately
be burned and the cow kept as widely apart as possible from the land, and not
receive the bull that goes with them'. The contagious nature of this disease was
confirmed by Lehunt (1878) who demonstrated that infection could be
transmitted by transferring the vagina discharge, or a piece of placental tissue
from a cow which had aborted, to the vagina of a healthy pregnant one.

   The discovery of the causal pathogen was made in Denmark. Bang (1895), with
Stribold who did the laboratory work, found a small Gram-negative bacillus in the
thick yellow exudate between the wall of the uterus and the foetal membranes
following abortion, and the organism was isolated in pure culture. This success
was even more commendable since the organism required a partial oxygen pressure for its growth. Bang confirmed the aetiology of the disease by reproducing it
in healthy pregnant heifers using pure cultures of the bacillus (introduced into the
vagina). The organism was named *Bacillus abortus* and has been referred to as
Bang's bacillus. Bang's work was soon confirmed in America and other European
countries, and by M'Fadyean and Stockman (1909) in the UK. Up to 1918 contagious abortion of cattle was regarded as an entirely distinct disease from that

producing abortion in goats and sheep, and consequently human disease was not attributed to this source. For this reason cases of undulant fever in the UK were falsely attributed to Bruce's 'micrococcus', although there was no history of drinking raw goats' milk.

In 1912 Weil and Menard reported a human case of undulant fever with strong presumptive evidence that bovine milk was the source of the infection, and in 1914 Kennedy found agglutinins for brucellae in the milk and blood of cows. But the association between contagious abortion in cows and human infection was not confirmed until the work of Evans (1918) opened the door for others to prove that undulant fever in man could be caused either by Bang's bacillus or Bruce's 'micrococcus'.

The causal agent of brucellosis in swine was originally described in the United States by Traum (1914). He isolated the organism from the foetus of a sow. A similar organism, differing only in minor biochemical proportion, was isolated from swine in Denmark (1931).

3. Relationship of various strains

The close similarity of the strains from the three animal species (namely *Micrococcus melitensis* in the goat, *Bacillus abortus* in cattle, and the porcine strain) was first demonstrated by Evans (1918). She showed strong serological similarities between the three organisms and demonstrated that each had common rod-shaped morphology, despite the earlier description by Bruce that the goat strain was a 'coccus'. To avoid further confusion she proposed the generic name *Brucella* in honour of Bruce.

Three species are generally recognized; namely *Brucella melitensis, Br. abortus*, and *Br. suis*. Within each of the three species a number of biotypes are recognized, differing only in minor respects (Table 15.3). These are often associated with particular geographical localities; for instance biotypes of *Br. melitensis* found in Malta may differ from those found in France or Palestine. A similar distribution is found with biotypes of *Br. abortus* and *Br. suis*. There is good agreement between bacteriophage sensitivity and metabolic tests for each of the various biotypes.

4. The epidemiology of human brucellosis in more recent times

Outbreaks of human brucellosis in advanced countries occurs sporadically. Relatively few cases are reported annually and the number is on the decline (Figure 15.1). This, in part, must be due to the public health measures to ensure the pasteurization of milk. A relatively high incidence is still found in individuals whose occupations bring them into close contact with farm animals especially at the time of abortion. Many farm residents drink raw milk, and unpasteurized dairy products are still the source of some human cases. However, direct contact

Table 15.3. Differential characters of the three species of the genus *Brucella* and their biotypes (modified from Wilson and Miles, *Principles of Bacteriology, Virology and Immunity*, 1975, published by Edward Arnold)

Species	Type	CO₂ require-ment	H₂S pro-duction	Thionin a	Thionin b	Thionin c	Basic fuchsin b	Basic fuchsin c	Aggl. A	Aggl. M	Phage Tb at RTD	Glutamic acid	Orni-thine	Ribose	Lysine	Urease	Most common host reservoir
Br. melitensis	1	−	−	−	+	+	+	+	−	+	−	+	−	−	−	+ or ±	Sheep, goats
	2	−	−	−	+	+	+	+	+	−	−	+	−	−	−	+ or ±	Sheep, goats
	3	−	−	−	+	+	+	+	+	+	−	+	−	−	−	+ or ±	Sheep, goats
Br. abortus	1	+	+	−	−	−	+	+	+	−	+	+	+	+	−	±	Cattle
	2	+	+	−	−	−	−	−	+	−	+	+	+	+	−	±	Cattle
	3	±	+	−	−	+	+	+	+	−	+	+	+	+	−	±	Cattle
	4	±	+	−	−	−	+	+	−	+	+	+	+	+	−	±	Cattle
	5	−	−	−	+	+	+	+	−	+	+	+	+	+	−	±	Cattle
	6	−	−	−	+	+	+	+	+	−	+	+	+	+	−	±	Cattle
	7	−	±	−	+	+	+	+	+	+	+	+	+	+	−	±	Cattle
	8	+	−	−	+	+	+	+	+	−	+	+	+	+	−	±	Cattle
	9	±	+	−	−	+	+	+	−	+	+	+	+	+	−	±	Cattle
Br. suis	1	−	++	+	+	+	−	−	+	−	−	−	+	+	+	++	Pigs
	2	−	−	+	+	+	−	−	+	−	−	±	+	+	−	++	Pigs, hares
	3	−	−	+	+	+	+	+	+	−	−	±	+	+	+	++	Pigs
	4	−	−	+	+	+	+	+	+	+	−	±	+	+	+	++	Reindeer
	5	−	+	−	−	+	−	+	−	−	−	±	+	+	±	++	Cattle, sheep

a = 1/25,000; b = 1/50,000; c = 1/100,000.

A and M correspond to monospecific *abortus* and *melitensis* antiserum respectively.

Phage Tb = Tbilisi strain.

RTD = routine test dilution.

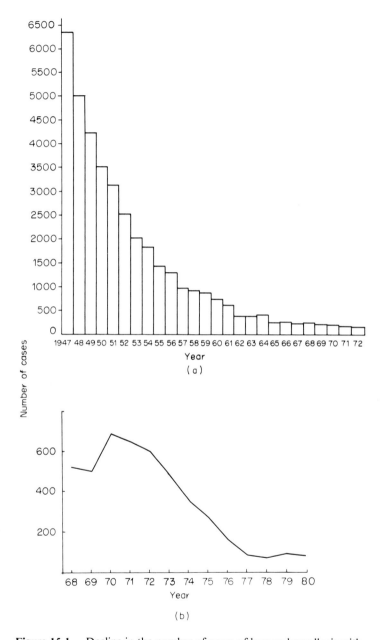

Figure 15.1. Decline in the number of cases of human brucellosis with time. (a) United States, 1947–72 (data from Centre for Disease Control, Atlanta, Report, January 1974); (b) England, Wales and Scotland 1968–80 (data from Communicable Disease Surveillance Centre, CDR, 25 April 1980)

with animals and their fluids now constitutes the main avenue of transmission to man and rarely, if ever, do person-to-person infections occur.

Recent annual returns in the USA and the UK indicate that brucellosis is a disease more frequently of males than females. The larger incidence of infection recorded in earlier years in the USA will be quoted, since these show more clearly the age and sex ranges of human cases. Data on human cases reported for a 7-year period (1940–46) show these differences very markedly. The cases have been analysed into two groups. Those in the first (315 males and 286 females) had no contact with farm animals. Presumably these contracted the infection from contaminated dairy products. A similar ratio of infection occurred between males and females and 11 per cent of all cases fell into the age range 20–49 years (Figure 15.2).

During the same 7-year period, a larger group of cases were known to have had contact with animals; out of a total of 1804 cases, 1582 were males and 220 females. By far the majority of cases occurred among men and, of these, the highest number occurred in the same age range as in the no-contact group (75 per cent in the age range 20–49 years). Since most farm residents, both males and females, frequently drink raw milk and cream, an explanation for the higher

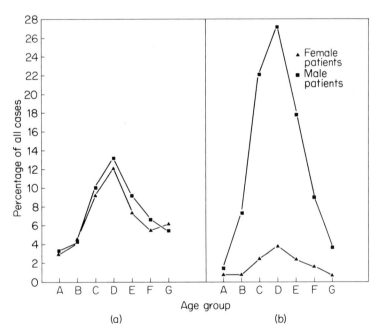

Figure 15.2. Distribution of cases of human brucellosis by age and sex reported in the USA (1940–46). (a) 601 cases in people not in contact with infected animals; (b) 1804 cases in people in contact with infected animals. A = 1–9 years; B = 10–19 years; C = 20–29 years; D = 30–39 years; E = 40–49 years; F = 50–59 years; G = 60 and over

incidence in men was sought. An analysis of the occupations of the males revealed that 73 per cent were farmworkers in direct contact with living animals. This may indicate that infection occurred through abrasions caused whilst restraining animals or, alternatively, the swishing of the tails of animals, contaminated with udder or uterine fluids, may sweep across the attendant's face and inoculate via the eye, a well-established route of infection (in-ocula).

5. Control of brucellosis in domestic animals

In the UK vaccination of animals against brucellosis has been practised for many years. Recently a scheme to eradicate the infection was commenced and is currently progressing.

(a) *Vaccination of animals*

Calves infected with *Br. abortus* at birth often throw off the infection within several weeks. It is the adult animal which is most susceptible, particularly when it becomes pregnant. Adult animals which have not been in contact with infection are most susceptible and may abort at the first and sometimes also the second pregnancies, thereafter usually carrying subsequent calves to full term. An immunity to infection obviously develops as a result of exposure to infection. Attempts to produce a similar degree of protection artificially have been tried. The most successful agents are the living S19 vaccine and the killed strain 45/20 of McEwen and Priestly.

The S19 strain was discovered as a naturally occurring isolate with very low virulence for guinea pigs and cattle. The strain has proved very stable relative to its degree of virulence and has good immunizing properties. It has been widely used for calfhood vaccination and, although the immunity is not absolute, protects the majority of young breeding stock through the period of their greatest susceptibility to the disease. Agglutinins appear in the animal's blood but in 90 per cent of animals the titre has fallen below a diagnostic level 12 months later.

The McEwen 45/20 vaccine consists of a rough strain of *Br. abortus*. It has the advantage over the S19 strain in not producing agglutinins against smooth antigens in the immunized animal and therefore avoids any confusion with naturally produced antibody. Initially used as a living vaccine, it is now administered as a killed vaccine, due to virulence instability. This vaccine was used mainly on adult animals, especially in situations where animals were exposed to heavy infection, but its use is now forbidden under the eradication scheme (see below).

(b) *Eradication of brucellosis in domestic animals*

Brucellosis is a menace both to human and animal health, and the cause of enormous economic losses by abortion and low milk yield. Over many years the control of human brucellosis has been limited to the pasteurization of milk. This failed to protect groups of workers in direct contact with infected animals and did nothing to safeguard susceptible animal stock. Vaccination had been the main means of control in domestic herds but the disease remained endemic. The more comprehensive approach was therefore to eradicate infection from animals and thereby eliminate the source of infection both for susceptible animals and man. Eradication of brucellosis has been successfully carried out in certain countries and a similar scheme is now under way in the UK. By virtue of being physically separated from the Continent, the UK has many advantages in any eradication procedure.

The success of any eradication scheme depends initially on developing a suitable test which can be done on every animal and which can detect infected animals with a high level of accuracy. The detection of brucella-infected animals depends on the ability to demonstrate specific agglutinins in the blood serum. Two tests are used. Initially all animals in a herd are screened qualitatively. A drop of neat serum is mixed with a drop of a stained suspension of brucellae on a white tile. This is called the Rose Bengal plate test (RBPT), the name indicating the dye used. Macroscopic clumping (agglutination) indicates a positive reactor. If at any time reactors are found, these are culled from the herd and the testing procedure on the herd has to be re-started from the beginning. The test is performed three times on each animal with several months gap between tests (Figure 15.3). A herd with no reactors after this period is then tested quantitatively by a tube serum agglutination test (SAT) and, if no reactors are found, is registered as brucella-free. The RBPT detects infection earlier than the SAT. It correlates well with other serological tests (e.g. the complement fixation test) in chronic infections and the test becomes negative sooner than the SAT in calves vaccinated with strain 19. The RBPT is considered to be of the order of 97 per cent reliable. Subsequent periodic testing after registration is essential to ensure that reinfection has not occurred.

Since the tests depend on detecting brucella agglutinins, any procedure which interferes with the test must be excluded. It is therefore forbidden to vaccinate adult animals against brucellosis in herds under test. At present calfhood vaccination with S19 vaccine is permitted since agglutinins stimulated by this vaccine disappear by the time the calf reaches adulthood. However, it is planned eventually to phase out the use of this vaccine to avoid any possibility of interference with the official tests.

To ensure that a herd is maintained brucella-free, the farm must be adequately fenced from neighbouring farms to keep the herd within the approved premises and prevent contact with other cattle. Also replacement animals may be purchased only from accredited stock and transport from one farm to another must conform to rigid control including disinfection of the vehicle.

The scheme in the UK is well under way. Areas of the country known to have a low incidence of infection were initially chosen. In these areas a voluntary scheme was first introduced. Incentives to encourage farmers to enter the scheme

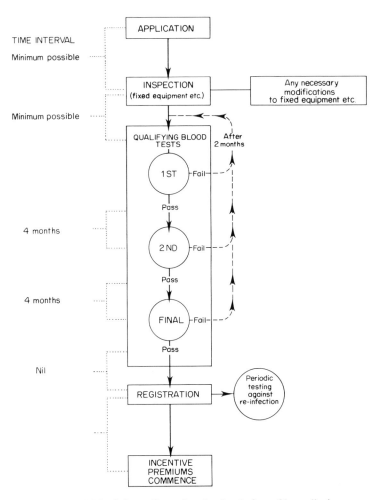

Figure 15.3. Scheme for registering herds free of brucellosis

voluntarily were introduced but all testing was carried out free. Eventually compulsory testing and eradication was imposed. Gradually the areas were extended (see Figure 15.4) and it is hoped that the whole country will be free of brucellae by the late 1980s. Once the national herd is free of infection, legislation governing the importation of stock and brucella-contaminated materials will need to be controlled rigidly to avoid reintroducing the infection into what will become a highly susceptible herd.

A close correlation has been demonstrated between progress with the eradication scheme and the number of laboratory confirmed human cases of brucellosis (Figure 15.5). In this figure the fall in human cases in each year is expressed as a

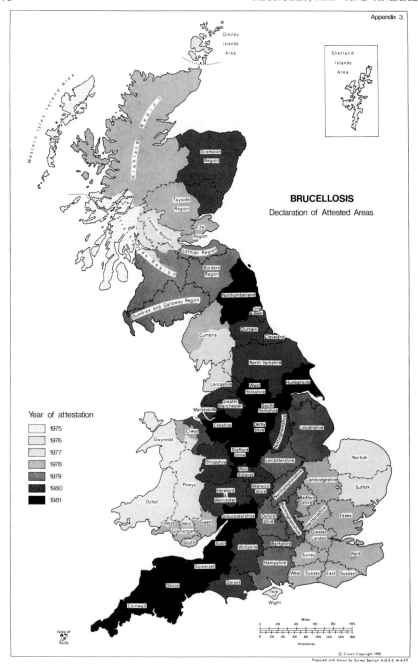

Figure 15.4. Brucellosis: attested and eradication areas, as at 1 November 1979 (from *Animal Disease Report* (1978), vol. 2, no. 1; ADAS. MAFF, Weybridge, England)

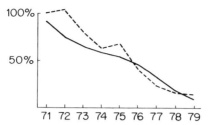

Figure 15.5. A comparison between the progress in eradicating brucellosis in cattle in England and the decline in human cases. ———, Human cases (measured by laboratory reports of infections) as a percentage of the 1971 total; - - - -, percentage of all herds not accredited brucella-free (from *Communicable Disease Report*, 1980; PHLS, no. 16. Reproduced by permission of the Communicable Disease Surveillance Centre, London)

percentage of all cases (220 in England alone) reported in 1971, the year in which compulsory orders leading to eradication were first made. Progress in the eradication scheme is demonstrated by plotting the percentage of all herds of cattle not accredited as brucella-free at the end of each year. It is clear that the number of human cases has declined since the scheme began; it has brought benefit therefore to both human and animal health.

FURTHER READING

Brucellosis (1950) A symposium under the joint auspices of the National Institutes of Health of the Public Health Service, the Federal Security Agency, the US Department of Agriculture, and the National Research Council, 1949. Bethesda, Maryland. American Association for the Advancement of Science.

Brucellosis (Eradication Areas) (England and Wales) Order (1974), No. 1151. HMSO, London.

Brucellosis Amendment (Eradication Areas) (England and Wales) (1976), No. 1853. HMSO, London.

Brucellosis Surveillance (1974 to date) Annual summaries. Center for Disease Control, US Department of Health, Education, and Welfare.

Communicable Disease Report (1980) *Human Brucellosis in Great Britain and Progress of Eradication in Cattle.* PHLS, CDR 80/16, pp. 3 and 4.

Dalrymple-Champneys, W. (1960) *Brucella Infection and Undulant Fever in Man.* Oxford University Press, London (196 pp.).

Ray, W. C. (1979) Brucellosis. In Steele, J. H. (chief ed.), *CRC Handbook Series in Zoonoses*, vol. I, pp. 99–183.

Wilson, G. S., and Miles, A. A. (1964, 1975) *Principles of Bacteriology and Immunity*, vols. I and II, 5th and 6th edns. Arnold, London.

Chapter 16

Waterborne infections

Waterborne infections result from drinking water polluted by human or animal faeces and, to a lesser extent, by urine. Fresh foods, such as salads, contaminated by polluted water and eaten uncooked, may also give rise to infections. The polluting pathogen must be able to survive or even multiply in the water. This depends not only on the nature of the pathogen but also on the pH of the water, its organic content, and the presence or absence of toxic substances such as ammonia.

Waterborne outbreaks are characterized by their explosive nature; this results in a steep rise in the epidemic curve. The decline may be equally steep if the source of the outbreak is quickly recognized and control measures implemented. The steepness of the falling curve may, however, be influenced by secondary cases which may arise by contact with susceptible persons. The source of a waterborne outbreak is usually not difficult to trace since the distribution of clinical cases indicates a common source. The ultimate cause of water pollution, however, may be much more difficult to establish.

Typical waterborne bacterial infections include typhoid fever and cholera; dysentery outbreaks occasionally have been waterborne. Hepatitis A is typical of a waterborne virus infection and giardiasis of a waterborne protozoan infection. An account of the epidemiology of cholera and cholera-like infections is presented to illustrate some of the features of waterborne outbreaks.

1. Asiatic cholera and cholera-like diseases

Cholera is an acute enteropathy, the violence of which seems out of proportion to the relatively normal appearance of the gut at post mortem. The causal organisms are localized in the gut lumen; there is no invasion of the blood stream, nor are the mesenteric lymph nodes affected. The vibrios multiply in the gut to reach enormous numbers; they can be seen readily in the 'rice-water' fluid which constitutes the gut excreta. Their arrangements in parallel, spiral rows, as seen in stained faecal smears, resemble fish in a stream.

It is now recognized that all the disease manifestations of cholera are due to the massive loss of fluids and electrolytes from the gut. This leads to a reduction in plasma volume resulting in shock and death. Loss of fluid from the gut may exceed 1 litre per hour and must be rapidly replaced. The rice-water faeces consist of isotonic fluid, low in protein and high in bicarbonate, and the leakage through the almost normal gut lining is attributed to the enterotoxin produced by the vibrios (Chapter 5).

Vibrios, although relatively delicate organisms, can survive outside the body for various periods of time. Survival is encouraged by moist, alkaline conditions; they can survive under conditions of minimal nutrients which would not support the viability of most bacterial pathogens.

Outbreaks are usually waterborne but spread may also occur from patient to patient due to poor sanitation. Infection can be spread from a sick to a healthy person by faecal contamination of food, either directly or by contaminated water, or on soiled linen. Contaminated clothing, if kept moist, has been shown to retain its infectivity for several weeks. Water supplies, contaminated with human excreta, account for explosive outbreaks in a circumscribed location. Spread to other areas is usually due to the movement of infected patients, a factor which has increased in importance with the speed of modern transport. This danger is potentiated by healthy carriers which excrete the organisms regularly or intermittently. Usually both waterborne and chain-spread infections (i.e. propagated from person to person) of cholera occur in an outbreak and this is illustrated by the Hamburg epidemic of 1892 (see later).

2. History of cholera

Endemic foci of Asiatic cholera have been known in parts of India for centuries and still exist today, but there is no historic evidence of spread to other parts of the world before 1817. The largest of these foci centred on Bengal in the delta regions of the Ganges and Brahmaputra rivers. Here were found extensive surface waters in densely populated flat areas where poor sanitation was practised. Since 1817, seven pandemics have been recorded, the last being as recent as the 1960s. Certain of these pandemics were more widespread than others. For instance, in the period 1832–33 the second of the pandemics spread first from India to Russia, then west to include the whole of Europe, subsequently reaching Canada, New York, and Cuba by immigrants fleeing, particularly from Ireland. Four thousand deaths occurred in London and 7000 in Paris. Müller, a Christian philanthropist of the nineteenth century, wrote from Bristol on 24 August 1832:

> the ravages of this disease are becoming daily more and more fearful. We have reason to believe that great numbers die daily in the city. Who may be the next, God alone knows. I have never realised so much the nearness of death. Except the Lord keep us this night, we shall be no more in the land of the living tomorrow. Just now, ten in the evening, the funeral bell is ringing, and has been ringing the greater part of this evening. It rings almost all day. 'Into Thine hands, O Lord, I commend myself'.

A later pandemic, which lasted for a longer period (1854–62), also reached Europe and 20,000 deaths were recorded in England alone. It was during this period that Snow made his famous observations on the waterborne nature of cholera (see later). The last pandemic, commencing in 1961 and persisting up to the present time, in a few countries, provided the impetus for intensive study of the disease and its epidemiology, and this has revealed the basic mechanisms of pathogenicity, not only of cholera, but of acute diarrhoeal diseases in general (Chapter 5).

From a world view the incidence of cholera is on the decrease, particularly in countries with adequate sanitary arrangements for the disposal of faecal waste and the provision of safe water supplies. It is probable that cholera will be confined in the future and cease to exist where reasonable health standards are maintained. Continued vigilance and health education, however, are necessary especially in deprived countries where foci of infection persist.

3. Cholera in Europe

Reference has already been made to the occurrence of cholera in Europe. It is thought that it was first introduced in 1831 by a ship from Hamburg carrying cholera cases. The infection spread slowly and eventually died out. Many cities suffered high mortality. The situation in Bristol has already been noted. The rapid growth of cities as a consequence of the Industrial Revolution soon overtook the already inadequate sewage and drainage systems, and waterworks, though much improved, were totally inadequate to supply the requirements of the people. Much of the water was drawn from wells and pumps or laboriously obtained by dipping it from rivers and streams (Figure 16.1). Often the water was collected below the town from rivers already heavily polluted (Figure 16.2).

Cholera was reintroduced into the UK in the summer of 1854 by a vessel of the Baltic fleet with a cholera patient on board. Many cases of cholera occurred in London, particularly in houses supplied with mains water by the Southwark and Vauxhall Water Company, one of four companies supplying water to the city. The piped water was obtained from the River Thames but the Southwark Company collected their water furthest downstream and was therefore the most heavily polluted. In the 4 weeks from 8 July to 5 August 1854, 563 deaths from cholera were reported in London. An epidemic was obviously already under way.

The notorious Broad Street outbreak was superimposed upon this already considerable epidemic. Although limited geographically and in duration, it was a great disaster. It started on 31 August 1854 and lasted only a few days into September. Of the 14,000 inhabitants living in the vicinity of the Broad Street pump, more than 600 died. This is an underestimate since many fled the district but died in the country or in hospital. In Broad Street itself 896 people lived in the grossly overcrowded 49 houses and of these 56 died in 3 days. Today it is almost

Figure 16.1. A common practice in the last century of taking water from the local river and distributing it by water cart for human consumption (from Shapter, Thomas (1849) *The History of Cholera in Exeter in 1832.* John Churchill, London (297 pp.))

impossible to conceive the fear and panic that prevailed, and it is little wonder that within a week of the commencement of the outbreak 80 per cent of the people fled. This was obviously an epidemic within an epidemic.

The outbreak was attributed to drinking water from the pump in Broad Street. The incidence of cholera in houses within the area which received mains water was no higher than in other districts receiving the same mains supply. On the other hand, people living elsewhere but who for one reason or another drank water from this pump, contracted cholera.

Dr John Snow, a London anaesthetist and amateur epidemiologist, had already produced evidence that cholera was a contagious disease. This was 30 years

Figure 16.2. The sources of drinking water were often polluted further upstream by washed soiled clothing and by sewage contamination. In this etching the house in which a cholera patient has died is an obvious source of contamination of the nearby mill-stream (from Shapter, Thomas (1849) *The History of Cholera in Exeter in 1832*. John Churchill, London (297 pp.))

before Koch discovered the cholera vibrio. During the Broad Street outbreak he documented each fatal case of cholera on a map of the area (Figure 16.3) and noted the concentration of cases in the area around the pump; this convinced him that water from the pump was the source of the outbreak. As a result of his observations the pump handle was removed on 8 September. Investigations revealed that the pump was supplied from a shallow well into which sewage percolated from a defective drain and cesspool serving the area. It was obvious that the pump water had been contaminated by sewage for years before the outbreak. Dr Snow's enquiries revealed that on 28 August a baby contracted cholera

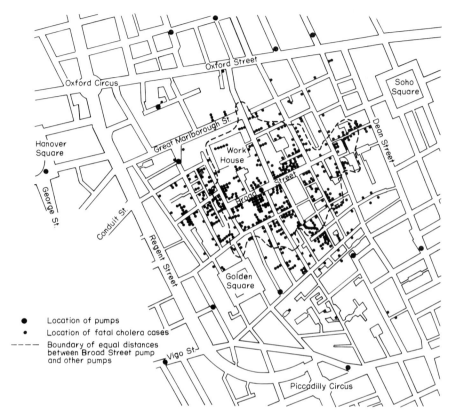

Figure 16.3.   Asiatic cholera and the Broad Street pump (from Snow, J. (1936) *Snow on Cholera*. The Commonwealth Fund, New York)

and the washings from soiled napkins found their way into the defective cesspool. The massive contamination of the well caused the sudden outbreak but how the baby became infected is not known. At a time when 'predispositions', 'miasmas', etc. were generally accepted causes of disease, these observations of Snow are the more creditable. For reasons not clear no further cases of cholera occurred even in people drinking the pump water after 3 September. One possible explanation for the disappearance of cholera vibrios from the water could have been its destruction by bacteriophage.

The Hamburg epidemic in 1892 was on a much larger scale than the Broad Street outbreak but in every respect vindicated the observations of Snow. There were 18,000 cases and 8200 deaths; at the height of the epidemic 1000 new cases fell ill each day. Both Hamburg and its sister town Altona drew water from the River Elbe. Although Altona drew water further downstream, where it was more polluted, it was passed through a simple sand filter. Hamburg suffered 34.3 cases per 1000 of the population but Altona suffered only 3 cases per 1000. It is believed that most of the cases in Altona in fact were infected in Hamburg. Where the border between the towns passed down the centre of one street, houses on the Hamburg side were affected while those on the Altona side were virtually free.

4. The causal agents of cholera

As far as is known the great pandemics of the nineteenth century were caused by *Vibrio cholerae*. With few exceptions cholera caused by this organism has now retreated to its Asian home. An epidemic occurred in Asia towards the end of the Second World War due to the mass movements of armies and refugees; in 1958 cholera broke out in countries neighbouring India and East Pakistan but subsided within 2 years. Even in the subcontinent of Indo-Pakistan the incidence has declined and it is a rarity at the present time.

A new situation arose in 1961 involving a cholera-like vibrio, *Vibrio El Tor*. This vibrio had been isolated in 1905 from pilgrims at the quarantine station El Tor. It caused mild diarrhoea in these patients, not nearly so severe as in cholera, and was disregarded from the Public Health point of view. Isolations of this organism were made from time to time but particularly from patients on the island of Sulawesi (Celebes) in the Dutch East Indies where it remained endemic from 1937. In 1961 infection spread to neighbouring islands, then east to many Asian countries and later west, including parts of Africa where cholera had not previously occurred and where it continues to be endemic.

On clinical grounds the illness is indistinguishable from classical cholera apart from severity. Prior to the occurrence of this pandemic, international regulations had not required member nations to impose quarantine and other regulations where El Tor disease (also called 'para cholera') arose. This may have contributed to the rapid spread in the 1961 pandemic. In view of the clinical and epidemiological observations made during this period, the World Health Organization decided that El Tor disease must be classified as cholera and not a different malady. It is unlikely that El Tor disease will gain a foothold in the future in developed countries but the outbreak in Naples in the 1970s, which revealed a disgraceful lack of basic sanitary facilities, is a reminder of the need for constant vigilance and maintenance of adequate standards. There is a continuing measure of anxiety regarding the vast crowds of pilgrims that congregate at Mecca and other religious sites, such as the River Ganges in India, where the provision of adequate sanitary facilities presents enormous problems.

5. The relationship between *V. cholerae* and *V. El Tor*

Biochemically the El Tor vibrios exhibit only minor differences from those of *V. cholerae* but both differ from non-cholera vibrios found frequently in faeces and natural waters, in their ability to ferment sucrose and maltose but not arabinose.

Table 16.1.   Some distinguishing characteristics* of *Vibrio cholerae* and *Vibrio El Tor*

	O sub-group	Haemolytic (sheep RBCs)	VP	Colistin sulphate	Phage IV
V. cholerae	O1	−	−	Sensitive	Sensitive
V. El Tor	O1	+	+	Resistant	Resistant

VP = Voges Prauskauer test.
*Some of these are variable but sensitivity to phage IV is an absolute distinguishing feature.

So similar are the two cholera vibrios that they have been named *V. cholerae* and *V. cholerae* biotype *El Tor* respectively. They both possess the O1 antigen which is definitive for *V. cholerae* (Table 16.1). The two pathogens were distinguished at one time by the haemolytic properties of El Tor vibrios but this property is now known to be variable; some strains are haemagglutinating rather than haemolytic. Only the activity of a specific phage (phage IV) distinguishes them unambiguously.

The pathophysiology of the two organisms is the same. In the past the severe dehydration, seen in classical cholera, occurred in only 10 per cent or less of cases infected with El Tor vibrio, and the mortality was only 2–5 per cent, compared with 40–75 per cent in untreated cases of classical cholera. However infections due to El Tor vibrio, especially in the recent pandemic, have been as severe as *V. cholerae* infections.

How the modern pathogen *V. El Tor* has arisen is not known. It has been found in the faeces of patients also excreting *V. cholerae*. Whether or not this indicates multiple infection, or is the consequence of some mechanism of transformation, is open to speculation. It is recognized that all strains of *V. El Tor* are lysogenic and it is possible that they may have arisen by lysogenization of classical vibrios.

The El Tor vibrios are, in some respects, more robust than classical vibrios. They are more resistant to drying, survive longer in food, are carried and excreted over a longer period from convalescent patients. These properties may account for the greater spread of the El Tor vibrio over the *V. cholerae*, in recent years.

6. Prevention and control

Modern treatment of cholera patients is directed almost entirely at replacing the enormous fluid and electrolyte loss. Where this is done sufficiently early the death rate can be reduced to below 1 per cent; probably a much lower figure is correct since mild and even subclinical cases should be included in the statistics. This treatment alone results in a disappearance of cholera symptoms, though the duration of the residual diarrhoea may be shortened by the use of tetracyclines. Recent reports of the emergence of tetracycline-resistant strains is a disquieting feature.

Vaccines have been widely used prophylactically (Chapter 10) but it is doubtful whether they give any protection and certainly they have no value in preventing the spread of cholera. Almost all vaccines used in the past have relied on somatic antigens consisting of killed suspensions or lysates of mixtures of *V. cholerae* and *V. El Tor* which stimulate specific antibodies of the IgG and IgM classes.

It is now known that all the symptoms of cholera are attributed to the enterotoxin which is produced in, and its functions confined to, the small intestine (Chapter 5). To be effective therefore a vaccine must aim at producing local protective immunoglobulins of the IgA class in the gut. The stimulation of this type of antibody should be the aim of any future vaccines. However, preventive measures and modern treatment by fluid replacement are far more effective than the use of vaccines.

Contrary to public opinion cholera is not a highly infectious disease. Preventive measures include the provision of safe water supplies, the avoidance of food contamination by sewage-polluted water, and strict personal hygiene. By these relatively simple expedients cholera can be avoided and controlled.

Where an outbreak has occurred a number of steps must be taken for its control and to prevent it spreading. The strategy for the control of any infectious disease is only possible where information on incidence and prevalence are made available. Countries which are members of the World Health Organization must notify within 24 hours every case of cholera, the number of deaths, and the measures taken to control the spread. An area is recorded as infected for 10 days after the last case has died, recovered, or been isolated.

Isolation of patients is desirable both in order that the patient may have proper treatment to aid survival and recovery, and to segregate them from the rest of the population. Contacts and carriers should also be isolated until they are clear of infection. During an outbreak movement of the population may be restricted and travel, including pilgrimages, may be forbidden to and from the infected area. Pilgrims travelling in their millions to Mecca are often advised to take prophylactic tetracyclines or sulphonamides.

Disinfection of infected excreta and articles of clothing, bedding, etc., is essential and water supplies should be safeguarded against contamination. Chlorinated lime has been used successfully as a general disinfectant.

Since cholera occurs more frequently among the poorer classes, health education becomes an essential requirement of any control programme.

FURTHER READING

Barua, D., and Burrows, W. (1974) *Cholera.* W. B. Saunders & Co., Philadelphia (458 pp.).
Elkin, I. I. (ed.) (1961) *A Course in Epidemiology*, pp. 247–55. Pergamon Press, Oxford.
Felsenfeld, O. (1964) Present status of the El Tor vibrio problem. *Bact. Rev.*, **28**, 72–86.
Felsenfeld, O. (1966) A review of recent trends in cholera research and control. *Bull. Wld. Hlth. Org.*, **34**, 161–95.

Gale, A. H. (1959) *Epidemic Diseases*. Penguin Books A456 (159 pp.).
Hobson, W. (1963) *World Health and History*. John Wright & Sons, Ltd., Bristol (252 pp.).
Pollitzer, R. (1959) *Cholera*. World Health Organization, Geneva, Switzerland (1019 pp.).
Wilson, G. S., and Miles, A. A. (1975) *Principles of Bacteriology and Immunity*, 6th edn., vol. 2. Arnold, London (2706 pp.).

Chapter 17

Airborne infections

A. RESPIRATORY INFECTIONS

In countries with a fairly high standard of living respiratory infections are more common than intestinal infections and, indeed, account for around half of all episodes of human illness. For example, in England and Wales in the decade 1967–77 there were 10 times per annum as many notified cases of measles, a common childhood respiratory pathogen, as there were enteric diseases of typhoid, paratyphoid, dysentery, and food-poisoning put together. In that period influenzal deaths amounted to 6 per cent of all the deaths attributable to infectious diseases. This was a greater proportion than normal since it included the influenza pandemic of the Hong Kong serotype in 1968. During the last 4 years of this period (1974–77) respiratory deaths amounted to 15 per cent of all non-violent deaths, with non-pandemic influenzal deaths accounting for one-fifth of these; in comparison, alimentary deaths amounted to 3 per cent and circulatory deaths to 0.5 per cent. The greater percentages of influenzal deaths occurred in men over the age of 70 years and women over 80 years.

1. The respiratory tract

The surface of the respiratory tract, from the trachea down to the larger bronchioles, is lined with a mucociliary epithelium which secretes mucus from the constituent goblet cells (Figure 17.1). The mucus is continually being expelled by the ciliary activity of the underlying ciliated cells and periodically aided by coughing. Mucus acts as a humectant and protective fluid for the underlying epithelium, and readily traps foreign particles in the inspired air which have escaped being filtered out at the turbinates or pharynx. Particles which penetrate as far as the trachea, bronchi, or broncioles range between 10 and 5 μm in size and are mostly precipitated on to the mucus layer at inspiration. Particles of less than 5 μm is size readily penetrate into the alveoli, where they become trapped on surface fluid and engulfed by the alveolar phagocytes.

Mucus varies in viscosity from the lower to the upper respiratory tract, being more watery in the lower regions, and the speed of outward flow varies inversely with the viscosity. The more fluid the mucus the more easily is it expelled by ciliary activity, and the more readily are entrapped particles expelled.

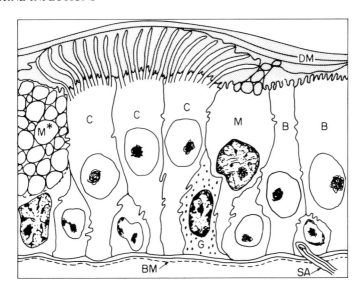

Figure 17.1. Mucociliary epithelium of the upper respiratory tract. B = brush-border cell; BM = basement membrane; C = ciliated cell; DM = mucus discharged on to surface of epithelium from the mucous cells; G = granule cell; M = mucous or goblet cell; SA = sensory axon. In the undischarged mucous cell (M*) mucus is in the form of numerous microdroplets. Brush-border cells are seen especially in the region of the larynx, and are often innervated there with sensory axons. (Redrawn from Weiss, L., and Greep, R. O. (1977) *Histology*, 4th edn. McGraw-Hill Book Co., New York)

The overall efficiency of the respiratory tract in removing airborne particles is remarkably high. Large particles are filtered out in the turbinates with an efficiency of around 85 per cent, the remainder being trapped by the respiratory mucus secretions. Few particles need to be dealt with by the alveolar phagocytes.

It has been estimated that an average of 10^5 airborne micro-organisms penetrate into the lower reaches of the tract per day. Approximately half of these are expelled within 1–2 hours of entrapment in mucus, and the remainder are expelled over longer periods of time (up to 24 weeks or more) according to the ease with which the alveolar phagocytes contend with them. Non-biological particles of similar sizes, such as carbon particles or asbestos fibres, which enter the alveoli may never be expelled.

2. Factors involved in infection of the respiratory tract, particularly by viruses

Viral infections of the respiratory tract are initiated by virus particles (virions). Virions of most respiratory viruses are around 150 nm in size and can penetrate to the alveoli at inspiration. Many have neuraminidase located on their surface which reacts with neuraminic acid in the mucus and so acts as a mucolytic enzyme. These features help the virus to gain rapid access to cells of the lower respiratory tract.

(a) *Target cells*

The cells to which respiratory viruses attach are difficult to identify in experimental studies. The goblet cells have been shown to constitute the major target cells for Newcastle disease virus (NDV), indicating that the virus multiplies principally in the upper and lower tubular parts of the tract, from the nasal epithelium to the secondary bronchi, but less so in the lung tissues. In a similar manner influenza is predominantly an upper respiratory tract infection, both in man and the ferret, with considerably less lower respiratory tract and lung involvement; indeed, the similar nature of the disease syndrome in the ferret makes it an excellent animal model for studying the disease.

Immunofluorescence studies have shown that influenza virus infects ciliated cells, goblet cells, and basal cells of the mucociliary epithelium, and adsorbs strongly to cilia and microvilli. The nasal turbinate tissues in the ferret produce around 1000 times more virus than the lungs. While influenza virus has a broader requirement than NDV for target cells located in the tubular parts of the respiratory tract it has a more prolific output in the head region, and these features constitute one important aspect of its virulence. In contrast, other respiratory viruses are more restricted in their cellular requirements; for example, Coxsackie virus A21 and Rhinovirus NIH 1734 are able to infect upper tract cells, while Adenovirus T4 infects lower tract cells.

(b) *Bacterial commensals*

Commensal micro-organisms in the respiratory tract may interfere with infection by respiratory viruses. Venezuelan equine encephalomyelitis virus has a better attack rate in pigeons treated with aureomycin, and this has been attributed to destruction of the respiratory commensals by the antibiotic; however no specific inhibitory bacterial products have yet been described.

(c) *Synergistic micro-organisms*

Viral infections are often succeeded by bacterial superinfection of the respiratory tract, frequently by a particular species of bacterium. These species must have specific colonizing ability for the virus-weakened respiratory tract or the virus and the bacterium may act synergistically. Sandford and colleagues have shown that influenza A virus acts synergistically with some bacteria. They found that there was specific adsorption by *Streptococcus agalactiae* and *Strep. sanguis* to the surface of influenza A-infected cells. Prior treatment of the cells with virus antibody prevented adsorption by the streptococci. It is suggested that the molecular configuration of the viral proteins mimics those of the tissue proteins to which streptococci naturally attach. Such mimicry offers an advantage for the streptococci to multiply on virus-weakened tissues. Sandford's findings may explain the frequent superinfection of animals by various streptococci, staphylococci, or *Haemophilus* species, following a primary influenzal infection.

White blood cells. Macrophages provide a first line of defence against viral pathogens. However, the macrophages which phagocytose the virus particles can penetrate into deeper tissues, both inside and outside the respiratory tract, thereby acting as transporters of the virus. African swine fever and rinderpest viruses promptly appear in the mediastinal lymph nodes following their inhalation, to which they have been carried in white blood cells; similarly ectromelia virus is known to be carried from the alveoli of the mouse by infected macrophages. In man both influenza and measles rapidly become bloodborne in infected lymphocytes, though there is no clear evidence that macrophages are involved.

(d) *Immune responses to virus infection*

Most experimental work has been carried out in mice. In these animals humoral antibody is formed against both surface and internal antigens of the virus. IgG is certainly protective against pulmonary infection but the role of IgA is much more speculative. While IgG is therefore of major importance in the mouse it is not known whether its principal action is by neutralization of the virus, inhibition of virus release, cooperation with T (killer) cells, or opsonization of virions prior to macrophage activity. In man it is known that the presence or absence of levels of specific IgG may impede or enhance the spread of pandemic strains of influenza A virus, and the speed of antigenic shifts and drifts exhibited by the virus (see below) may be regulated by the degree of immunity possessed by a population. The secretory IgA constitutes a greater threat to the survival of respiratory pathogens than the mucociliary barrier. Its adsorption to the surface of the pathogen renders bacteria more amenable to phagocytic action and neutralizes virus infectivity. There is need therefore for sufficient IgA molecules to be present in the secretions to initiate a rapid interaction with the pathogen.

Cell-mediated immunity does not play as important a role in virus infections of the respiratory tract. However, mouse thymus-derived T-cells show a specificity for the stimulating virus and also for the major mouse histocompatibility (H2) genes. Involvement of the histocompatibility genes in the T-cell response suggests a cell-to-cell (cytocletic) triggering process in which T-cell 'targets' appear to be determined both by the H2 genes and the viral genes. When mice are immunized by influenza A the T-cell response is able to cross-react with *all* influenza A serotypes but not with influenza B serotypes. This indicates that the cell surface-located H and N antigens of the virus are not intimately involved with the T-cell response, and, indeed, this is corroborated by the finding that viral antigens added to cytotoxic T-cells do not reduce their cytotoxicity. It has been suggested that the T-cell receptor is in fact an 'altered self' antigen expressed as a result of the influenza infection. In man HLA antigens (human lymphocyte antigens used in tissue typing) are known to be similarly involved in the specificity of the T-cell response.

(e) *Fever*

The production of fever during the course of a virus infection is a natural reaction of the body. Fever can contribute significantly towards reducing the rate of replication of the virus, and hence survival of the infected animal, although viruses with elevated ceiling (maximum permissible replication) temperatures are better able to withstand it.

B.   INFLUENZAVIRUS

At present the family Orthomyxoviridae contains one recognized genus, *Influenzavirus*, and within the genus are types A and B influenzaviruses. A second (yet unnamed) genus may be provided by type C influenza virus. These viruses constitute three morphologically similar but antigenically heterologous serotypes which do not share any virus-coded antigens. Complement fixation tests conducted with their haemagglutinin and neuraminidase antigens serve to subtype the viruses. Neutralization tests indicate that antigenic variation is most common with type A viruses, less common with type B viruses, and does not occur with type C viruses. Genetic recombination between the principal serotypes has not been substantiated, and the serotypes therefore deserve recognition as separate viral species.

1. The nature of influenza A virus

Influenza A virus, more correctly termed influenzavirus type A, is undoubtedly the best-studied respiratory tract virus. It will be considered here as a prime example of an infectious agent which is transmitted from person to person via the respiratory route, and which owes its survival in the population especially to its propensity for antigenic variation.

(a) *Animal host range*

Influenza virus infections occur in swine, horses, and a variety of avian species as well as in man. Stability and antigenic variability are qualities which can enhance virus survival and transmission in a population, and it is found that type A viruses are especially implicated in human pandemics, type B in epidemics, and type C in sporadic respiratory infections of man. It may be deduced from this that viruses capable of antigenic variation are more successful epidemiologically.

(b) *Functional morphology*

Most of the biochemical and structural studies on influenza virus have been undertaken with type A virions and the following description of morphology refers especially to these. The *virion* (Figure 17.2) consists of centrally located, delicate ribonucleoprotein (RNP) surrounded by an ether-sensitive bilipid envelope set with spikes. The RNP displays helical symmetry and contains single-stranded RNA in association with the soluble, complement-fixing nucleoprotein (NP) antigen which is common to all members of the serotype. The RNP has a diameter of 9 nm, which serves to distinguish the Orthomyxoviridae from the

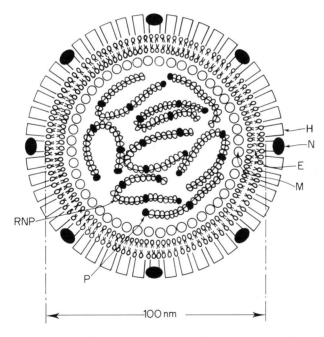

Figure 17.2. The morphology of influenza A virus. H = haemagglutinin spike; N = neuraminidase spike; E = envelope; M = membrane protein; RNP = ribonucleoprotein complex, made up of RNA with complexed nucleoprotein, some of which has a polymerase (P) function. The ratio of the two varieties of spikes H:N is 5:1. There are eight separate lengths of RNP within the mature infectious virion. The diagram was constructed from data of Matthews, R. E. F. (1979) Classification and nomenclature of viruses. *Intervirology*, **12** (3–5), 219

Paramyxoviridae (15–18 nm diameter). It exists in eight unequal lengths or segments which total around 800 nm overall, and they may be conjoined in the virion with further NP. It is the divided genome of type A viruses which supplies the facility for genetic recombination, multiplicity reactivation, and the von Magnus phenomenon (the production of genomically incomplete virions).

The bilipid virion envelope contains and protects the RNP. It is formed at the time when new virions are being released from the cell by a process of exocytosis or 'budding-off', and is essentially virus-modified host cell plasma membrane. The envelope is lined internally by a non-immunogenic structural M (variously *m*embrane, *m*atrix, or *m*ajor) protein, the function of which is to align the newly synthesized RNP beneath those areas on the plasma membrane of impending exocytosis. Protruding radially through the outer lipid layer of the envelope are 700–900 immunogenic haemagglutinin (H) and neuraminidase (N) spikes, the 'feet' of which may not extend into the inner lipid layer.

The more numerous H spikes consist of two H-bonded glycoprotein molecules, each molecule consisting of two disulphide-linked glycopeptides. The outer, H_1, and inner, H_2, glycoproteins are obtained by post-translational proteolytic cleavage of an entire precursor H_0 glycoprotein when still in the cytoplasm. While this proteolytic cleavage is not essential for virion maturation and release from the surface of the host cell, it is vital for infectivity. Reduced cleavage means reduced infectivity, and in some instances reduced infectivity of the virions may be restored by trypsin treatment which will generate the necessary 'sticky' outer end on the H_1 spike. The outer end of H_1 spike has a strong affinity for sialic acid (*N*-acetylneuraminic acid) 'receptor' molecules which are commonly found constituents of the plasma membrane of animal cells and which therefore enable the virions to adsorb strongly to the cell surface. It can be seen that infectivity and hence pathogenicity is vested in the adsorption capabilities of the virion, which in turn are dependent upon adequate post-translational cleavage occurring within the host cell. Because of the successful pandemic capabilities of influenza A subtypes it is obvious that cells of the respiratory tract are capable of providing this cleavage. Anti-haemagglutinin neutralizes virion infectivity.

Arising out of Hirst's fortuitous discovery of haemagglutinin the diagnostic haemagglutination and haemagglutination-inhibition tests have been developed. These assay for the influenza virus and neutralizing antibodies respectively and have proved indispensable for diagnostic and epidemiological studies with influenza.

The N spikes number around one-fifth of the number of H spikes. They consist of two disulphide-linked glycoprotein molecules (foot and stalk) which are non-covalently linked to a further glycopeptide neuraminidase 'head' on the spike. The enzyme can split off sialic acid from a variety of glycoproteins in a similar way to the 'receptor-destroying enzyme' of *Vibrio cholerae* and *Clostridium perfringens*. It will thus break down the receptors to which adjacent H spikes have adsorbed. It is considered that the enzyme serves a dual purpose at the time of infection: first, it acts as a mucolytic agent to permit the virion to 'burrow' through the surface mucus in the respiratory tract to the underlying cells; second, it permits the virion to detach from the cell in the event of it not being speedily pinocytosed, and to further 'browse' the epithelium until a receptor cell is found. Both functions enhance virion infectivity considerably as well as aiding the virion to escape any premature removal from the respiratory tract. Neuraminidase also appears to play a role in the release of progeny virions from the host cell, since anti-neuraminidase has been shown to inhibit release of mature virions from the cell surface, and this suggests that the enzyme also helps to prevent the anomalous readsorption of virions back on to the cell which generated them. This function similarly enhances virion infectivity and spread in the respiratory tract.

(c) Nomenclature

In the early years each influenza isolate was named by its investigator; in some instances the initials or names of the patients from which they were isolated were used, e.g. WS, DSP, LEE; in others the place of isolation was given, e.g. PR8, MEL, PHILA, while in other instances laboratory numbers were given, e.g. 1233. It was not uncommon to find additional names added to designate special features of adapted or modified strains, e.g., NEURO WS. Such *ad hoc* nomenclature was clearly unsatisfactory in view of the increasing number of isolates, and it was left to the WHO to evolve a satisfactory, flexible, coded nomenclature for universal adoption. In this code the first statement denotes the serotype, the second the natural host (usually omitted if man), the third the place of isolation, the fourth the number of the isolate and year of isolation, and the fifth the H and N subtypes (Table 17.1). The coded identity for type A Asian influenza isolated in Singapore

Table 17.1. Pandemic subtypes of influenza A

Start of pandemic	H/N subtype	Source of subtype (and popular name)	Interval since last pandemic (years)
1874	Heq2Neq2*	Russia?	—
1890	H2N2*	Siberia (Asiatic flu)	16
1899	H3N2*	England	9
1918	Hsw1N1	France (Spanish flu)	28
1933	H0N1	London	15
1946	H1N1	Melbourne	16
1957	H2N2	Kweichow Province, China (Asian flu)	11
1968	H3N2	South-eastern China (Hong Kong flu)	11
1976	Hsw1N1-like	Fort Dix, USA; expected pandemic failed to occur	—
1977	H1N1	China	9

Major antigenic shifts occurred in 1957 and 1968. The mean interval for the listed pandemics is 14 years.
*Serological type deduced from epidemiological evidence.

in 1957 is thus A/Singapore/1-57/H2N2. Additional statements can be added readily to such a code as more detailed characterization of serotypes becomes available.

2. The strategy of antigenic variation by influenza A virus

The majority of infectious micro-organisms are antigenically stable throughout the course of infection in both a single host and a succession of hosts. Even in persistent or chronic infections antigenic stability is maintained despite prolonged exposure to the immune defence mechanisms of the host. Survival of pathogens depends on their ability to resist these mechanisms by, for example, intracellular multiplication sites, or rapid transmission to fresh hosts. There are exceptions to this generalization, and *Borrelia*, trypanosomes, malaria, and equine infectious anaemia virus undergo antigenic variation to maintain their hold upon the host in a manner similar to influenza virus.

Influenza virus presents a unique strategy of antigenic variation. The prevalence of a single subtype within a population over a restricted period of time is followed by its prompt disappearance and the simultaneous appearance of a new subtype. With influenza, therefore, the epidemiological pattern is one of successive pandemics produced by a succession of new subtypes which arise by a process of antigenic variation. Of considerable epidemiological interest is the unusual subtype A/USSR/77/HINI which appears to have persisted for 27 years without significant antigenic variation (see later).

(a) *Antigens involved in antigenic variation*

In influenza A, thirteen H and eight N antigens have been identified in strains isolated from man and animals (Table 17.2). Sixteen further major antigens are

Table 17.2.   The H and N antigens of influenza A subtypes in man and animals

Animal species	H antigens	N antigens	Possible combinations
Human	H0 to H4	N1, N2	8
Swine	Hsw1	N1, N2	2
Equine	Heq1, Heq2	Neq1, Neq2	4
Avian	Hav1 to Hav10	Nav1 to Nav6	40

Recent studies by Schild *et al.* (1980) have indicated that the following H and N subtypes are related: H0, H1, and Hsw1; H3, Heq2, and Hav7; Heq1 and Hav1; Nav1 and Nav6; Nav2 and Nav3. There are therefore 13 distinct H subtypes and eight N distinct subtypes, and it is proposed to number these accordingly.

known to exist in type A strains, and these represent a variety of major and minor structural and non-structural proteins associated with each virion as well as the complement-fixing (CF) antigen. Type A strains share these antigens in varying amounts and these, together with the qualitative variations obtainable with the H and N antigens, account for the antigenic variations between subtypes.

While antigenic variation in influenza has been studied extensively with the H and N antigens, recent genetic evidence indicates that variation can occur in all of the segments of the genome. Major variations are termed 'antigenic shift'; minor variations, 'antigenic drift'. New pandemics break out as a result of a shift and this places entire populations at risk due to lack of immunity. Drifts tend to be far less spectacular in their effects upon a population, but allow the virus to survive in the inter-pandemic periods. When a shift has been detected epidemiologists therefore need to consider the existing state of immunity of a population in order to speculate on the severity of any ensuing pandemic. The continuing antigenic shifts and drifts favour the worldwide persistence of influenza.

(i) *Antigenic shifts*. These are associated with type A virus and occur when major changes in H, and less often N, antigens take place. A shift is therefore heralded by the emergence of a new subtype. It is considered that shifts occur by a process of recombination or other genetic intermixing rather than by mutation, and involve major changes in the RNA segments coding for one or both of the H and N antigens as well as in other segments associated with virulence and transmissibility. When it occurs in human populations animals may assist in the process by providing one of the recombinants and acting as the intermediate host, but new pandemic subtypes in man can be derived by animal-to-man transmission of animal subtypes. It is considered that the pandemic spread of a recombinant is aided by the immunity of the population against the previous serotypes. In the laboratory antigenic shifts have been obtained by recombination techniques.

Avian hosts appear less likely to subscribe to recombinational processes since animal and avian viruses do not readily cross-infect between the species. The possibilities of avian participation are nevertheless intriguing in view of the con-siderable numbers of influenza A subtypes isolated from them.

The periodicity of the shifts has occurred about every 15 years and this suggests that 1983 may see the emergence of a major pandemic subtype. A simple arithmetical calculation with the figures in Table 17.2 will indicate that, with the five H and four N antigens available (ignoring the avian antigens), 20 different H and N combinations and hence subtypes are possible. It is clear that type A virus will continue to plague mankind with fresh pandemics for many more decades to come, since the survival advantage of antigenic shifts is so enormous.

The lack of shifts in type B virus is probably related to a lack of animal strains.

(ii) *Antigenic drifts*. These are associated with both type A and B viruses, and occur when minor changes, usually but not exclusively of N antigen, gradually take place. Drifts occur within a family of virus strains which are related in respect of internal and external structural antigens. Mice inoculated with one strain bearing, say, N2 antigen, will resist challenge with another strain bearing homologous N2 antigens better than challenge with a strain bearing heterologous N2 antigen. Drifts therefore provide a survival advantage in the presence of existing antibody though it is a lesser advantage compared with antigenic shift. A recent example of drift is provided by the H3N2 serotype which first appeared in 1968 as Hong Kong flu (Table 17.1) and exhibited drift in the successive England/42-72, Port Chalmers/1-73, Scotland/840-74, Victoria/3-75 and Texas/1-77 isolates. In the laboratory drift is initiated by culturing virus in the presence of antibody in order to induce mutations in the strain-specific determinants of the H antigen without necessarily producing simultaneous changes in the cross-reactive determinants. Such drift entails the substitution of new amino acids in the reactive regions of the antigenic determinants. Drifts take place within an immune human population without animal participation; the respiratory tract of man is the prime initiation site of drift.

(iii) *A/USSR/77/H1N1*. This subtype was isolated in Anshan Shenyang, China, in May 1977 and was subsequently implicated in a pandemic in young persons. Laboratory tests suggested the virus was closely similar to the A/Fort Warren/1-50/H1N1 subtype isolated in 1950. The question, 'how was an influenza A subtype maintained in a population for 27 years without antigenic variation?' has not been answered, and the virus must be considered an anomaly.

3. Sources of fresh pandemic serotypes of influenza A

The rapid establishment of a new pandemic serotype with the equally rapid loss of the preceding serotype suggests the existence of a delicately balanced state of equilibrium between virus and host. Loss of the preceding subtypes may be attributed to the rise in antibody levels within a population but this is unlikely to be the full explanation since many other viruses survive in the face of rising antibody levels without resort to antigenic variation. Antibody levels alone cannot be deemed responsible for the survival or non-survival of particular viral subtypes, and a better, possibly multifactorial, explanation needs to be sought. This must encompass the epidemiological features of the virus, which include its rapidity of spread within populations (in man aided by a short 48-hour incubation period and an equally short 48-hour period of shedding infectious virions, which indicates its highly infectious nature and efficient transmission); its rapid stimulation of antibody production with an equally rapid loss of susceptible hosts within a population; its apparent failure to establish chronic (carrier) infections; its poor survival outside the body; and its genetic mutability which ensures a continuing hold on the population.

The most pertinent epidemiological question which can be asked is, 'what are the most probable sources of fresh pandemic subtypes of influenza A?' These may result from genetic recombinations which generate novel subtypes continually, immunoselection of new subtypes from an already existing worldwide pool, or a combination of both. These alternatives in turn depend on the existence of animal reservoirs, and geographical epicentres.

(a) *Genetic recombination*

All fresh pandemic subtypes arise from genetic recombinations of already existing antecedent subtypes, and the successfull recombinant will be the virus against which the population immunity is low. Because the number of known H and N antigens is limited (Table 17.2) there is only a relatively small number of possible recombinants, and hence of shifts, which can occur, and recycling of subtypes must eventually take place. In the laboratory influenza A displays genetic stability under constant cultivation conditions; as stated above cultivation in the presence of homotypic antibody will induce antigenic drift but not shift, and antigenic shift can be achieved only by genetic manipulation such as occurs when genetic recombination of two 'parental' subtypes occurs. In genetic recombination it is necessary that an exchange of similar genomic segments takes place between the 'parental' subtypes in order to allow an entirely new recombinant progeny genome to be produced. By such exchanges it is possible to create *in vitro* any desired subtype.

Genetic recombination has been shown to occur in swine inoculated with a mixture of Hsw1N1 swine influenza and Hav1Neq1 viruses which yielded both of the possible Hsw1Neq1 and Hav1N1 recombinants, and in turkeys inoculated with a mixture of Hav6N2 turkey influenza and Hav1Neq1 fowl plague viruses which yielded both of the possible Hav6Neq1 and Hav1N2 recombinants.

Genetic recombination within animal hosts is now a proven phenomenon; therefore natural recombination can take place in the host respiratory tract following simultaneous infection with two influenza A serotypes. This is a finding of considerable epidemiological importance. For the human population it means that the introduction of two different influenza A serotypes may produce recombinants capable of initiating a fresh influenza pandemic subject to existing low levels of immunity in the population. Where human and animal populations exist in close proximity it is likely that cross-infections between the species may generate recombinants, which again will be dependent for subsequent spread upon existing immunity levels in both populations.

(b) *Immunoselection*

Natural infections of influenza induce a more efficient host immunity than vaccination with inactivated virus, no doubt because of the greater variety of viral antigens produced in infected cells. Following vaccination with 'live' virus it has been demonstrated that influenzal immunity persists for around 2 years. This immunity is less closely related to respiratory antibodies of the IgA variety than to the serum antibodies of the IgM and IgG varieties. Repeated infections by succeeding serotypes provide a broadening of antibody response; hence adults are more immune than young persons.

The immunoselection pressures exerted by a population upon virus subtypes alters during the course of a pandemic. At the commencement the pressure increases towards the suppression of the initiating subtype but fails to suppress new subtypes against which no immunity exists. The immune level in a population against a previously encountered subtype slowly falls with time. This is partly due to the decay in immunological 'memory' and partly due to death of the older members of the population. In consequence earlier subtypes appear again and recycle in the population. For H antigens the evidence (Table 17.1) suggests a recycling time of between 30 and 70 years. For example, it was found that H2N2 (1957 Asian flu) and H3N2 (1968 Hong Kong flu) vaccines induced a secondary homologous response in some elderly persons who had previous experience of the subtypes. By inference it is considered that H2N2 and H3N2 had circulated in the past and probably caused the pandemics of 1890 and 1899 respectively. The recycling times of both serotypes might therefore be placed reasonably at 68 years. The reappearance of swine influenza in a totally localized outbreak in the USA in 1976 suggests a recycling time of 57 years, while the reappearance of the H1N1 serotype in 1977 suggests a recycling time of 31 years.

The variability of recycling times cannot be explained in simple terms and undoubtedly reflects an interaction of a number of biological variables, the most important of which would be a reduction in population immunity leading to an increased ability of influenza to colonize the respiratory tract.

(c) *Animal reservoirs*

In 1941 Shope reported (though this work is now disputed) that domestic swine could act as a reservoir for swine influenza (Hsw1N1) for long periods of time, aided by the mechanical transmission of virus in the lungworm (*Metastrongylus* spp.) and the earthworm (*Lumbricus* spp.). Because Hsw1N1 at the time was a prime candidate to be the causal agent of the 1919 pandemic it was logical to speculate that swine, and possibly the horse and various avian species known to be hosts for other influenza A subtypes, might act as animal reservoirs for the various subtypes infecting man. Of these animals, swine and horses were considered the more likely reservoirs in view of the known difficulty of infecting avian species with mammalian viruses. However, despite frequent and close contact of man and swine, human infections with swine influenza have been difficult to demonstrate. In the 1960s it was demonstrated that persons in close contact with swine (for example, veterinary surgeons, abattoir workers, and farmers) had higher antibody levels against swine influenza virus than persons out of contact with them, which suggested some degree of transmission. The true situation now appears to be that infection is subclinical, due to the low virulence of the virus for man, and may pass unnoticed. Persons in close contact with swine frequently develop high levels of antibody against human influenza viruses, which indicates that contact with swine does not dramatically alter immune responses against human viruses. Possibly the relative stability of Hsw1N1 compared with the human subtypes partly explains this. Repeated exposure to an unvariable subtype will generate high homologous levels but will not compromise the ability to respond fully against other variable subtypes, even when these share common antigens.

Concern about swine influenza was again raised in 1976 following the localized outbreak of A/New Jersey/swine-like (Hsw1N1-like) influenza in 11 persons, including Army recruits at Fort Dix, USA (see Table 17.1). Epidemiological investigations failed to establish that the virus was derived from swine, and person-to-person spread was the expected mode of transmission. The death of one recruit recalled the devastating Hsw1N1 pandemic of 1918–19, and initiated large-scale production of vaccine, but no community spread ensued.

Equine influenza Heq1Neq1 and Heq2Neq2 subtypes are also relatively stable compared with human subtypes. The viruses are widespread in horses, and antibodies against Heq2Neq2 have been detected in them in Outer Mongolia. The close association of man and horse must presuppose some degree of horse-to-man transmission and antibodies against Heq2Neq2 have been detected in aged men in the UK, USA, Holland, and Czechoslovakia in the absence of pandemics. A similar subclinical or mild influenzal infection occurs in man with this virus as with swine influenza virus.

Genetic recombination between swine and avian subtypes has been demonstrated *in vivo* and *in vitro*, while serological surveys have suggested recombination between human and equine subtypes *in vivo* (see above). In the light of these observations it has been suggested that animal and avian influenzaviruses are the direct result of domestication of the various species.

The evidence points to human influenzavirus subtypes having been the progenitors of animal subtypes and, if human pandemics are not caused by any reverse transmissions of animal subtypes, continuing man-to-animal transmissions *may* provide the key to the origins of new pandemic subtypes. For this idea to be worthy of more detailed examination it will be necessary to demonstrate shared antigens of human and animal subtypes in the pandemic strains, but this has so far not been done. On the other hand, the subtypes isolated from domestic ducks in southern China in 1978 included the H2N2 subtype, which disappeared from man in 1968 (Table 17.1), and the recombinant H2Nav6. These isolates reinforce the idea of man-to-bird transmission, of recombination occurring subsequently and of the *potential* of recombinant subtypes for later back-transmission to man.

In wild avian species there is certainly a much larger pool of influenza A viruses than might be expected, mostly of avian subtypes. Avian influenza is mainly enteric in nature, though the respiratory tract is also infected. The increasing number of isolations made in the last decade have increased speculation about avian species being a major influenza reservoir and attention has been directed towards all domesticated species. Both feral and domesticated ducks are natural hosts to many influenza A subtypes and all of the pandemic subtypes (Hsw1N1, H0N1, H1N1, H2N2, and H3N2) will replicate in their respiratory tracts. While transmission from ducks to man has yet to be documented, despite the transmission advantage offered by enteric replication over respiratory, this species holds promise of novel epidemiological findings in the future. For these reasons regular surveillance of domestic poultry is indicated to determine the nature and extent of future epidemics for man.

Few other animals have been shown to harbour Orthomyxoviridae. This is in distinct contrast to other families of viruses, such as the Poxviridae, Adenoviridae, Herpetoviridae, and Picornaviridae, which have representative strains in a wide variety of animal species. The reason for this has not been satisfactorily explained. Amongst the poikilothermic animals, amphibia and reptiles possess receptors and their lack of subtypes has been ascribed to their lower body temperatures. Unexpectedly, the marine whale is known to harbour the virus and here the epidemiology must be particularly unique. It is clear that while domesticated animals occupy a unique position in the epidemiological strategy of influenzavirus type A, wild animals are much less implicated.

(d) *Geographical epicentres*

The difficulty encountered in detecting influenzavirus between pandemics led eventually to the idea of animal reservoirs and to geographical epicentres or 'breeding grounds'. Central Asia, with its considerable animal (especially swine) and peasant populations, has been seriously considered to be the epicentre of new pandemic subtypes and claims have been made that *all* pandemics may be traced to that area. However, there are a number of epidemiological facts which suggest this is not so. These are: the 1890–96 pandemic commenced in Bokhara, Greenland, and Canada simultaneously, and the 1918–19 pandemic occurred in France in April and in China in the following May–June. While the exact area of origin of the 1946 pandemic is not known the first isolations were made in Australia in that year and in the USSR in 1949. The re-emergence of the 1918 subtype occurred in 1976 in the USA.

It is highly unlikely that there is any single geographical epicentre of new pandemic subtypes but rather that the correct combination of existing influenza subtypes, population immunity, population nutrition, and animal reservoir will encourage the emergence of new subtypes whatever the geographical location.

(e) *Reactivation of latent subtypes*

Many of the unanswered epidemiological puzzles would be solved if influenza latency *in vivo* could be proved. It is an intriguing problem. Aside from the known longevity of influenza virus in swine and the demonstrated persistence of influenzal antigen in postencephalitic Parkinsonism there is no other evidence at present to indicate *in vivo* latency or persistence of infection.

In summary influenzavirus must be considered to be one of the most successful viruses. With high infectivity, moderate to low virulence, good survival in and between hosts, and genetic adaptability or mutability, influenzavirus is likely to remain a continuing problem in man and animals for many years to come.

FURTHER READING

Kilbourne, E. D. (ed.) (1975) *The Influenza Viruses and Influenza*. Academic Press, New York.

Schild, G. C., and Oxford, J. S. (1976) The immunology and epidemiology of influenza virus. In Oxford, J. S., and Williams, J. D. (eds), *Chemotherapy and Control of Influenza*. Academic Press, London, pp. 3–8.

Selby, P. (ed.) (1976) *Influenza: Virus, Vaccines and Strategy*. Academic Press, London.

Youmons, G. P., Paterson, P. Y., and Sommers, H. M. (1980) *The Biologic and Clinical Basis of Infectious Diseases*, especially Section III (pp. 177–305) and Section IV (pp. 309–446). W. B. Saunders & Co., Philadelphia.

Chapter 18

Gonorrhoea—an example of a contact infection

The Gonococcus is a necessary but not sufficient cause of gonorrhoea
(M. G. McEntegart)

Gonorrhoea is an infection by *Neisseria gonorrhoeae*. This delicate and fastidious Gram-negative diplococcus primarily attacks the columnar cell-lined mucous membranes of the genital tracts of men and women. It is usually transferred from one person to another during the act of sexual intercourse. There is no animal reservoir of infection. Certain non-human primates have been infected experimentally but other animals appear totally refractory to infection by ordinary routes. Local spread of infection to the rectal mucosa in women is common, usually as a result of spread of infected secretions from the vagina. Infections of the throat of both sexes and of the male rectum can also occur. The conjunctival sac of adults can become infected due to autoinoculation, and the eyes of babies may be infected from the cervix of an infected mother during birth. This last infection is named gonococcal ophthalmia neonatorum.

1. Infection and pathology

Infection is acquired by the deposition of viable gonococci on an area of susceptible columnar epithelium. The organism is non-motile but once it has been planted on an appropriate site it will spread from cell to cell through the tissues. It may also spread by floating in inflammatory exudates by which it can invade areas of the genito-urinary tract distant from the site of original infection.

Local invasion of the mucous membrane follows attachment. Attachment of deposited organisms on the urethra surface is a very necessary first step if the organism is to withstand the very effective flushing action of urine. Pili have been observed on the surface of virulent gonococci. These have non-polar ends which probably attach to appropriate target areas on the genital epithelial cells. Electron micrographs have shown gonococci effectively located by pili, and strains of gonococci lacking pili are not able to set up infection in man. However, pili are not the only virulence factor required for establishing infection and an ability to produce an appropriate 'gap junction' is also necessary.

Once successfully attached the gonococcus seems to enter the cell passively by phagocytosis. This process is known as pinocytosis and it can be prevented in the laboratory (and thus infection prevented) by inhibitors of cell surface motility such as cytochalasin. Once inside the cell the gonococcus multiplies and eventually causes death and rupture of the cell with the passage of the organisms into the submucosa producing much inflammatory discharge.

Clumps of gonococci can frequently be found inside polymorphonuclear neutrophil phagocytic cells in the inflammatory exudate; indeed the demonstration of such organism in cells has long been recognized as a reliable criterion for diagnosis of the infection. Earlier theories that the organisms were actually multiplying in these cells have given way to the realization that those cells are actively phagocytosing the bacteria. There is some evidence that virulent piliated gonococci will survive better, and even multiply, inside phagocytes, but it is more likely that most of the organisms are dying and represent a victory of the host's defences. There is some evidence that the phagocytosed gonococci will sometimes kill the polymorphonuclear cell which in turn is phagocytosed by a macrophage. It is possible that multiplication in this cell system may occur whilst being protected from other cells, antibiotics, and antibodies which may be present in the rest of the environment.

Infection commences in the lower genito-urinary tract but may spread in the female to the Fallopian tubes and in the male to the epididymus. Invasion of the submucosa and destruction of the mucosa is the basic pathological process of gonococcal infections and results in the production of an inflammatory discharge. It may give rise to immediate local symptoms of discomfort and discharge, and can produce fever and pain if the infection spreads to the upper genital tract. The fever and pain of salpingitis can be quite severe. Subsequent healing may lead to permanent loss of ciliated cells and structural disturbances due to scarring. These can result in urinary retention due to obstruction of the lower urinary tract, and to sterility in the male from blocking of the vas deferens inside the epididymis. In the female damage to the cilia and tubular stenosis or blockage can result in sterility or to impaction of a fertilized ovum in the tube instead of the uterus—with a resulting ectopic pregnancy—an acute and life-threatening surgical emergency. It has been calculated that in 1972 there were approximately 175,000 hospital admissions in the USA which were a result of gonococcal infections, totalling 1.2 million hospital days, requiring 102,500 surgical procedures, and costing $212,000,000. In certain parts of Africa where gonorrhoea is endemic its effects cannot be so precisely measured but, even more significantly, fertility has become so low that depopulation is becoming a problem.

Spread of a local infection can occur to distant sites. This is called disseminated gonococcal infection (DGI). Spread to the heart valves with fatal endocarditis was a not uncommon occurrence in the days before antibiotics were available, and spread to the blood stream—and thence to skin and joints—is still being recognized frequently today. It seems that many factors play their part in this. It occurs in about 1 per cent of infections but more commonly in the female than the male.

Pregnant women and those menstruating seem most prone to this complication, but women taking oral contraceptives are also more prone than those not using them. There may well be an association with the fall in cervical tryptic activity and local antibody production which occur in these conditions. Certain families also have a higher incidence than normal of disseminated neisserial infection, and an association with this and congenital absence of the later-acting components of the complement cascade has been established. However, not only are host factors involved, but it seems that certain strains of gonococci are more likely to disseminate than others. Typing of gonococci is not a precise science but it is clear that a significant proportion of strains producing disseminated gonococcal infection tend to produce very small colonies on culture media, are exquisitely sensitive to penicillin, and belong to a limited number of nutrient-requiring biotypes (auxotypes). Thus it would seem that variation, both of the soil and its seed, can produce situations conducive to dissemination of the infection.

2. History of gonorrhoea

The gonococcus was first visualized under the microscope in 1879 by Albert Neisser and was cultured by Bumm a few years later, but the disease has probably been known for several millennia. Unlike syphilis, which was not known until it appeared in Europe in the fifteenth century at the time when Vasco da Gama was opening the sea routes around the world, it seems to have been present in Ancient Egypt. Moses was probably trying to control it when he ordered the execution of all women who had 'known man by lying with him'. Boswell is also considered to have suffered from gonorrhoea and it was rife in Victorian England.

3. Survival outside the body

As already stated, this organism is very sensitive. It dies rapidly when exposed to air and allowed to dry. This, together with its predilection for multiplication in columnar epithelium-lined organs, dictates the epidemiology of the disease. It must be spread by direct inoculation of the organisms from an infected site to another susceptible site. This can only occur during sexual intercourse as only then are suitable areas of susceptible epithelium sufficiently closely approximated. Infections of non-genital sites (perhaps with the exception of the rectum) rarely act as anything but end-stage infections or end-host sites since infectious secretions from them are unlikely to find their way to susceptible mucosae on another person.

4. Social factors

A social factor is also essential for this disease to spread in the community. An infected person must have sexual contact with more than one individual. If only one sex partner exists then the infection will not spread beyond that second individual, and the commununity will be protected. Thus the behavioural pattern of the population can provide a substantial degree of herd immunity. If, however, an infected person has sexual contact with more than one partner, and these partners behave similarly, then an infection can rapidly become widely disseminated. Such relationships are usually termed promiscuous. Not everyone in an infection situation need have more than one partner and a single promiscuous female can be responsible for a whole chain of cases. An alteration in the population's behavioural pattern can rapidly erode the herd's immunity and render a community highly vulnerable to epidemics of gonococcal infection.

Increased vulnerability has been traditionally associated with increasing urbanization, wars, population mobility, and times of social unrest and change. These factors all lead to disruption of the stable family unit and have been associated with increased gonococcal infection rates. Prostitution is another traditional cause of much gonorrhoea. In the Far East it is undoubtedly a major factor but in the West currently a severe epidemic of gonorrhoea is being experienced and here it does not play a major role. This increase in rates—which is only now being contained by the control services in most countries—is due to a change in the social behaviour patterns of modern communities. Previous forces which tended to reduce promiscuity and to reinforce single-partner relationships have diminished. We are in a time of altered values and of greater intellectual and sexual freedom. Previous constraints of religion, fear of conception, and fear of untreatable disease have to a large extent ceased to hold due to changes in

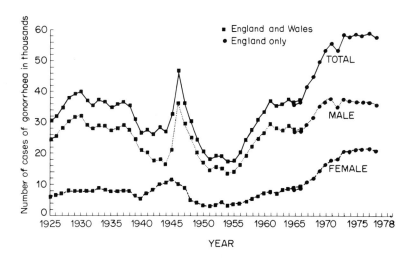

Figure 18.1. Number of cases of gonorrhoea reported, 1925–78 (compiled from annual reports of the Chief Medical Officer on the state of public health)

philosophy, and the advent of the Pill and of effective chemotherapeutic agents. This, added to the traditional factors of increasing urbanization and diminution of the strengths of the family unit, together with an unparalleled increase in the affluence and leisure of the bulk of the population, have all led to a lessening of the so-called 'morals' of the communities. There has been a concomitant increase in the number of sexual partners and a decrease in the stability of these partnerships for individuals in most countries. This has led to the current situation. The graph depicting this (Figure 18.1) also demonstrates the decrease in the male-to-female case ratio. This again is a new phenomenon and reflects the changing population at risk.

5. Control of gonorrhoea

Clearly it is of paramount importance that this situation is brought under control. There are two ways this can be achieved; either by increasing the herd immunity or by rapidly removing all sources of infection from the community.

(a) *Control of sources of infection*

Currently removal of sources is relied upon. This is most effectively achieved by rapid detection and curative treatment of as many cases and carriers of the disease as possible. Until recently it was assumed that women tended to have fewer symptoms and that men always had urethral discharge and discomfort when infected, but it has now been realized that there are significant numbers of asymptomatic carrier men as well as women. Thus the sheet anchor of present control is detection and effective treatment of cases as they present, and of the cases and carriers detected as a result of contact tracing of these original cases. When this is carried out really effectively then good control may be expected. It should be remembered that tuberculosis, brucellosis, and foot-and-mouth disease—which are all highly contagious diseases—have virtually been eradicated from the animal population of this country by similar means. It is not too much to hope for a similar success with gonorrhoea but other means of combating the disease should also be considered.

Herd immunity could be increased by a return to more stable sex partnership patterns. History suggests that there will eventually be a reversal of the present trend but most of those involved in public health are pessimistic about influencing its arrival, and beyond efforts to educate the public to the risks of casual encounters ('if she's available she's got it'), are seeking other methods to reduce the likelihood of successful transmission. The most effective measure currently known is the use of the condom. This must present a mechanical barrier to infection. A recent publicity campaign in Sweden advocating the use of this device has coincided with the first decline in infection rates in that country for many years. However condoms are more acceptable in the more stable partnerships—these are less likely to spread the infection anyway—and the effect of the campaign in Sweden seems already to be falling off.

(b) *Vaccination*

A vaccine could also improve herd and individual immunity. If a significant proportion of the population were immune to the disease then the infection chain would be broken and the disease would tend to die out. Numerous attempts at producing a vaccine have been made. These have sometimes resulted in the production of measurable levels of antibodies in vaccinees but clinical trials have shown no protection. This is not surprising as clinical infection itself does not lead to immunity, and it is unlikely that very good protection against this organism will become available with anything but a highly sophisticated vaccine. One carefully investigated assessment of a vaccine has been reported where a great increase in the dose required to set up an infection was demonstrated after a course of pilus vaccine. However the vaccine in its present state has unacceptable local toxicity. Limited protection has also been demonstrated in chimpanzees. Further developments in this field are awaited in the hope that a non-toxic but effective preparation will eventually be found. When it is found, the benefits of its use will have to be weighed against any risk of inducing an undetected carrier state in any subsequently infected vaccinee with all its implications, and also against any increase in promiscuity which such vaccination programmes may induce, thereby undoing the good which may have been achieved.

(c) *Chemotherapy*

Both treatment of cases and the strategy for control of the disease (in which effective therapy plays a major part) are based on the deployment of non-toxic, cheap, and effective antibiotic agents. The gonococcus has revealed an ability to develop resistance to chemotherapeutic agents and the pattern of control in the various areas of the world today correlates well with the degree of success in keeping this in check.

In 1939 sulphonamide drugs were first used on cases of gonorrhoea with dramatic effect, and for the first time physicians were given an effective curative agent. The drug was widely prescribed but unfortunately this produced a selection pressure for resistant mutants so that the gonococci in circulation rapidly developed an increasing resistance. By the end of the Second World War sulphonamides were proving largely ineffective, nearly all strains encountered were clinically resistant, and the use of sulphonamides had to be abandoned. The selection pressure then immediately eased and it is interesting to note that the great majority of gonococcal strains encountered today are again fully sensitive to sulphonamides.

It was fortunate that as the effectiveness of the sulphonamides faded away so another antibiotic became available. This was penicillin. It was virtually non-toxic and was cheap and highly effective. A single dose of 300,000 units proved curative. Gonococci had minimal inhibitory concentrations (MICs) of 0.002 mg per litre or less and no resistant strains seemed to exist. Rapid development of resistance, as had occurred with the sulphonamides, did not arise and for a honeymoon period of 10 years all seemed well.

6. Antibiotic resistance in gonococci

At the end of 10 years of successful treatment it became apparent that the previously effective dose of penicillin was failing to cure a significant proportion of cases. This stimulated surveys into the sensitivities of the then currently circulating strains, and it was realized with dismay that the variation in resistance between the most sensitive and most resistant strains had widened from its previous 10-fold variation to one of 200-fold, with the most resistant strains, now 1000 times more resistant than the most sensitive, having MICs in the range of 0.1 mg per litre and sometimes even higher.

This situation arose because inadequate doses of penicillin had been used, resulting in blood levels below those necessary to eradicate the organism. The levels achieved would often be such that they would inhibit the growth of a fully sensitive organism but any slightly more resistant mutant which happened to be present in the population of gonococci within the patient would not be affected, and would soon outgrow and replace the previously fully sensitive parent strain. This resistant organism would then maintain the infection in the treated patient—producing a so-called treatment failure—and more importantly would then be available to infect the patient's next sex partners. If these next persons were treated with large doses of penicillin then their infections would respond, but if they too received doses resulting in blood levels minimally higher than that of the index case then a second mutation occurring in one of these patients could allow a yet slightly more resistant gonococcus an advantage in the face of the selection pressure, resulting in the replacement of the first mutant population with a second slightly more resistant one. It was the summation of small increases in resistance which resulted in the notably large increase in resistance ultimately recognized.

Factors tending to produce these low blood levels (and thus engendering resistance) included:

(a) The then current use of long-acting depot preparations of penicillin, such as benzanthine penicillin. These gave good therapeutic blood levels followed by very long 'tails' of low blood levels which were ideal for the selection of resistant 'persisters'.
(b) Inadequate penicillin dosage. This resulted from initial success with very low

doses of penicillin and the practical problems of arranging for more than the traditional 'one-shot' course of therapy at a VD clinic. It also followed self-treatment with antibiotics bought without medical supervision and the sharing of drugs prescribed for a case with consorts.

These factors were recognized and in many areas of the world it proved possible to slow down the increase in resistance by increasing dosage to one producing satisfactory blood levels. This was possible because the organisms were still sensitive to higher levels of penicillin—being less sensitive rather than fully resistant. Indeed in localities where effective chemotherapy has been used it has been possible to observe an actual fall in the resistance levels since minor penicillin resistance has ceased to be a survival advantage for the gonococcus in these areas.

Unfortunately in many areas of the world rational and effective antibiotic therapy has not been available and totally uncontrolled and often inadequate quantities of penicillin have been used to treat infections, often prophylactically. This has led to a great increase in the proportion of relatively resistant strains in these areas. Thus in 1957, whilst in the UK and Sweden—with good VD services—17 per cent of strains of gonococci were resistant to 0.1 mg per litre, reports from Africa revealed 70 per cent were resistant. The stress of the Vietnam war produced the classic situation for an increase in gonorrhoea with a large population of medically uncared-for prostitutes passing infection to huge numbers of American servicemen. In this situation an explosion of gonorrhoea was to be expected but, in this instance, the added ingredient of antibacterial agents being available to the local population on a totally unsupervised and uncontrolled basis resulted in much inadequate dosage, with large numbers of the at-risk population experiencing sub-therapeutic (resistant-mutant-selecting) blood levels of antibacterials over prolonged periods. The inevitable result was the selection of ever-increasingly resistant gonococci, until organisms with MICs of between 2 and 4 mg per litre were being recognized. These resistant strains were brought back to Western countries by returning servicemen and other travellers, and the resistance of strains in circulation in the Western hemisphere increased.

The situation was paralleled to a greater or lesser extent in other areas of the Far East. In many areas of Africa, too, gonorrhoea is rife and the inadequate medical facilities there have resulted in haphazard and frequently self-administered treatment with inadequate dosage of antibacterials creating selective pressures for the resistant gonococci which have predictably emerged.

The mechanisms for some of these resistances have been worked out at the cellular level. Mutation loci have been named Pen A, MTR, and Pen B. The effects of these mutations are cumulative and a strain bearing all three mutations will have an MIC of 100 times that of its parent non-mutant type. Pen A affects the penicillin resistance only, but MTR and Pen B loci affect certain other antibiotics also. Resistance to streptomycin, rifampicin, and spectinomycin appears (as in other organisms) to be due to large-step (go–no-go) ribosomal mutations which occur independently of the small penicillin changes.

Even the most resistant strains encountered up until 1976 were actually sensitive to penicillin, provided that a sufficiently high concentration could be achieved, though this would not be possible with a one-shot regime. However in 1977 the advent of truly penicillin-resistant gonococci was recognized. These strains were different in that they had the ability to actually destroy penicillin. This had arisen—as it had been predicted that it would—in the areas of profligate and indiscriminate antibiotic usage in the so-called Third World. They were first recognized by Percival in strains imported from West Africa and were recognized in America almost simultaneously in cases returning from the Far East. It is possible that the organisms obtained their resistance at about the same time. The resistance is due to possession of R-plasmids which code for the production of a β-lactamase enzyme which will destroy penicillin. How these plasmids were acquired is not known, but they bear similarities with those found in *E. coli* and *H. influenzae* and it is likely that they were transferred from one of these organisms. β-Lactamase-producing *E. coli* and *H. influenzae* have been common in areas of high antibiotic usage for some years because selection pressures have existed for these organisms too. It seems that it was only a matter of time before a suitable conjugation transferred these highly potent R-plasmids to the gonococcus in an environment of high antibiotic usage which gave the recipient the advantage needed to multiply and disseminate. The resulting gonococci are totally resistant to penicillin therapy and alternative antibiotics must be employed. This is a major problem for those countries experiencing the strains because the alternative antibiotic—spectinomycin—is a much more expensive drug and must be given by injection—both of which factors will reduce the number of sufferers who can be effectively treated by health services of limited resources. Ironically in the more affluent West these strains are less of a problem. Alternative antibiotics are available at a price which can be afforded and the apparent 'drain on the resources' of the organisms resulting from its having to support a large episome seem to present it with a biological handicap which may be preventing it from spreading as effectively as other non-episome-bearing strains. The African strains seem to show a high spontaneous loss rate for the episome and, in the UK where a high selection pressure due to widespread abuse of penicillin is not a factor, it was hoped that the organism would not gain a permanent foothold. At first even in the USA secondary spread of infection within the country was the exception rather than the rule. These strains are a cause of great concern, however, and at last appear to be gaining a foothold in Holland and to a lesser extent in the USA. Even in the UK—where careful monitoring and effective therapy of all such cases is the rule—chains of infection, which cannot be traced back to one importation, are beginning to be recognized and threaten us with a new dimension in this disease.

7. Summary

The gonococcus presents a picture of a parasite which is enjoying great success. To achieve this it has developed a very narrow host- and tissue-specificity and it seems to have gained the ability to parasitize areas that other bacteria find difficult to attack. Once it has arrived there are no other bacterial competitors which might make its environment less habitable. An organism which can colonize in these moist tissues so successfully finds itself well protected from dehydration and oxidation. Thus its great sensitivity to these is no disadvantage, except to dictate that its spread must be by a venereal means, and its success in parasitizing a community must depend to a very significant extent on the degree of sexual activity, degree of multiple partnerships, and rapidity of partner-change in the community involved. The conditions of moral values and social change, of increasing affluence and leisure are all currently tending to promote conditions conducive to the effective dissemination of the organism. Further, indiscriminate use of antibiotics has created selection pressures ensuring that those gonococci which do spread are becoming more and more resistant.

Efforts to contain and eradicate this infection are based on rapid and effective detection and treatment of cases and carriers. This is still possible despite the emergence of resistant strains of gonococci, and is the cornerstone of current control practice strategies where antibiotic prophylaxis and vaccine deployment currently have little place. Antibiotics disseminated widely in the community cannot but engender increasing resistance, and vaccines, even when they can be developed, will still pose cultural problems and may even induce the carrier state. The most effective means of controlling the spread of infection would be to alter the behavioural pattern of the community—and it is likely that until this does change the gonococcus will be a match for all man's wiles against it.

FURTHER READING

Morton, R. S. (1977) *Gonorrhoea*. Vol. 9 in *Major Problems in Dermatology*. W. B. Saunders & Co. Ltd., London.

Roberts, R. R. (1977) *The Gonococcus*. John Wiley & Sons, Chichester.

Skinner, F. A., Walker, P. D., and Smith, H. (eds). (1977) *Gonorrhoea—Epidemiology and Pathogenesis*. Academic Press, New York.

WHO Scientific Group (1978) *Neisseria gonorrhoeae and Gonococcal Infections*. World Health Organization Technical Report Series No. 616. WHO, Geneva.

Chapter 19

Arthropod vector-transmitted diseases

Many infectious diseases are transmitted by arthropods; most are caused by viruses, some by protozoa and spirochaetes, but relatively few by bacteria (Table 19.1). Among prokaryotes rickettsiae are an exception since all rickettsial diseases are arthropod-transmitted; otherwise only the bacteria causing plague and tularaemia are vector-transmitted. Some vector-transmitted diseases are limited to one animal species which constitutes the reservoir of infection; others have multiple animal reservoirs of which man may be but one. To illustrate some of the features of vector-transmitted diseases the epidemiology of rickettsial diseases will be considered.

Table 19.1. Examples of vector-transmitted infectious diseases

Micro-organismal group	Disease	Vector	Reservoir
Viruses	Yellow fever	Mosquito	Man, monkey
	Dengue fever	Mosquito	Man, jungle mammals
	Encephalitis	Mosquito	Wild birds
	Rabies	Dog, jackal	Dog, jackal
	Paralytic rabies	Vampire bat	Vampire bat
Bacteria	Plague	Flea	Rat, other rodents
Spirochaetes	Relapsing fever 1	Louse	Man
	Relapsing fever 2	Tick	Rodents, ticks
Rickettsiae	Epidemic typhus	Louse	Man
	Endemic typhus	Flea	Rat
	Scrub typhus	Mite	Field mouse, small rodents
	Rocky Mountain spotted fever	Tick	Wild rodents
Protozoa	Malaria	Mosquito	Man, monkey
	African sleeping sickness	Tsetse fly	Man, antelopes

A. Rickettsial diseases

Rickettsiae are very small bacteria, found naturally in the alimentary canal of arthropods. Most of them are adapted very well to this environment and do no harm to the arthropod. In some the association is much stronger, the rickettsiae being transmitted vertically from generation to generation via the eggs. Where this occurs the arthropod can transmit the rickettsial infection without coming into contact with an infected host.

Rickettsiae will not grow on inanimate media (Table 19.2). Living cells must be present, even for the most independent rickettsiae, indicating that their metabolism is to a certain degree dependent on that of the host's living cells. Hence, *in vitro* growth of rickettsiae is possible in tissue cultures although even these are not very reliable; growth is more consistent on the chorio-allantoic membrane or in the yolk sac of the chicken embryro, or in living animals.

For vector transmission to take place, rickettsiae are taken up by the vector in a blood meal taken from an infected host and transported to another susceptible host. This will occur only if the numbers of rickettsiae present in the small volume of blood ingested constitutes an infective dose for the vector. Where the number of particles of the agent falls below a minimum threshold the host becomes noninfectious. The rickettsiae multiply in the vector before being passed on; this latent period may vary from a few days to a few weeks. The length of time will depend on the ambient temperature which influences the temperature within the vector. The vector, in most instances, transmits the infectious agent by depositing it with its saliva, oesophageal contents or faeces into the wound produced by its bite. Alternatively the rickettsiae, which remain alive for a long time in faeces from the vector, may enter a susceptible host by other routes. The arthropod faeces may

Table 19.2. A comparison of selected properties of bacteria, rickettsia and viruses

	Bacteria	Rickettsiae	Viruses
Approximate size	1 μm diameter	300 nm	<300 nm (many < 150 nm)
Nucleic acids	DNA and RNA	DNA and RNA	DNA or RNA
Muramic acid	+	+	−
Growth (+) on artificial media	+	−	−
Intracellular replication (+)	−	+	+
Mode of replication	Binary fission	Binary fission	Intracellular assembly of constituent parts (nuclei acid, protein sub-units, etc.)
Sensitivity to therapeutic antibiotics (see Table 11.2)	+	+	−

dry into dust and the rickettsiae-laden dust may be inhaled, setting up a primary lesion in the lungs; it may be blown into the eyes and produce a primary lesion in the conjunctiva; or, more often, it may be introduced into an abrasion consequent upon scratching by the host.

Most lesions in animal hosts, including man, occur in the endothelial cells of the blood vessels and the serous cavities, and the damage produced is the consequence of the copious multiplication of rickettsiae in these cells.

Rickettsial diseases have played a dramatic role in human history. Without question epidemic typhus fever has been one of the great world scourges. It has occurred frequently where large numbers of people were confined under insanitary conditions, as in gaols (gaol fever) and in times of war and famine. The fate of great military campaigns has often been decided not by armaments but by outbreaks of this fever. Our knowledge of the epidemiology of the disease, however, was realized much later than that of most other bacterial diseases, doubtless because of the minute size of the causal agent and the need to discover its transmission by vectors. By the First World War other rickettsial diseases, such as Rocky Mountain spotted fever (RMSF) and trench fever, were recognized. The distinctions and similarities between epidemic typhus and the endemic typhus fevers of Malaya awaited serological differentiation, but once this was realized extensive search in various parts of the world revealed the broader extent of the rickettsial fevers.

Rickettsial diseases may be divided according to whether they are 'demic', that is, transmitted from man to man, or 'zootic', that is, transmitted from animals to man (Table 19.3). This subdivision also provides a means of classifying the causal agents.

1. *Demic rickettsioses*

(a) *Epidemic typhus.* This is caused by *Rickettsia prowazekii* and is transmitted by lice from man to man. Though not common, it is still found in Eastern Europe. During and after the First World War it caused extensive outbreaks in Eastern and Central Europe, involving both the fighting forces and civilians. Hardest hit was the USSR. It is thought that 25–30 million people fell ill there, with more than 3 million deaths. Outbreaks occurred as far west as the Netherlands and Portugal, but these remained localized.

Of all the vectors of rickettsioses, the louse (*Pediculus humanus corporis*) which transmits the causal agent of typhus fever, is the only vector killed by a rickettsial agent. The rickettsiae taken up by the louse in a blood meal from an infected person invade the epithelial cells of the louse gut where they multiply copiously, making the cells swell until they burst; eventually, insufficient epithelium is left for digestion of food, and the louse dies.

In the human host *R. prowazekii* parasitize the endothelium of the small blood vessels of the brain, producing headache and other cerebral manifestations;

Table 19.3. Classification of some rickettsiae

Characterization	Species	Site in cell	Agglutination with *Proteus*		
			X2	X19	XK
A. *Demic*, from man to man by louse					
1. Epidemic typhus (classical typhus) Brill's disease (mild, sporadic disease in NY)	*R. prowazekii*	Cytoplasm	+	+++	−
2. Trench fever	*R. quintana*	Extracellular	?	?	?
B. *Zootic*, from animals to man by arthropod vectors					
1. *Flea-borne:* Endemic or murine typhus (shop typhus of Malaya—from man to man by rat-flea	*R. mooseri*	Cytoplasm	+	+++	−
2. *Tick-borne* (a) Rocky Mountain spotted fever. Tick-borne fever of South Africa (reservoir in rodents, dogs and sheep)	*R. rickettsii*	Intranuclear	+	+	+
(b) Q Fever—sometimes tick-borne (reservoir in opossums, bandicoots, dogs, cattle, sheep, and goats)	*R. burnetii*	Cytoplasm (some extra-cellular)	−	−	−
3. *Mite-borne:* Tsutsugamuchi, scrub typhus (from rats, mice, voles, to man)	*R. orientalis* (*R. nipponica*)	Cytoplasm	−	−	+++
Rickettsial pox	*R. akari*	Intranuclear	−	−	−

together with fever, these are characteristic of epidemic typhus. The rickettsiae also settle in the capillaries of the skin, producing a rash; their effect on the small blood vessels of fingers and toes may lead to gangrene of the extremities. When the infected endothelial cells of the human host burst, rickettsiae are liberated into the blood. Although many are phagocytosed, sufficient are present in the circulation to be taken up by lice taking a blood meal. Lice tend to leave a host with high fever and search for another with normal temperature. Here they excrete the rickettsiae in their faeces—they defaecate as they bite—and the rickettsiae are involuntarily scratched into the skin by the new host. The fever subsides as the rickettsiae are removed from the circulation.

The question which puzzled typhus workers for a long time was where the rickettsiae persisted in the community between epidemics and in the absence of clinical disease. It was thought that these time-intervals were bridged by inapparent or mild infections. For instance, *R. prowazekii* was isolated in small numbers from the blood of patients suffering with a mild febrile illness called Brill's disease. This occurred only in immigrants from Eastern Europe with a history of past infection. Some of these immigrants settled in and around New York and a few in Ireland and Switzerland. The body louse was often absent. The possibility that Brill's disease was a new infection was considered, but the more likely explanation was a flare-up of an old infection in persons whose immunity had decreased. Further evidence was provided when rickettsiae were isolated from the lymph nodes of two healthy persons who had had typhus fever a long time before, in their country of origin, and who were still serologically positive. There is therefore no doubt that the rickettsiae remain alive in their human hosts for many years and may occasionally, for unknown reasons, pass into the circulation and produce symptoms of Brill's disease. However the numbers in the blood are too few to infect a louse and so cannot initiate a new outbreak. Therefore another explanation for their survival between outbreaks must be found. This may be provided by recent findings that *R. prowazekii* have been demonstrated in the blood of sheep, goats, and zebu in Ethiopia and in ticks living on these animals. This reveals a cycle in which *R. prowazekii* is well adapted. It suggests that the cycle man→louse→man may be a secondary one which provides no advantage to the rickettsiae since the louse invariably succumbs, and sometimes man. No doubt another primary cycle will be demonstrated in Eastern Europe to explain how the causal agent of typhus fever lingers on.

2. *Zootic rickettsioses*

(a) *Endemic, murine, flea typhus.* The infectious agent is *Rickettsia mooseri* and its distribution is worldwide, causing disease under a variety of names: Mexican typhus or Tabardillo in Mexico, Toulon typhus around the Mediterranean, shop typhus (as it occurs mainly in urban areas) in the Far East. It also occurs in the south-east of the USA, in South Africa, in some small foci in Russia, and in Australia. In Australia it is an occupational disease of people handling rat-infested grains.

R. mooseri causes a mild endemic disease of rats; sporadic cases in man occur in persons brought into close contact with infected rats, infection being transferred by the rat flea *Xenopsylla cheopis*. The infection chain is usually rat→rat flea→rat, but occasionally rat→rat flea→man.

The clinical picture is very like that of classical typhus, with a rash and headache, but less severe, the patients remain quite orientated and the lethality is low. *R. prowazeki* and *R. mooseri* are closely related and there is cross-immunity after infection though not after vaccination with killed rickettsiae; hence, although antigenically related, they are not identical and do not share all their immunogenic antigens.

(b) *Tick typhus.* Diseases transmitted to man by ticks include a group of closely related rickettsiae: *Rickettsia rickettsii, R. conorii,* and *R. sibirica.* All these are passed on horizontally from male to female ticks with the sperm, and vertically from generation to generation transovarially. In addition, non-infected ticks may become infected by taking blood from an infected host and, by this means, new lines of infected ticks are created.

R. rickettsii causes Rocky Mountain spotted fever (RMSF) which occurs not only in the Rocky Mountains but generally on the Pacific side of both North and South America and, as San Paolo typhus, on the Atlantic side of South America. It is transmitted from rodents, sheep, and dogs to man by ticks, such as the wood tick *Dermacentor andersoni.* An animal reservoir is, however, unnecessary since the ticks themselves serve both as reservoirs and vectors. RMSF is a severe infection which, before the advent of antibiotics, carried a lethality of 20 per cent, a little higher even than epidemic typhus, and caused all sorts of neurological–paralytic complications. *R. rickettsii* morphologically resembles the other typhus rickettsiae but it alone invades the nucleus as well as the cytoplasm of parasitized host cells (Table 19.3).

R. conorii is the cause of 'fievre boutonneuse', which occurs around the Mediterranean, the Black Sea, and the Caspian Sea. The site of the primary lesion, the tick bite, is usually a little ulcer on the leg with a black crust, the 'tache noire', but not infrequently, dried tick faeces are blown or rubbed into the eyes and then the primary lesion is an inflammation of the conjuctiva. The course of the disease is typical, rather like endemic typhus but without cerebral symptoms except headache, and the prognosis is generally good. Experimentally infected guinea pigs show a scrotal reaction as in endemic typhus. *R. sibirica* caused the Sibirian tick bite fever which is clinically indistinguishable from fievre boutonneuse; it also produces a 'tache noire'.

All these rickettsiae are antigenically related and give cross-immunity after infection but not after vaccination with killed rickettsiae. This is similar to that obtaining with *R. prowazekii* and *R. mooseri* and it is thought, therefore, that all the tick-borne rickettsiae stem from a common ancestor and diverged in evolution. Immunologically related to this group is the rickettsia which produces rickettsial pox (see below).

(c) *Mite typhus.* Mite-borne rickettsiae include *Rickettsia orientalis,* the cause of mite fever, tsutsugamushi, or scrub typhus respectively; the last name distinguishes it from shop or murine typhus which occurs in the same region. It is transmitted to man by mites such as *Trombicula akamuchi* var. *diliensis.* The disease is found in the Far East only. There are a number of strains which differ antigenically and which produce outbreaks of varying severity; the lethality of particular outbreaks may be as low as 1 per cent or as high as 60 per cent.

Small, wild rodents form the reservoir from which the rickettsiae are transmitted to man by the larval stage of mites of the genus *Trombicula* and produce a disease with a primary sore at the site of the bite, a fever, a rash, and—in severe cases—cerebral symptoms. As in typhus, there are foci of infiltration around the small blood vessels, and occasionally a haemorrhagic pneumonia.

The rickettsiae persist in the mite through many generations. In fact, this is their only way of survival. The larvae of the mites suck blood but only on a single occasion; then they fall off the host and the next stages are vegetarian. In man the rickettsiae persist for many years in the lymph glands, in the same way as *R. prowazekii*. Although no illness corresponding to Brill's disease is known, the rickettsiae seem to be shed into the circulation periodically without causing symptoms. Some years ago, a healthy Japanese student donated blood, and three people who received his blood fell ill with the disease tsutsugamushi. Previous donations of his blood had been administered without incident.

(d) *Rickettsial pox*. This disease first appeared in 1946, on a housing estate near New York, when about 100 people fell ill with fever and a vesicular rash. There were no complications or deaths. The rickettsiae were recovered from patients' blood by injection into guinea pigs, mice, and the chicken embryo. The causative agent was *Rickettsia akari* and the vector was a mite, *Allodermanyssus sanguineus*. Mites infected with the rickettsiae were found in large numbers on the mice that infested the estate. Like those which transmit scrub typhus, this mite also could become infected either by sucking the blood of patients or transovarially. Since 1946 rickettsial pox has been reported from many places, for instance Russia and South Africa, and it seems to have a worldwide distribution. Often mistaken for chickenpox its rickettsial nature was unsuspected until 1946.

(e) *Q fever*. Unlike other rickettsial diseases, the rickettsia of Q fever is in part arthropod-borne and in part dust-borne. First described in Australia it is believed to have a worldwide distribution. Derrick (1935) distinguished it on clinical grounds from other fevers of man and called it Q fever pending an internationally agreed name. Fuller knowledge has not produced a better name and the useful term Q fever is retained. Q fever in man is an acute, febrile disease with severe headache and, not infrequently, pneumonia but, in contrast to most rickettsial diseases, no rash develops. Similar diseases reported a few years later in the USA and in troops in the Balkans and Italy during the Second World War, were also identified as Q fever. Infections of cattle, and in turn man, have been reported in the UK. The causal organism has been named *Rickettsia burnetii*.

Human infections are usually derived from domestic animals, principally sheep and cows. There is an absence of symptoms in these animals despite the presence of large numbers of rickettsiae in the milk and foetal membranes; the latter probably constitutes the main source of infection for other domesticated animals and indirectly for man. There is a well-marked seasonal incidence of human infections derived from sheep and associated with the lambing season. Man is mainly infected via the respiratory route. Inhalation of infected dust from dried animal materials is considered to be more important than the drinking of milk from an infected animal. The rickettsiae are very resistant to drying. Thus laundry workers handling contaminated clothing from slaughterhouse personnel have been infected, as well as workers in direct contact with the animals.

Domestic animals in Australia derive their infection from ticks which also parasitize wild animals such as bandicoots, opossums, and kangaroos. Ticks infected with rickettsiae may be brought onto a farm by dogs running in the open country.

R. burnetii produces agglutinins in infected hosts but these do not agglutinate Proteus OX2, OX19 or OXK (Table 19.3). Using the specific antigen propagated in the yolk sac of developing chick embryos it is possible to demonstrate the presence of agglutinins by a complement-fixation test. The presence of the organism in milk or other animal tissues can be demonstrated by intraperitoneal inoculation into guinea pigs; agglutinins may be demonstrated subsequently in their sera.

3. The antigenic structure and serology of rickettsiae and the Weil–Felix reaction

A simple serological diagnostic test for epidemic typhus has been available since the First World War. Edmund Weil found that the serum of typhus patients agglutinated non-flagellate Proteus spp. which he had isolated from the urine and faeces of typhus patients. He isolated more than twenty of these strains, called them Proteus X and numbered them in the order of their isolation. The two strains which gave the most pronounced agglutination with sera from typhus patients were X_2 and X_{19}. This agglutination reaction is called the Weil–Felix reaction, after Weil and his collaborator, Felix. Although not fully specific, it is still the routine laboratory method for the diagnosis of typhus and the strains of X_2 and X_{19}, isolated in the First World War, are still being used.

It seemed rather strange, at first, that a patient's serum should contain antibodies to a high titre against an organism without there being any causal relationship between the organism and the illness. Various theories to explain this have been proposed but it is assumed that the relationship is coincidental, based on a similarity between a rickettsial antigen and a somatic antigen of the Proteus. Amongst the heat-stable antigens of Proteus is an alkali-stable one, loosely called the X-factor, which is responsible for the reaction with rickettsial antibodies. As this antigen is somatic, non-flagellate variants of X_2 and OX_{19} give the clearest reactions.

After the First World War, workers studying the endemic typhus in Malaysia observed that the OX19 strain reacted with the sera of some patients but not with the sera of others clinically ill with typhus fever. Thinking that their strain was losing its agglutinability, the laboratory tested a different strain of *Proteus* OX19. This had been brought out to Malaya by Kingsbury and, to distinguish it from the old strain, was labelled OXK after him. Both strains were used in parallel for a time and quite unexpectedly gave complementary results, i.e. the sera which agglutinated with the original OX19 strain did not react with the OXK strain, and vice-versa. It turned out that strain OX19 was agglutinated by the sera of patients with shop typhus and strain OXK by the sera of patients with scrub typhus, thereby distinguishing two groups of typhus fever found in the same locality.

There is something unique about the way in which micro-organisms present in typhus patients appear to acquire the antigenic properties of the rickettsiae. *Pseudomonas aeruginosa*, and even some strains of *Escherichia coli* isolated from the faeces of typhus patients, in addition to *Proteus* strains, have been shown to agglutinate with the sera of patients. Perhaps in the present climate of advanced genetic manipulation, the acquisition of characters by the mediation of bacteriophage or plasmids does not seem so strange as in earlier days. A *Proteus* variant, specific for the diagnosis of tick-borne typhus, has yet to be found.

The complement-fixation test is the most reliable and specific of all serological tests used for the diagnosis of rickettsial diseases. In parts of the world where typhus-like diseases may be due to either of several possible causes (for instance, in America where Rocky Mountain spotted fever, classical and murine typhus may occur in the same area) the complement-fixation test is able to distinguish between them. The specific rickettsiae used as antigens in this test are grown in the yolk sac of the developing chick embyro, a method discovered just before the Second World War and used both for the preparation of antigens in serodiagnosis and also for the production of vaccines.

4. *Control of rickettsial diseases*

Three means of control have been practised; namely, the use of insecticides to control the vectors, of vaccines, and the treatment of sick patients.

Epidemic typhus can be controlled by the use of DDT on humans and their clothing. The outbreak in Naples among civilians during the Second World War was controlled by this means, and is the pattern should future outbreaks occur. For the control of endemic typhus DDT is also used but it is generally sufficient to sprinkle the insecticide where rats regularly run with a view to reducing the number of fleas they carry.

No account of the control of typhus would be complete without mention of the concentration camps of the Second World War in Germany. As the allied forces reached these camps serious foci of typhus were revealed. By the sheer horror of its conditions the Belsen camp in the British sector was the most notorious. This camp epitomizes the extremes of human misery and suffering imposed on civilian populations as a consequence of war and pestilence. Typhus was introduced into the camp in January 1945, and the British army entered in April of the same year. It is estimated that 20,000 died of typhus during those 4 months. On arrival the British found 10,000 unburied dead and 45,000 living, many of whom were beyond hope. Of these 13,000 died later from typhus fever, tuberculosis, and dysentery.

The first task was to control the outbreak. This included thorough delousing programmes with DDT, washing and clothing of patients, and transport of survivors to a nearby tank training barracks commandeered as a hospital. By the end of April, 30,000 living persons had been dusted with DDT, as had also the corpses and 'human kennels' prior to burial and destruction. By these measures the last case of typhus fever occurred on 14 May. There can be no doubt that the success of this programme was due to the use of DDT powder and, had this not been available, Western Europe, debilitated by war and shortage of food, and with many displaced persons, would have suffered a major epidemic.

Vaccine production requires the propagation of the specific rickettsiae but their intracellular nature necessitates an *in vivo* system for their propagation. At the time of the First World War this involved infection of lice from which the gut was removed and suspended to make vaccine. At first the lice were infected either directly, by feeding them on typhus patients, or indirectly, by injecting a patient's blood into guinea pigs and then feeding the lice on these. The louse was particularly suitable for this technique, the gut being normally sterile and, after inoculation, acted as a living culture medium. In 1917 Weigl developed a technique of infecting lice directly into the tail end of the gut. For 25 years the value of the phenolized louse-gut vaccine was not disputed. The fact that about 10 million people were vaccinated with this material was an enormous feat, considering that about 150 louse guts were needed to vaccinate one person; the enormity of the louse-breeding programme can be visualized. However, the immunity obtained by it was not absolute, but generally infection, when it occurred in vaccinated people, was milder and the numbers of rickettsiae in the blood too small to infect lice feeding on that patient.

Only the pressure of the Second World War made it imperative to look for methods that would allow vaccine production on a commercial scale. It is strange that this development took place only in the West. In Central Europe, even in the Second World War, typhus vaccination continued to be organized on a louse-breeding and louse-feeding scale. The rectal injection of the lice, extraction of the louse-gut and purification of the rickettsial suspension were, to some extent, mechanized but the vaccine was still prepared in lice.

In the West, various attempts were made to break away from the arthropod material for vaccines. French workers had tried to use a living vaccine in the form of the dried brains of infected guinea pigs; this vaccine produced cases of typhus and had to be abandoned. In 1938, Cox in the USA perfected a vaccine devised a year before by Barykine. Rickettsiae were inoculated into the yolk sac of the chicken embryo. The Cox vaccine, which became the standard material for vaccination, consisted of chick tissue ground up and phenolized. It did not, however, work for scrub typhus. Despite its widespread use, especially in war zones, the immunogenic components of the rickettsiae are still unknown and its protective value still in doubt.

The use of broad-spectrum antibiotics has transformed the clinical management of the rickettsial group of diseases. Chloramphenicol in particular has been widely and successfully used in the therapy of Rocky Mountain spotted fever, the typhus fevers, rickettsial pox and Q fever, both in reducing the mortality rate and the length of the febrile period.

FURTHER READING

Report (1979) *Rickettsial Diseases Surveillance, Summary 1975–1978*. Center for Disease Control, US Department of Health, Education, and Welfare, Atlanta, Georgia.

Steele, J. H. (ed.) (1980) *Rickettsial Diseases*. In Steele, J. H. (chief ed.), *Handbook Series in Zoonoses*, vol. 2. CRC Press Inc., Florida, pp. 279–434.

Wilson, C. S., and Miles, A. A. (1975) *Principles of Bacteriology and Immunity*, 6th edn, vols I and II, pp. 1195–1205 and 2343–63.

Chapter 20

Wound infections

Any breach in the integrity of the skin increases a host's susceptibility to certain infections to which it may be exposed. Wounds may range from minor abrasions (e.g. scratch wounds), through minor puncture wounds (e.g. the bite of an insect—some have been considered in the previous chapter), to major wounds which may occur accidentally or as a consequence of surgical interference. A number of important bacterial infections are mechanically introduced into wounds. These include infections by many of the clostridial pathogens, e.g. *Clostridium tetani* and *Cl. perfringens* and by *Bacillus anthracis*. In this chapter a further example of a wound infection is considered: the rabies virus, commonly introduced by the bite of a rabid animal.

Rabies

Rabies is an infectious disease characterized by acute and profound dysfunction of the central nervous system. It is caused by a medium-sized (180 × 60 nm), rod-shaped virus of the *Lyssavirus* genus. The virus exists as the wild-unattenuated form or 'street' virus and as the laboratory-attenuated or 'fixed' virus. Recent studies have shown that a number of subtypes exist throughout the world which differ in the antigenicity of their surface glycoprotein 'spikes'. In addition, serologically distinct rabies-related viruses have also been described, viz., the Lagos bat virus (Nigeria, 1970) and the Duvenhage bat virus (South Africa, 1971). No doubt other rabies-related viruses also exist in tropical areas.

1. *History of rabies*

Rabies is a disease of antiquity; it was stated by Ahuja to be present in India 5000 years ago and reported in classical times by Aristotle and Celsus. In Western Europe historical accounts of fox epizootics date back to AD 1271 and the opening of Atlantic trade routes enabled it to spread to North America in 1753 and to South America in 1803. In 1964 a WHO survey indicated that the disease was worldwide with few territorial exceptions, notably those countries which had natural water barriers separating them from neighbouring countries, such as Great Britain (in which the disease was eradicated in 1922), Australia (eradicated in 1876), New Zealand, Tasmania, New Guinea, the Hawaiian archipelago,

297

Polynesia, Iceland, and some Caribbean islands. In Western Europe the disease has spread since 1945 in a westerly direction from Poland and has now reached mid-France. The alarm felt about rabies reaching the French Channel ports has not so far been realized.

2. *Reservoirs of infection*

The natural reservoir of the virus is the wild mammal, especially the carnivore, and the various species of bats. Since reservoir animals hold the key to ultimate eradication of the disease, their identification is always a prime objective. The role of wild rodents has formed the basis of a number of independent studies since the first reported transmission of rabies in Trinidad from a fieldmouse to man in 1929. Since then wild earth-burrowing rodents in Yugoslavia have been implicated as reservoirs based on the coincidence of fox enzootics with periodic surges in wild rodent populations. The inference has been that the fox, a natural predator of the rodent, is more likely to contract the disease when rodents are plentiful. However, other studies do not indicate that rodents function as reservoirs and so the initiation of European fox enzootics may depend on other factors.

In the mammal the virus multiplies at the site of entry, and may produce a transient viraemia before infecting the central nervous system via the peripheral nerves, and the salivary glands. Infection of the salivary glands, and the subsequent production of infected saliva in both silent and acute infections, is of prime importance in transmission. In acute infections infected saliva is produced about 3 days before classical symptoms commence and the virus enters the dermal or subdermal tissues of the next host via licks or bites. Respiratory or ocular entry, due to saliva aerosols, also occurs. The eating of infected tissues by carnivores is considered to be an alternative transmission route.

The bites of vampire bats transmit infection, indicating that infectious saliva is found in these animals also. Bat-to-bat spread is considered to occur during periods of social grooming and similar activities. The blood-licking bats, of the family Chiroptera, are an important reservoir in the USA. In Europe the bat is rarely implicated. Rabies transmitted by vampire bats is usually of the rarer paralytic form ('dumb' rabies).

3. *Human rabies*

Man becomes infected through close contact with an infected animal and in Western Europe the classical sequence of transmission is shown in Figure 20.1. Man cannot function as a reservoir of the disease since rabies is invariably fatal and to date no carriers have been recorded. The normal disposal of human dead by burial or cremation further precludes transmission back to animals.

The risk of working with rabies virus in the laboratory needs little emphasis. In 1977 an American worker became infected when preparing aerosols under conditions of imperfect containment and survived with severe mental defect. The more

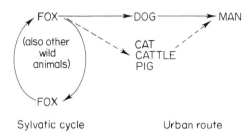

Sylvatic cycle   Urban route

Figure 20.1. Classical transmission routes
for rabies

unusual and fatal cases of adventitious rabies occurred in 1979 following corneal transplantations from unexpectedly infected donors and serve to illustrate the considerable problems attendant upon screening tissue-donors.

Human rabies nearly always results from contact with a rabid animal and therefore there is need for close surveillance of the disease in wild and domesticated animals. Recent European figures indicate the fox as the reservoir of greatest importance. In Germany, 80 per cent of the country in 1969 had a rabid wild animal population and 83,636 cases of animal rabies were reported in the period 1950–76, the fox accounting for 63 per cent of these (Table 20.1). Local fox control measures significantly reduced the incidence of feral disease and the cessation of this control in 1964 resulted in a subsequent rise in the incidence of rabies in the fox. In France, in 1968, rabies crossed the border from Saarland into the Department of Moselle. From here the disease maintained a spread of 30–60 kilometres per year in a south-westerly direction. In the first half of 1977 the disease crossed into Switzerland from France. The incidence rate (Table 20.1) is similar to that for Germany and the fox again stands out as the prime reservoir as in past European epizootics.

In Great Britain rabies was first eradicated in 1902, and the country remained free of the disease until its reintroduction by an illegally landed dog in 1918. 328 cases in animals occurred up to the time it was eradicated in 1922. In the period 1922–69 there were 27 cases in quarantined animals but none outside the quarantine premises. In 1969, in the Camberwell incident, the disease was accidentally reintroduced in a quarantine dog due to an unusually long incubation period which exceeded the 6 months quarantine by 10 days. In 1970 a similar incident occurred with a second quarantined bitch in Newmarket in which the incubation period was 9 months. Neither animal provided secondary cases and they were not related in any way. No further cases in animals have occurred since, but the entry of Great Britain into the EEC in 1973 prompted the swift drafting of legislation as a precautionary measure.

In the USA, in the period 1969–73, skunks headed the list of wild animal reservoirs, with foxes and bats in second and third places (Table 20.1). In the skunk the production of infected saliva precedes clinical symptoms by 14 days,

Table 20.1. Percentage distribution of rabies amongst animals submitted for diagnosis in Germany, France, and the USA

Animal species	Germany 1950–76	France 1968–75	France 1977	USA 1969–73
Dogs	3.9	2.3	2.6	5.6
Cats	5.5	3.2	4.4	4.4
Cattle	7.3	10.2	8.2	10.1
Other domestics (sheep, goats, pigs)	7.3	2.5	7.0	1.8
Foxes	63.3	78.4	77.0	17.8
Badgers	1.2	1.5	NA	NA
Other mustelidae	2.6	1.9	NA	NA
Wild deer	8.9	NA	0.8	NA
Skunks	—	—	—	43.3
Racoons	—	—	—	4.7
Bats	NA	NA	NA	10.5
Total animal cases	83,636	9,467	NA	19,307

NA = Not available.

compared with the 2–4 days in the dog, which makes the skunk a very much more dangerous transmitter of the disease. There is, therefore, an obvious difference in the principal wild animal reservoir in the USA compared with Europe, although the relative positions occupied by dogs, cats, and cattle are roughly similar. Human infections in the USA, having decreased from a mean of 22 cases per annum, in the period 1946–50, to 1–5 cases per annum in the period 1960–80. A similar decrease has been recorded in the confirmed cases amongst domestic animals; in the dog, for example, the number of cases has fallen from 8000 in 1946 to 122 in 1978, which has drastically reduced the likelihood of human exposure to these animals even though dog (and bat) bites still constitute the principal reason for initiating prophylactic anti-rabies treatment.

In American wildlife, especially in skunks, foxes, racoons, and bats, there has been an upward trend in the numbers of confirmed cases in recent years, from 3000 cases in 1976, 1977, and 1978 to 5000 cases in 1979. This is thought to represent a true increase of spread in the principal reservoirs rather than any increase of awareness and surveillance by the public health authorities. Since 1968 wild animals have accounted for 70 per cent of all confirmed cases of animal rabies and so they remain the prime source of infection for both man and his domesticated animals throughout the United States.

It should be borne in mind that in addition to natural geographical shifts in the principal animal reservoirs shifts may be induced as a direct outcome of eradication programmes. For example, until the institution of an eradication programme in the early 1950s the jackal was the principal reservoir in Israel. The subsequent drastic reduction in both jackal and stray dog populations saw the fox emerge as principal reservoir.

In Western Europe the fox and the dog stand out as main links in the chain of transmission to man, the fox being the classical reservoir and the dog being man's principal animal contact (Figure 20.1). Since AD 1271, when it was first recorded that a human epidemic coincided with the deaths of large numbers of foxes in southern France and Germany, the fox has been the principal reservoir in European history. Subsequently, other urban epidemics similarly coincided with outbreaks of the disease in foxes, though the linking of the fox disease with the human disease remained to be demonstrated. In 1804, 1807, 1823–28, and 1847–48 there were coincidental fox and human outbreaks of the disease. In more modern times, in Germany in 1940, Yugoslavia in 1950, Vermont in 1965, and France in 1968, outbreaks in foxes and human populations have taken place simultaneously. More recent data for Europe confirm the fox as the principal reservoir (Table 20.2).

It may be asked why the fox is implicated when *all* mammals are susceptible to the disease; the simple answer is that the fox is very highly susceptible to the virus compared with other animals, needing only 10 mouse LD_{50} doses to infect (Table 20.3). It has been claimed that there is a reciprocal relationship between sensitivity and the level of serum ribonuclease in these animals, suggesting that a protective effect is exerted by the enzyme. It is also claimed that bites from rabid wolves are more serious because of the higher level of hyaluronidase (spreading factor) in the saliva. The wolf is a notable reservoir in Eastern Europe and the Middle East.

4. *The control of rabies reservoirs*

It follows that control programmes must be directed initially against the fox. Fox control is not as simple an operation as might first appear. Fox-to-fox transmission can be reduced to negligible proportions when the fox population is reduced

Table 20.2. Rabies in Europe in 1979 and 1980

	1979	1980
Number of cases		
Human rabies	4	3
Animal rabies	17,073	18,603
Percentage in		
Domestic animals	20.6	23.4
Wild animals	79.4	76.6
(Foxes	70.1	68.6)

Data adapted from the *Rabies Bulletin in Europe*, 4/80, 1980.

Table 20.3. Order of susceptibility to rabies

Fox	
Cattle	
Skunk	Ascending order
Man	of sensitivity
Racoon	
Opossum	

to around 1 per 700 acres (283 hectares). At this low density food is more available and the proportion of female:male cubs in new litters rises from 50:50 to
80:20 as a natural means of compensation. The population then quickly increases
to a point where a further reduction in numbers is needed.

By far the most important biting mammal for man in all rabid areas of the
world is the dog. In addition to dogs which display signs of active infection,
infected saliva has also been detected in clinically healthy dogs; these animals
function as silent carriers and constitute a subtle threat to human safety. In many
areas of the world the relatively large number of stray dogs provide the crux of the
transmission problem; young dogs tend to be more vicious than older dogs in their
biting habits.

To control the numbers of rabid dogs prophylactic immunization of registered
dogs, together with culling of strays, needs to be undertaken. Immunization in the
USA considerably reduced the frequency of canine rabies in the period 1946–78
(see below) and this action, together with more efficient human treatment,
successfully reduced the human cases within the same period from 20 per annum
to 2.

Reservoir control is possible only when the reservoir animals can be eradicated
readily, or vaccinated. The greatest successes have been achieved against foxes
(especially in Europe) and dogs (in other rabies areas of the world). For instance,
in Taiwan in 1951 there were 238 human cases of rabies, and an anti-rabies
campaign directed at the dog population was commenced in 1952 with the culling
of strays and the compulsory vaccination of the remainder. By 1959 human cases
had been reduced to zero after a total cull of 1961 strays and the vaccination of a
total of 4670 others in the period.

5. *Control of rabies by prophylaxis*

Over and above such control measures there is the need for adequate prophylactic
treatment of the human contacts which inevitably occur in rabid areas of the
world. The chance of a person developing clinical symptoms of the disease after
being bitten by a rabid animal is only 1 in 5, and it may be questioned what makes
rabies *so* feared that major efforts are directed towards the control of animal

rabies and towards the development of efficient prophylaxis? The answer is to be found in the fatal outcome of the disease once symptoms are manifest and the horrific progression of the disease in the period preceding death. In both man and animals there is a prodromal period of non-specific upsets, such as fever, respiratory and gastrointestinal upsets, insomnia, and anxiety which are hardly pathognomonic. This period is followed by one of excitement ('furious' rabies) in which are displayed changed behaviour, cyclical arousal, muscular spasms, photophobia, salivation, and manic behaviour; it is in this period that hydrophobia (the inability to swallow due to spasm of the pharyngeal muscles) is observed in the patient. A subsequent period of depression ensues ('dumb' rabies), with diminished or absent vocalization, apathy, increasing loss of voluntary muscle movement, stupor, and coma being exhibited before the interruption of vital functions—respiration and heart rhythm—causing death. Herbivores infrequently exhibit 'furious' rabies; in dogs the proportion of animals which exhibit 'furious' rabies is 75 per cent, while in cats the proportion is 25 per cent; in many other animals the 'furious' stage may not occur. In man the 'furious' stage is more common and the 'dumb' stage rare. The patient is usually aware of the progression of the disease up to the time at which coma intervenes, the period varying from days to weeks. It is this awareness of the disease, its progression and fatal outcome, which makes it so feared.

In the entire recorded history of the disease there is only one fully documented case of survival following the display of clinical symptoms (Hattwick *et al.*, 1972), and only one recorded case of prolonged survival for 19 weeks (Emmons *et al.*, 1973), in which the presence of the disease was incontrovertible. With such an ephemeral chance of survival once the symptoms become manifest it is imperative that prophylactic treatment is commenced immediately following exposure, and here the usually extended incubation period of the disease may be used to the patient's advantage. While the incubation period may extend from 1–2 weeks to 18 months (as documented for man and dog), the nearer the virus enters the tissues to the central nervous system, the shorter will be the ensuing incubation period. This is one reason why young children, who are inevitably bitten or scratched on or near the face or head, usually succumb rapidly to the disease despite prophylactic treatment.

Today exposed persons are given prophylactic treatment which involves the vigorous cleansing of bites, scratches, or saliva-contaminated skin with soap and water, and the administration of an efficient vaccine. Immune serum is also administered at the same time as the first vaccine. The choice of vaccine is based principally on efficiency and safety, and the human diploid-cell (WI-38) vaccine and human immunoglobulin are first choices because of lower risks of adverse reactions in the patient.

 Modern rabies prophylaxis, using the human diploid-cell vaccine, may be considered to be a breakthrough in the control of rabies since the failure rate is now approaching zero. It remains for a similar breakthrough to be realized in the control of feral rabies before this zoonosis loses its fear for man.

FURTHER READING

Communicable Disease Report (1981) *Rabies in Europe 1980*. No. 16. PHLS, Colindale, London.

Emmons, R. W., Leonard, L. L., De Genaro, F., Photas, E. S., Bazeley, P. L., Giammona, S. T., and Sturckow, K. (1973) A case of human rabies with prolonged survival. *Intervirology*, **1**, 60–72.

Hattwick, M. A. W., Wies, T. T., Stechschulte, C. J., Baer, G. M., and Gregg, M. B. (1972) Recovery from rabies. A case report. *Ann. Intern. Med.*, **76**, 931–42. (The 5-year follow up on this case was documented in *C.D.C. Vet. Public Hlth. Notes*, December, 1975. Center for Disease Control, US Department of Health, Education, and Welfare, Atlanta, Georgia.)

Kaplan, C. (1977) *Rabies: The Facts*. Oxford University Press, London.

Macdonald, D. W. (1980) *Rabies and Wildlife: A Biologist's Perspective*. Oxford University Press, London.

MAFF Handbook (1976) *Rabies and its Controls: a selected bibliography of English-language publications*. MAFF, Surbiton.

Chapter 21

The epidemiology of fungal disease

Recognition of the importance of fungi as pathogens of man and animals is comparatively recent. In the early days of mycopathology at the turn of the century, although oral thrush was well known, interest centred mainly on ringworm, especially scalp ringworm in children, which was rife in Europe at the time and was very obviously a contagious disease. By the 1930s the major systemic mycoses (fungal infections of the internal organs) of man in the Americas had been firmly attributed to fungi, and a search was being made for the sources of infection, which were far less obvious than in the case of ringworm. Two or three decades later it had been accepted that not only the systemic mycoses, but also the subcutaneous diseases found in many parts of the world, are caused by fungi which grow as saprobes (saprophytes) in the environment, and that infection is contracted only from this source.

Meanwhile, as medical procedures became more sophisticated and the use of broad-spectrum antibiotics, steroids, immunosuppressive drugs, and major surgery became commonplace, increasingly frequent reports were appearing of a whole new range of systemic, subcutaneous, and other infections of man, attributable to fungi known previously only as causing mild or rare conditions. The era of the iatrogenic mycoses due to opportunistic fungi had begun. In animals iatrogenic infections are rare because animals are expendable, but opportunistic mycoses resulting from poor husbandry are well known.

1. Nature and pathogenicity of fungi

Fungi are heterotrophic organisms. Because they lack chlorophyll they are unable to synthesize their own carbohydrates but must depend for these essential nutrients on materials already synthesized by plants and animals. Saprobic fungi utilize the residues after the death of plants and animals, commensals grow in sufficiently close contact with living organisms to enable them to use excreted organic substances, and parasites extract their sustenance from the living cells of the host. Stages in the life cycles of fungal pathogens of man and animals are found in all these categories, and many pass from one category to another as the opportunity arises.

The great majority of fungi favour substrata of vegetable origin. These organisms are largely responsible for the continuous recycling of plant remains which occurs in nature, and are also the agents of the majority of plant diseases, including the most devastating plagues of man's crops. In contrast, fungi play only a small part in the breakdown of animal residues, with the exception of keratinized materials such as hair and horn, and are much less common as agents of animal disease than are bacteria. This may account for the comparatively recent realization of the importance of mycoses in medical and veterinary practice.

Many plant diseases are caused by fungi which are so well adapted to their role that they are either obligate parasites or can only survive in a resting phase in the absence of their preferred host. But this is only true for a very small proportion of animal pathogens most of which, in the normal course of events, are harmless saprobes or commensals for which pathogenicity is an accidental and unnecessary part of the life cycle. Some of these fungi are *primary pathogens* which are able to attack a healthy host, but many are *secondary* and can only invade an already compromised host, i.e. one in which the natural defences are impaired. Fungi causing secondary infections are classed as *opportunists* and this group must include, too, the agents of most subcutaneous mycoses, since these are harmless unless inoculated into the body through a wound. The small group of fungi which are 'obligate' pathogens, in the sense that they are soon overwhelmed by other environmental organisms when growing as saprobes, are all primary pathogens.

2. Source of infection in fungal disease

In fungal disease the pathogen may reach the host in different forms and by different routes, depending on its life cycle and on how essential the pathogenic phase is in that cycle. There are three distinct patterns of behaviour:

(1) Some potentially pathogenic fungi are primarily saprobes, inhabiting a particular environmental niche and participating in the normal way in the breakdown and recycling of constituents of its substrate. In these organisms the spores produced in the saprobic phase are the source of infection; cross-infection between animal hosts does not take place directly or indirectly. This type of life cycle is characteristic of the causal organisms of all the major systemic and subcutaneous mycoses.

(2) Some fungi, although able to grow in pure culture, cannot survive as saprobes in a natural environment where they are in competition with other micro-organisms. As there is no naturally occurring saprobic phase, dissemination to new animal hosts is by direct or indirect cross-infection. Many of the dermatophytes are included here.

(3) A few fungi, notably *Candida albicans* and species of *Malassezia*, are normal commensals of the human or animal body and have no other saprobic phase. These may become actively pathogenic if changes in the circumstances of the host result in conditions especially favourable to the proliferation of the

fungus. Disease, therefore, is usually the result of endogenous infection. Clearly cross-infection is a regular occurrence, but the outcome is more often the beginning of a commensal relationship with the new host than of a diseased state.

These three variations on the natural history of mycoses form a convenient basis on which to study the epidemiology of the diseases.

3. Epidemiology of mycoses caused by saprobic fungi

(a) *Fungal spores and their dispersal*

At some time in their life cycle most fungi sporulate. The spores produced may be of asexual origin or they may be formed following the sexual fusion of two complementary nuclei. They are of infinite morphological variety and some fungi have life cycles which include up to five different types. The main function of spores is to enable the fungus to colonize virgin sites beyond the range of its mycelium which extends only through the adjacent substrate, but they may also, in unfavourable environmental conditions, pass into a resting state, remaining dormant until conditions improve. Fungi causing plant disease sporulate at the surface of the host during the pathogenic phase, but animal pathogens remain within the tissues and sporulation is rare in this phase, mainly because of the inadequate supply of oxygen.

Spore dispersal may be active, passive, or by each means in succession, and both the spores themselves and the sporogenous apparatus are variously and often beautifully adapted to a particular means of dispersal. In active dispersal the spore is either forcibly projected from the tip of the sporogenous hypha, e.g. the basidiospore of the toadstools, or extruded from the mouth of the small flask-shaped fruiting bodies which are common in the large group of Ascomycetes. Passive dispersal may be by means of air currents, films or droplets of water, running water, the legs of insects, etc.

From the epidemiological point of view a very fundamental distinction is that between 'wet' and 'dry' spores, for this governs the means by which they are dispersed and hence their route of entry into the host and the site of the resulting infection. Dry spores have water-repellent walls, are very light and easily airborne, and are dispersed by air currents, sometimes for many miles. They enter the host by the respiratory route, either giving rise to pulmonary disease or being carried from the lungs by the blood stream to other organs. Wet spores, on the other hand, either have hygroscopic walls and are dispersed in moving films or droplets of water, or are coated with a layer of mucus which, when moist, enables them to stick to another object for transport to a new substrate, and also to remain firmly attached to the new substrate until germination and penetration can occur. Such fungi, in order to pass the barrier of the skin, must be inoculated into the animal host, resulting in subcutaneous infections.

(b) *Geographical distribution*

Some of the potentially pathogenic, saprobic fungi, for instance *Aspergillus fumigatus* and *Cryptococcus neoformans*, are of worldwide distribution; some, such as *Histoplasma capsulatum*, though they are found in every continent, occur in strictly localized pockets where their special environmental requirements are met; a few, such as *Coccidioides immitis*, are very strictly localized geographically. Clearly the diseases of which they are the agents must be limited to the same geographical areas as their causal fungi. Tables 21.1 and 21.2 show the distribution of the major systemic and subcutaneous mycoses of man and animals together with the preferred substrate of the causal fungus if this is known. However, cases are not distributed uniformly within the endemic areas because of the intervention of other, overriding, limiting factors.

The epidemiology of some representative systemic and subcutaneous infections will be considered in the light of the biology of their causal fungi.

(c) *Systemic mycoses*

Histoplasmosis is a primary, pulmonary mycosis which was first described in man. Ninety-five per cent of patients are asymptomatic, the disease being detectable only by means of X-rays and a skin test with an extract of the causal fungus, *Histoplasma capsulatum*. Sometimes, however, clinical lung infection is apparent and very occasionally the disease progresses to a disseminated and usually fatal form.

The epidemiology of histoplasmosis has been most thoroughly investigated in the USA where, in the focal district in the eastern central states, it has been shown that 80 per cent of the population has been infected. The further away from these states that observations are made, the lower the incidence becomes. Skin testing surveys in other countries have shown a high incidence of positive reactors in Mexico and Panama; a low incidence in parts of South America, Europe, and Australia; and none in Alaska. In endemic areas most species of wild and domestic animals may be infected but overt disease is as rare as in man. In the laboratory it has been shown that for optimal growth the saprobic phase of *Histoplasma capsulatum* requires a high humidity and a temperature of between 20 and 30°C, and it is possible that the geographical distribution of the fungus is partially governed by these factors.

As a saprobe *Histoplasma capsulatum* is a filamentous fungus bearing large, round macroconidia with thick, tuberculate walls, and small round to pyriform microconidia (Figure 21.1). It thrives in soil in competition with the normal microflora provided that its special nutritional requirements are met, and although at present these requirements cannot be exactly defined, it is abundantly clear that they are supplied by the droppings of bats, chickens, and many other bird species. The fungus can be isolated readily from dried deposits of these

Table 21.1. Geographical distribution of the major systemic mycoses of man and animals

Disease	Host	Causal fungus	Preferred substrate	Geographical range
Aspergillosis	MA	*Aspergillus fumigatus*	Decaying vegetable matter	Worldwide
Cryptococcosis	MA	*Cryptococcus neoformans*	Pigeon droppings	Worldwide
Histoplasmosis	MA	*Histoplasma capsulatum*	Excreta of birds and bats	Worldwide in pockets
Coccidioidomycosis	MA	*Coccidioides immitis*	Desert soils	Central America, northern South America
Blastomycosis	MA	*Blastomyces dermatitidis*	?	North America
Paracoccidioidomycosis	M	*Paracoccidioides brasiliensis*	?	South America
Adiaspiromycosis	A (rodents)	*Emmonsia crescens*	Soil	Worldwide
Mycotic abortion	A (cattle)	Miscellaneous fungi	Mouldy forage, etc.	Worldwide

M = man; A = animals.

Table 21.2. Geographical distribution of the major subcutaneous mycoses of man and animals

Disease	Host	Causal fungus	Preferred substrate	Geographical range
Sporotrichosis	MA	*Sporothrix schenckii*	Timber, vegetation	Worldwide
Subcutaneous histoplasmosis	M	*Histoplasma duboisii*	?	Africa
Epizootic lymphangitis	A (horses)	*Histoplasma farci*	?	Europe, Asia, Africa
Mycetoma	M	Fungus according to geographical origin	Various	Worldwide
Chromomycosis	M	*Phialophora* spp.	Vegetation	Tropics
Subcutaneous phycomycosis	M	*Basidiobolus haptosporus*	Rotting vegetation	Tropics
Hyphomycosis	A (horses)	*Hyphomyces destruens*	?	Asia, North America
Rhinosporidiosis	MA (horses)	*Rhinosporidium seeberi*	?	India and elsewhere

M = man; A = animals.

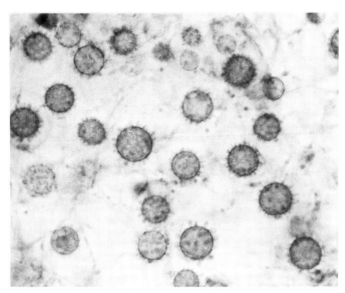

Figure 21.1.   Saprobic phase of *Histoplasma capsulatum*. Tuberculate macroconidia ($\times 450$)

materials the disturbance of which—by cavers, chicken farmers, and others—precedes outbreaks of histoplasmosis.

Cryptcoccus neoformans is also a coprophilous (dung-loving) fungus pathogenic for man and animals, but it is less catholic in its tastes than *Histoplasma capsulatum*, apparently being restricted to the excreta of pigeons. It has been postulated that the creatinine content and low pH of pigeon droppings may make them a selective medium for the fungus, which has been found in this substrate in most countries where a search has been made. In man, cryptococcosis usually presents as an infection of the central nervous system and it is presumed that this results from haematogenous spread of the organism from minimal primary lung lesions. It may be diagnosed in a previously healthy patient shortly after known exposure to a source of the fungus (primary cryptococcosis), or in a patient suffering from some other grave or debilitating illness (secondary cryptococcosis). In the latter case any association with pigeon droppings is obscure, but it is presumed that latent lung lesions become activated as the patient's immunological status is lowered. Until recently only the heavily capsulated yeast form of *Cryptococcus neoformans* (Figure 21.2) was known and it was assumed, in spite of the large size of the units, that it was in this form in which the fungus was dispersed aerially from the dried droppings. A more probable means of infection would seem to be the recently discovered basidiospore, produced by the perfect state of the fungus *Filobasidiella neoformans*. These

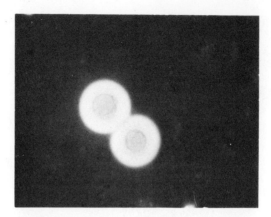

Figure 21.2. *Cryptococcus neoformans* in cerebrospinal fluid. Mounted in Indian ink to show
the large capsules (×450)

spores are catenulate, are only 1.8–2.5 μm in diameter, and would be much more
easily airborne than the yeast phase.

Unlike the two preceding pathogens, the preferred substrate of *Aspergillus
fumigatus* in its saprobic phase is vegetable debris of almost any type. As well as
colonizing such material at normal environmental temperatures it is tolerant of
temperatures of 40°C and upwards such as are found at the centre of a compost
heap. Geographically the fungus has been isolated in most parts of the world. The
small conidia, produced in huge numbers on the mop-like sporing heads (Figure
21.3), are easily airborne and are ubiquitous in the air of town and country alike,
yet a healthy host exposed to an average dose of spores runs no risk of contracting pulmonary aspergillosis. Infection is dependent on an overwhelming dose of
spores, on a depressed state of immunity in the host due to physical disease, drug
therapy or stress, or on a combination of these circumstances. In domestic
animals, and more particularly birds, exposure to huge concentrations of spores
from mouldy bedding or feed, especially within the confines of the buildings in
which the animals are housed, can precipitate massive outbreaks of aspergillosis.
Man is less often exposed to high doses of spores in this way, but cavities
produced in the lungs by tuberculosis and other diseases, asthma in atopic
patients, transplant and other major surgery, are all predisposing conditions
which are exploited by *Aspergillus fumigatus*. Fatal pulmonary aspergillosis in
wild birds and animals newly taken into captivity would seem to be an example of
the combination of stress and spore dose as predisposing factors.

Mycotic abortion in cattle is the cause of considerable financial loss to farmers
and has therefore been the subject of much research effort. However, although it
is known that the condition can be caused by many different species of fungi, it
has yet to be reproduced experimentally by any but entirely artificial means. The
majority of causal organisms, which include *Aspergillus fumigatus* and a number

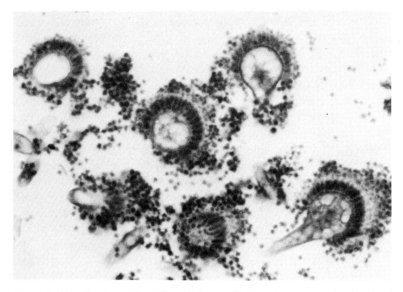

Figure 21.3. Section of the infected air sac of a hen showing sporing heads of *Aspergillus fumigatus* (× 450)

of Phycomycete species, are common in mouldy hay and straw, and the concentration of their spores in the atmosphere of cowsheds may be very high indeed. As most cases of mycotic abortion occur in the winter months when cattle are being fed on hay it can be assumed that this is the source of infection. In the absence of direct evidence, it seems probable that infection of the placenta follows haematogenous spread of the fungus from the lungs after inhalation of the airborne spores.

Understandably, most mycoses of mammals have been reported from domesticated species, but adiaspiromycosis is an exception. This disease, caused by *Emmonsia crescens* and *Emmonsia parva*, is known almost exclusively as a pulmonary infection of wild, burrowing animals, and is of worldwide distribution. In their saprobic phase in soil *Emmonsia* spp. are mycelial fungi reproducing by means of small conidia which are, presumably, inhaled by the animals during their subterranean activities, and develop in the lungs into the large, round adiaspores from which the disease is named. The fact that adiaspiromycosis is seldom reported from animal species which do not inhabit burrows suggests that the spores of the saprobic phase of the fungus are not dispersed in the atmosphere in any quantity.

(d) *Subcutaneous mycoses*

These diseases, caused by a wide variety of fungi and presenting in clinically varied forms, have in common that damage to the skin is a prerequisite to the entry of the fungus. The limbs, being the parts of the body most subject to casual traumatic injuries such as grazes and scratches, are the sites most usually affected. The fungus, usually a wet-spored species, may grow on the object causing the injury, e.g. tree-bark, a thorn, soil; or the object may merely be contaminated with the spores of the fungus.

The clinical condition known as mycetoma may be induced by aerobic and anaerobic actinomycetes as well as by a large number of different fungi, the causal organism in a given case being to some extent governed by the geographic origin of the patient. For instance, the fungus responsible for most cases of eumycetoma in Africa is *Madurella mycetomi*, but *Madurella grisea* is the predominant organism in South America, and *Monosporium apiospermum* in Europe and the USA.

Unlike mycetoma, only one fungus, *Sporothrix schenckii*, is responsible for the clinical entity, sporotrichosis. The disease has been reported from all continents and from both tropical and temperate zones, and when the saprobic phase of the fungus has been traced it has always been found on tree-bark and other plant surfaces. Grazes sustained from such objects permit subcutaneous inoculation of the fungus, ulcers form at the site, and there is secondary spread along the lymphatics draining the area. Large outbreaks of this disease, such as occurred in the gold mines of the Rand in 1942, and are discussed in more detail later, follow massive contamination of the environment by the fungus. Each infection is freshly contracted from the saprobic phase; cross-infection does not take place.

Subcutaneous phycomycosis is confined to the tropics where the causal fungus, *Basidiobolus haptosporus*, is common in leaf mould and other decaying vegetable matter and can be isolated from both the dung and intestinal tract of frogs. The disease occurs most often in children and young people and may affect the trunk or limbs. It would seem that slight injuries to scantily clad bodies, as children play in any available space near their villages, could facilitate entry of the fungus from its preferred substrate of rotting leaves and other plant material.

(e) *Mycoses caused by occasional pathogens*

Theoretically any fungus capable of growth at the body temperature of an animal host is a potential pathogen of that host, and many normally saprobic fungi have been recorded as human or animal pathogens on a single occasion only, or on a very few occasions. For instance, one case of mycetoma in man due to *Pyrenochaeta romeroi* has been reported, and a severe, indurated lesion of the face has been reliably attributed to *Cercospora apii*. Such infections are almost always iatrogenic or secondary to some predisposing trauma or underlying disease, and their onset depends on the chance conjunction of the spores of a fungus growing in the environment with a host in a suitably vulnerable condition. Infections vary in severity from fatal systemic conditions in which a saprobic fungus has gained entry to a vital organ during a major operation, to the colonization of the traumatized toenails of elderly persons by a wide variety of common and uncommon moulds. Recognized pathogens can also gain access to unusual sites in similar circumstances. For instance, a cardiac infection by *Aspergillus fumigatus* precipitated the death of Britain's first heart transplant patient; sub-cutaneous infections by *Cryptococcus neoformans* in man and animals following local trauma are well recognized; and there have been two reports of mycetoma in cats due to *Microsporum canis*, the fungus already present in ringworm lesions being introduced into the subcutaneous tissues when the animals received routine protective inoculations against other diseases.

(f) *Prevention*

There are at present no vaccines against the group of diseases under discussion. Any attempt to control their incidence must, therefore, be aimed at reducing exposure to the saprobic phase of the fungus, especially when the condition of the host is such as to render it particularly susceptible to infection. This would be possible either by eliminating the fungus from its natural habitat, which is practic-able only when that habitat is circumscribed and easily recognized, or by prevent-ing the potential victim from coming into contact with the fungus.

High concentrations of *Histoplasma capsulatum* are associated with the roost-ing sites of birds and bats, and with old fowl houses. These are well-defined areas and, if the soil has to be disturbed, it is quite practicable to soak it with water beforehand and so reduce the danger of dispersal of the airborne spores of the fungus. In contrast, *Coccidioides immitis* (Figure 21.4) is widely disseminated in the soil in the districts in which it is endemic, and attempts to control it are impracticable. There can be few people who do not inhale the spores of *Aspergillus fumigatus* daily, and it would be impossible to take precautions against this, but it would obviously be wise for those with any pre-existing lung condition to avoid handling mouldy hay, or turning over the compost heap, thereby exposing themselves to excessive concentrations of spores. Similarly, if mycotic abortion and pulmonary aspergillosis in farm animals and poultry are to be prevented, only good-quality straw and hay should be used on the farm, especially when the animals are housed.

Many of the fungi causing subcutaneous mycoses are widespread in nature, and the source of others has yet to be discovered: the control of any of them under ordinary circumstances is impossible, but adequate clothing and footwear provide excellent protection against the wounds and scratches by means of which they are inoculated into the body. Only in very localized outbreaks of sporotrichosis, such as the South African epidemic mentioned above, can the saprobic phase of the fungus be effectively attacked. In that instance the wooden pit-props, which were eventually identified as the source of the fungus, were sprayed with a fungicide, after which no new infections occurred.

4. Epidemiology of diseases due to 'obligate' pathogens—the dermatophytes

The dermatophytes belong to three closely related genera of fungi: *Microsporum*, *Trichophyton* and *Epidermophyton* (Figure 21.5a, b), all of which are able to lyse

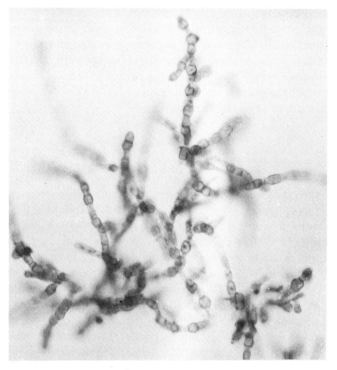

Figure 21.4.   Saprobic phase of *Coccidioides immitis*. Chains of
thick-walled arthrospores (×450)

keratin *in vitro*. In the sense that all grow in culture, none is an obligate parasite, yet the saprobic phases of the majority of the zoophilic species and all the anthropophilic species are unable to exist in competition with the normal microflora of the soil so that pathogenicity is essential for their survival. Epidemiologically, as there is in practice no saprobic phase, dissemination of these species depends entirely on direct or indirect contact between infected and uninfected hosts. However, cross-infection is not indiscriminate, for each dermatophyte has definite host preferences. Anthropophilic species are confined to man, infection of animals being exceedingly rare, while zoophilic species—with the exception of *Trichophyton mentagrophytes*—are restricted to one or a few animal hosts although all are more or less frequently transmitted to man.

The genera *Microsporum* and *Trichophyton* contain non-pathogenic species some of which are common in soil, where they presumably play an important part in the decomposition of keratinized animal residues. Some of these geophilic dermatophytes, the members of the *Microsporum gypseum* complex, occasionally cause ringworm in both man and animals. They present a very different epidemiological problem from the anthropophilic and zoophilic species since

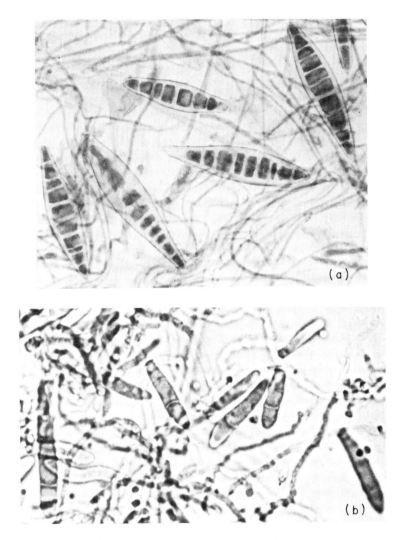

Figure 21.5. Macroconidia produced by dermatophyte fungi in the saprobic phase: (a) *Microsporum*; (b) *Trichophyton* (× 450)

infection is contracted directly from the saprobic phase in the soil, very rarely by cross-infection in the pathogenic phase.

The life cycles and epidemiology of geophilic, zoophilic, and anthropophilic dermatophytes are summarized in Figure 21.6, in which the increasing host dependence and specificity of the three groups is clearly shown. The geophilic species reproduce freely both sexually and asexually, but as dependence on the host increases in the zoophilic and anthropophilic groups, first sexual reproduc-

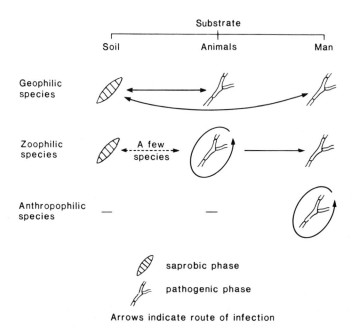

Figure 21.6. Relationship of saprobic and pathogenic states in dermatophytes (arrows indicate route of infection)

tion and then—in a few anthropophilic species—asexual sporulation as well, is lost from the saprobic phase. Such species cannot be other than dependent on cross-infection for the survival of their race.

The zoophilic dermatophytes affecting domestic animals are listed, together with their preferred hosts, in Table 21.3. Geographically the distribution of most species coincides with that of the host, although *Microsporum nanum*, *Trichophyton gallinae* and *Trichophyton simii* are of very local occurrence. *Trichophyton verrucosum* (Figure 21.7) and *Microsporum canis* have a particular affinity for man. Among wild animals, rodents (*Trichophyton mentagrophytes*), hedgehogs (*Trichophyton erinacei*) and voles (*Microsporum persicolor*) are often lesion-free carriers of their respective dermatophyte species. Cross-infection to man may be limited by his infrequent contact with the hosts.

Not only do the anthropophilic dermatophytes have a very strong specificity for the human host, but each species shows a distinct and constant preference for a particular body site on that host. These primary sites are the scalp, trunk, groin, and feet, and from them secondary spread to other predictable areas may or may not take place, again depending on the species of dermatophyte involved. The anthropophilic dermatophytes, with their preferred sites of infection, their geographical range, and the economic status of the communities in which they are normally found, are listed in Table 21.4. It is evident that there is a sharp contrast in

Table 21.3. Dermatophytes affecting domestic animals

Fungus	Host
Microsporum canis	Cats, dogs
Microsporum equinum	Horses
Microsporum gallinae	Chickens
Microsporum nanum (also geophilic)	Pigs
Trichophyton mentagrophytes	Rodents, etc.
Trichophyton equinum	Horses
Trichophyton simii	Chickens
Trichophyton verrucosum	Cattle

the epidemiological behaviour of species causing scalp and body ringworm compared with those causing foot infections.

To judge by surviving clinical descriptions, scalp and body ringworm have been known since ancient times, but 'cheiropompholyx', an eruption of the feet, was first described in the medical literature only in the nineteenth century and was not attributed to fungal infection until 1908. The story of the appearance and spread of tinea pedis is an instructive one epidemiologically. The two cases reported in 1908 were caused by *Epidermophyton floccosum*, a fungus already well known at the time as the agent of primary groin ringworm. The occasional foot infection was probably secondary to groin lesions, for up to that time bathrooms in houses were by no means universal, and public swimming baths and similar facilities were rare, so that opportunities for host-to-host transmission of foot infections were severely restricted. In 1914 the first cases of tinea pedis due to a previously unknown fungus, *Trichophyton interdigitale*, were reported from Europe. This organism closely resembles *Trichophyton mentagrophytes* and is probably a downy mutant from that granular, zoophilic species, the physiological requirements of which are better satisfied by the environment offered by a shod foot than by smooth skin. Later, in 1921, tinea pedis due to *Trichophyton rubrum* was reported from the USA. Again the fungus appeared to be downy mutant of a common, granular form, this time of an anthropophilic species causing body ringworm in the Far East. It was presumably imported to the West by white colonizers and travellers. From 1916 onwards both European and American medical literature documents an explosion in the incidence of tinea pedis mainly due to the two 'new' dermatophytes, *Trichophyton interdigitale* and the downy form of *Trichophyton rubrum*. Until today in developed countries tinea pedis is the most common form of ringworm. The establishment and rapid spread of these two species occurred as domestic bathrooms and communal sports and bathing facilities became available to all, so providing the fungi—already adapted physiologically to conditions on the shod foot—with the necessary facilities for transfer from host to host during the brief periods in which footwear is discarded.

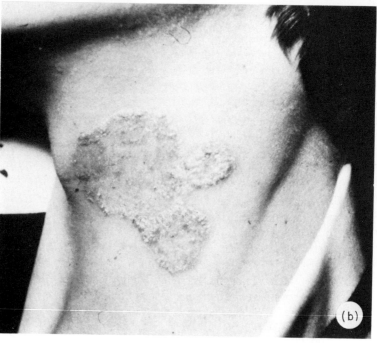

Table 21.4. Distribution and preferred sites of anthropophilic dermatophytes

Fungus	Preferred site	Geographical range
Associated with affluence		
Trichophyton interdigitale	Feet	
Trichophyton rubrum (downy)	Feet	Cosmopolitan
Epidermophyton floccosum	Groin, feet	
Associated with deprivation		
Microsporum audouinii	Scalp	
Trichophyton tonsurans	Scalp	
Trichophyton violaceum	Scalp, beard	Cosmopolitan
Trichophyton schoenleinii	Scalp	
Trichophyton rubrum (granular)	Trunk	
Microsporum ferrugineum	Scalp	Far East, West Africa
Trichophyton concentricum	Trunk	Far East, Pacific area
Trichophyton megninii	Beard	Western Europe, North Africa
Trichophyton soudanense	Scalp	
Trichophyton gourvilii	Scalp	Africa
Trichophyton yaoundii	Scalp	

(a) *Control*

Successful control of both human and animal ringworm must be directed towards breaking the chain of host-to-host transmission, and it is important that the measures taken should be based specifically on the habits of the particular fungus involved.

When human or animal hair is invaded by dermatophytes the hair becomes very brittle and the fungal hyphae break up into huge quantities of arthrospores (Figure 21.8a, b) which are easily disseminated if disturbed and will settle on any horizontal surface. Being light in weight, arthrospores may be airborne for quite long distances. When skin is invaded, however, dermatophyte hyphae are embedded in the substrate (Figure 21.8a, b) and dispersal is by means of infected squames which, being heavy, tend to fall rather than to become airborne but which, like arthrospores, may adhere to uninfected skin, clothes, and towels. Hyphae in skin squames are unaffected by brief exposure to disinfectants for they are well protected by the keratin in which they are embedded. Whether as free arthrospores or embedded in fragments of skin, dermatophytes can survive for long periods in a dry environment but are quickly killed by competing micro-organisms in moist conditions. These factors must be borne in mind when devising control methods for ringworm outbreaks.

Figure 21.7. Ringworm lesions in (a) a calf; (b) man, due to the zoophilic dermatophyte *Trichophyton verrucosum*

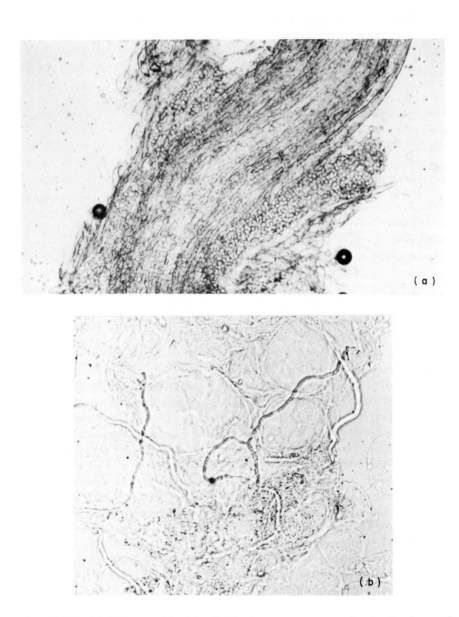

Figure 21.8. (a) Hair parasitized by *Trichophyton verrucosum*, showing hyphae and arthrospores of the fungus (×75); (b) hyphae of a dermatophyte in human skin. KOH squash preparation (×100)

Ringworm in domestic pets is not too great a problem, as only one or a few animals are kept in each household. The classic work of La Touche (1955) outlines the steps to be taken should there be a large outbreak of *Microsporum canis* infection among cata in a residential district. The situation is much more serious in breeding establishments, farms, and large stables where numerous animals of the same species are housed together. In such conditions the introduction of a single infected animal can speedily result in a large-scale epidemic. Because hairs infected by *Microsporum canis* fluoresce under Wood's (ultraviolet) light, it is possible to control ringworm in catteries and kennels simply by screening all newcomers before introducing them into the establishment, together with regular precautionary screening of inmates. Yet this practice is extremely rare among breeders. In stables during outbreaks of *Microsporum equinum* and *Trichophyton equinum* infection, each horse should have its own set of tack and grooming kit, never used for any other animal, and a blow-lamp used on walls and partitions before whitewashing helps to eliminate infective particles. Outbreaks of cattle ringworm, especially in calves, are so commonplace that frequently no attempt is made to control them. However, in the USSR and Eastern Europe a vaccine against *Trichophyton verrucosum* is in use on a large scale and is reputed to be very effective, but is not at present obtainable in the West.

Although most zoophilic dermatophytes can be transmitted to man by their animal hosts, further transmision from human to human is uncommon. Control of animal ringworm in man is therefore directed primarily at the animal reservoir (English, 1972).

Human scalp and body ringworm due to anthropophilic dermatophytes have always been, and still are, associated with a low standard of living and will only be eliminated as the living standards of the communities concerned are improved. Tinea pedis, however, as has already been indicated, is a disease of affluence, and a satisfactory means of controlling it has yet to be devised. In the home certain obvious precautions against spread can be taken, especially in the bathroom, but swimming baths and other communal bathing facilities present a problem which has so far proved insoluble. Disinfectants are of little use for cleaning the floors of swimming baths, for the reasons given above. Cleaning is best carried out by thorough brushing or hosing down to remove infective particles from circulation. Footbaths for bathers are also useless and should be abandoned, as they instill a false sense of security. The most hopeful approach is that suggested by Gentles, Evans, and Jones (1974) who, by supplying the users of a public swimming bath with a prophylactic powder for use after bathing, were able to reduce the incidence of tinea pedis from 8.5 to 2.1 per cent in $3\frac{1}{2}$ years.

5. Epidemiology of diseases due to normal body commensals

The incidence of *Candida albicans* in its commensal phase in the alimentary tracts of normal persons varies from population to population, but 30–50 per cent would not be unusual. The fungus is also associated with the lung and alimentary tract of normal animals and birds. Although it has been reported on a few occasions from soil and vegetation it is rare in such situations and there is little doubt that its normal environmental niche is that of a commensal of the human or animal body, when it occurs in the yeast phase. Because of the absence of an independent saprobic phase transmission of *Candida albicans* to a fresh host, whether as commensal or pathogen, is dependent on direct or indirect contact between hosts. In fact, successful transmission is unlikely in the first instance to result in disease, but merely in the setting up of a commensal relationship with the new host. Only two conditions are commonly a direct result of a new infection; oral thrush of the newborn is contracted following delivery of the baby through an infected birth canal, and vaginal thrush may sometimes be transmitted by the sexual partner.

Many cases of vaginal thrush, and almost all other *Candida albicans* infections, are endogenous in origin and are precipitated by some local or general disturbance of the checks and balances normally operating between the living host and the fungus, such disturbances frequently being iatrogenic in origin. From its commensal headquarters in the alimentary tract, *Candida albicans* is well placed to reach any part of the host's body of which the normal defences may be temporarily lowered. Subsequent colonization of the site should present no difficulty to a fungus which, unlike the saprobes, is already partially adapted to the physiological and nutritional environment offered by the living host. *Candida albicans* is thus an extraordinarily versatile fungus, capable of the prompt exploitation of abnormal circumstances arising in any body site (Figure 21.9). In Table 21.5 some of the manifold forms of candidiasis (candidosis) of endogenous origin are listed, together with the special factors predisposing to them. The major hazard is clearly modern medical practice including drug therapy, catheterization, and major surgery. Control measures include constant vigilance, possibly involving elimination of the fungus from the alimentary tract, and avoidance where possible of predisposing factors other than essential medical treatment.

In domestic animals *Candida albicans* rarely presents problems. Although occasional outbreaks of infection of the crop and upper alimentary tract are reported in flocks of chickens, turkeys, and geese the source of the yeast is not clear and the association of disease with specific predisposing factors has not been ascertained.

Malassezia furfur, also known as *Pityrosporum orbiculare*, resembles *Candida albicans* in that it has yeast-like and mycelial phases which are often found together. It is a lipophilic organism without keratinolytic capability, which is responsible for pityriasis versicolor, a very superficial infection of the human skin. Until quite recently the fungus was known only as a pathogen and the epidemiology of the disease was considered to be similar to that of ringworm caused by anthropophilic dermatophytes, although certain predisposing factors

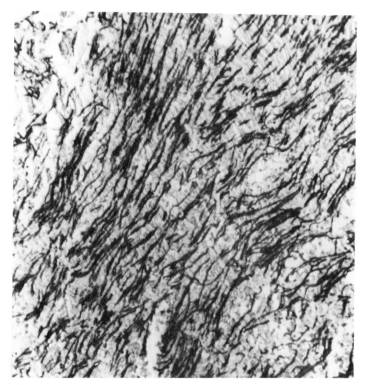

Figure 21.9.   Infection of the oesophagus by *Candida albicans*. The invasive phase in strongly mycelial. Grocott silver stain ($\times 75$)

Table 21.5.   Candidiasis of endogenous origin

Form	Precipitating factors
Oral thrush (adults)	Ill-fitting dentures
Oesophageal candidiasis	Immunosuppressive therapy
Mycotic keratitis	Trauma, local steroids
Otomycosis	?
Intertrigo	Diabetes, steroids
Chronic paronychia	Wet work
Vaginal thrush	Pregnancy, the Pill
Systemic candidiasis	Immunosuppressive drugs, antibacterial antibiotics, surgical procedures, indwelling catheters, drug addiction
Mucocutaneous candidiasis	Congenital immune deficiencies, poor iron absorption

such as excessive sweating and genetic susceptibility were acknowledged. However, Roberts (1969) showed that the fungus is carried in small quantities in its yeast phase in the skin of the trunk, flexures, and face of over 90 per cent of normal subjects and is, therefore, a skin commensal. The change to pathogenicity takes place when certain known and unknown predisposing factors produce suitable conditions in the skin, while the pathogenic state in the fungus is marked, as in *Candida albicans* infections, by the production of mycelium. Although cross-infection undoubtedly occurs, particularly in conjugal cases, the source of the fungus is usually endogenous, and preventive measures must be aimed at correcting the underlying precipitating factors rather than controlling cross-infection.

6. Economic status and fungal disease in man

(a) *Ringworm*

It need not be further emphasized that scalp and body ringworm are diseases of poverty and underprivilege, whether in Third World countries or in deprived communities in more affluent lands, whereas tinea pedis is only common in developed countries and well-to-do communities. As has been pointed out earlier, only the well-off can afford the comparative luxury of a bathroom in every home and communal bathing and sports facilities readily accessible to all, which are the precondition for the propagation of the disease.

(b) *Systemic and subcutaneous mycoses*

Like scalp and body ringworm the long-established primary systemic and subcutaneous infections are almost all diseases of the Third World or of deprived communities in developed countries. They are rare in affluent countries and communities even when the causal fungi are known to occur in the environment, although *Histoplasma capsulatum* is exceptional in affecting rich and poor alike. Lack of medical facilities probably plays only a small part in this situation for, until recently, no effective antifungal chemotherapy has been available except in the case of sporotrichosis. Rather, the virtual elimination of these diseases, at least in their disseminated forms, from the wealthier sections of humanity is due to the vastly improved social conditions that are achieved with increased wealth. The robust health conferred by an adequate diet results in a high degree of resistance to all infectious diseases, including the mycoses, and especially helps to prevent the progress of the systemic mycoses from primary, subclinical lung lesions. Then too, the time spent on agricultural work under primitive conditions is drastically reduced, with a consequent massive reduction in exposure to airborne fungal spores, for instance those of *Coccidioides immitis*. Finally, it becomes possible to afford adequate clothing and footwear to protect the limbs from the superficial wounds which initiate the subcutaneous mycoses.

Candida albicans and *Aspergillus fumigatus* are commonly isolated in medical laboratories and there is a tendency to consider them of little clinical significance. Consequently when, some 30–40 years ago, in laboratories in the West, these fungi began to appear on culture plates with increasing frequency, there was some doubt as to the interpretation to be put upon them, since many forms of candidiasis and aspergillosis are notoriously difficult to diagnose clinically. However, the increasing sophistication of serological diagnostic techniques has confirmed that the isolation of these organisms is meaningful in a significant number of cases and that the incidence of candidiasis and aspergillosis, as well as of secondary cryptococcosis, is rising rapidly. This phenomenon is only occurring, however, in the technologically advanced countries where patients who, in the past, would have faced an early death from grave constitutional diseases, can now be kept alive for many years by techniques which, often as a side-effect, bring about a great reduction in immune competence. Patients subjected to such regimens fall an easy prey to weak but ubiquitous pathogens such as those responsible for the secondary mycoses, which have little opportunity for mischief in populations whose members tend to die young of primary diseases.

FURTHER READING

Ainsworth, G. C., and Austwick, P. K. C. (1973) *Fungal Diseases of Animals.* Commonwealth Agricultural Bureaux, Farnham Royal.

Emmons, C. W., Binford, C. H., Utz, J. P., and Kwon-Chung, K. J. (1977) *Medical Mycology.* Lea & Febiger, Philadelphia.

English M. P. (1972) The epidemiology of animal ringworm in man. *Br. J. Dermatol.*, **86**, 78–87.

English, M. P. (1980). *Medical Mycology.* Arnold, London.

Gentles, J. C., Evans, E. G. V., and Jones, G. R. (1974). Control of tinea pedis in a swimming bath. *Br. Med. J.*, **2**, 577–80.

La Touche, G. J. (1955) The importance of the animal reservoir of infection in the epidemiology of animal-type ringworm in man. *Vet. Rec.*, **67**, 666–9.

Odds, F. C. (1979) *Candida and Candidosis.* Leicester University Press.

Roberts, S. O. B. (1969) Pityrosporum orbiculare: incidence and distribution on clinically normal skin. *Br. J. Dermatol.*, **81**, 264–9.

Index

abortion, 4, 9, 14, 17, 18, 225, 237, 309
acetylcholine, 88
N-acetylglucosamine, 104
N-acetylneuraminic acid, 115, 262, 266
Acinetobacter, spp., 186
actinomycin, 174
active immunity, 24 *et seq.*, 156 *et seq.*
adenine arabinoside, 175
Adenoviridae, 274
adenylate cyclase, 97, 98, 101
adiospiromycosis, 309, 313
adjuvants, 24, 110, 158, 162, 164
aerosols, 152
aflatoxicosis, 223
African sleeping sickness, 286
agar gel diffusion test, 28, 219
age, a factor in disease, 5 *et seq.*, 244
aggressins, 66, 67, 68, 78
airborne infections, control of, 151 *et seq.*
air-conditioning systems, 187, 197, 198
air disinfection, 152, 153
alimentary tract, structure in different
  species, 38
alkaline disease, 220
allergy, 22, 29 *et seq.*
Allodermanyssus sanguines, 292
alveolar phagocytes, 260, 261
Amanita phalloides, 222
Amanita virosa, 222
amantadine, 175
amatoxins, 222
amidase, 172, 177
amikacin, 175
aminoglycosides, 178
amodiaquine, 155
aminoacyl transferase II, 83, 84
amoebiasis, 207
amoebic dysentery, 148
amphoteracin, 174, 175, 177
ampicillin, 172, 174, 175
amoxycillin, 175
amylolytic bacteria, 46

anamnestic response, 24
anaphylaxis, 30, 156, 162
Anopheles bellator, 169
anthrax, 4, 10, 13, 54, 62, 77 *et seq.*, 134,
  137, 138, 139, 146, 207
anthrax toxin, assay of, 80
  mode of action, 80, 81
  production of, 78 *et seq.*, 80
antibiosis, 171 *et seq.*
antibiotics, 6, 12, 171 *et seq.*, 183, 184,
  188, 194 *et seq.*, 200
  influence on gut flora, 39, 41, 42 *et seq.*
  mode of action, 176 *et seq.*
antibiotic-resistance, 41, 43, 44, 149, 159,
  163, 180 *et seq.*, 195, 233
antibody, *see* immunoglobulins
antisepsis, 193, 194
Aristotle, 208
Arthus reaction, 30, 111
aseptic techniques, 190, 192
aspergillosis, 309
Aspergillus flavis, 223
Aspergillus fumigatus, 308, 309, 312, 313,
  315, 327
ataxia-telangiectasia, 127
athlete's foot, 134, 137, 319 *et seq.*
autoimmunity, 29
avirulence, definition of, 49

Bacillus anthracis, 4, 19, 21, 52, 56, 64,
  67, 73, 77 *et seq.*, 92, 134, 136, 146,
  157, 207, 297
Bacillus cereus, 92, 217, 221 *et seq.*
Bacillus melaninogenicus, 35
Bacillus mycoides, 92
bacitracin, 173, 174, 175
bacterial adhesion, 32, 62 *et seq.*, 276
bactericides, tissue, 20, 66
bacteriocins, 41, 94
Bacteroides, spp., 39, 40, 42, 178, 179
Bacteroides amylophilus, 46
Bacteroides succinogenes, 46

host defences, alimentary tract, 19
  anatomical, 19
  skin, 19
  upper respiratory tract, 19
hot–cold haemolysins, 95
human lymphocyte antigens (HLA), 121, 264
humidifiers, 185
humoral antibody, *see* immunoglobulins
hyaluronidase, 52, 73, 75, 76, 92, 94
hypersensitivity, 22, 29 *et seq.*, 111, 180
hypersensitivity, delayed, 30
hyphomycosis, 310

iatrogenic diseases, 11, 33, 305
iatrogenic mycoses, 305
idoxuridine, 175
immune complex disease, 125
immune responses, 24 *et seq.*
immunity, acquired, 23 *et seq.*
  active, 156 *et seq.*
  passive, 5, 7, 156 *et seq.*
immunocytes, 23
immunodiffusion test, 28, 84, 219
immunoglobulins, 7, 8, 25 *et seq.*
  complement-fixing, 29
  IgA, 25 *et seq.*, 124, 164, 165, 263
  IgE, 25 *et seq.*, 124
  IgG, 25 *et seq.*, 124, 263
  IgM, 25 *et seq.*, 124
  neutralizing, 27, 28, 124
  pooled, 156
  precipitating, 28
  protective, 27, 158
immunosuppressive drugs, 5, 167, 183, 197
impedins, 66, 78, 93
incubation period, 135, 136
indigenous flora, 32 *et seq.*
infection, the process of, 61 *et seq.*
infectious bovine rhinotracheitis, 167
infections, endogenous, 14, 184, 187
  exogenous, 13
infectious hepatitis, 147
infectious laryngotracheitis, 167
infectious mononucleosis, 138
inflammatory response, 6, 21 *et seq.*
influenza, 131, 136, 139, 143, 187
  animal reservoirs, 272
  antigenic drift, 269
  antigenic shift, 269
  antigenic variation, 268

genetic mutability, 270
genetic recombination, 266, 271
geographical epicentres, 274
haemagglutinin (H) spike, 266
Hong Kong flu, 270
immunoselection of serotypes, 271
M-protein, 266
multiplicity reactivation, 266
neuraminidase (N) spike, 266
nomenclature of subtypes, 267
reactivation of latent subtypes, 275
recycling time, subtypes, 272
shift periodicity, 269
von Magnus phenomenon, 266
vaccines, 165
innate resistance, 3
insect repellants, 154
insect vectors, 143
  control of, 153 *et seq.*
  eradication of, 154
interferon, 11, 123
intravenous drips, 106, 108, 190
irradiation, 6, 7, 10, 11, 183, 190
isolation, of infected hosts, 146, 197, 200
isoniazid, 175
Isospora, spp., 207
intoxication, 87, 217 *et seq.*
invasiveness, 68 *et seq.*

Jenner, 159
Johne's disease, 4, 8, 14, 134

karyokinesis, 117
K88 antigen, 39, 63, 100, 182
K99 antigen, 63, 100, 182
kanamycin, 44, 173, 174, 175
Kennedy, 241
kidney, infections of, 16 *et seq.*
Kingsbury, 294
Kitasato, 71
Klebs, 71
Klebsiella, spp., 184, 186, 195
Koch, 71, 208
Kupffer cells, 122
kuru, 119, 126, 127

β-lactamase, 93, 172, 177, 284
lactobacilli, 37, 40, 56
Lactobacillus acidophilus, 35
lamb dysentery, 7, 8, 88, 89, 92
lame sickness, 221